THIRD EDITION

Biology with Computers

Biology Experiments Using Vernier Sensors

David Masterman
Scott Holman

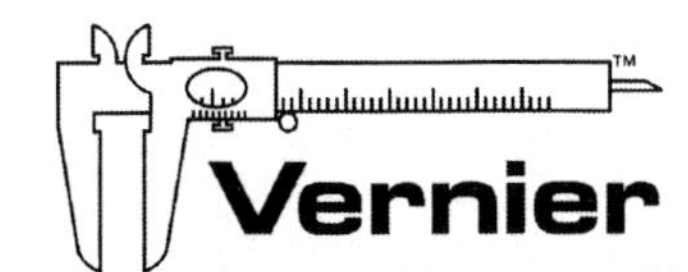

Vernier Software & Technology
13979 S.W. Millikan Way • Beaverton, OR 97005-2886
Toll Free 888-837-6437 • 503-277-2299 • FAX 503-277-2440
info@vernier.com • www.vernier.com

THIRD EDITION

Biology with Computers

Biology Experiments Using Vernier Sensors

Published by
Vernier Software & Technology
13979 SW Millikan Way
Beaverton, OR 97005-2886
Toll Free 888-837-6437
Ph. 503-277-2299
Fax 503-277-2440
info@vernier.com
www.vernier.com

ISBN 1-929075-26-X
Third Edition
Second Printing
Printed in the United States of America

About the Authors

David Masterman received his B.A. in Zoology and his M.A. and Ph.D. equivalency in Biology from UCLA. He taught at Mount Saint Mary's College in California, and at Jackson Hole High School in Wyoming for fifteen years before taking his current position at Lakeside School in Seattle, Washington. David has been active in the field of computer-assisted data collection for many years. Specifically, he

- conducted numerous workshops for the Woodrow Wilson National Fellowship Foundation, where he integrated computer-assisted data collection into the Chemistry curriculum.
- is the president of MasterComp, Inc. Previously, he designed and constructed interfacing software and hardware for Apple II computers. Glencoe/McGraw Hill is currently distributing his data analysis software as an integral component of their middle and high school science programs.
- was the principal investigator and instructor at the Center for Science Studies summer workshops for computer interfacing.
- was the computer instructor for the first Woodrow Wilson National Fellowship Foundation's 1991 Summer Biology Institute for High School Teachers.
- hosted a World Wide Web session sponsored by Genentech's Access Excellence on the subject of using computers, CBLs and probeware in the science lab. He is an Access Excellence Fellow and a returning Fellow.

David is a Dreyfus Master Teacher, Tandy Technology Scholar, a 1988 Presidential Awardee for Excellence in Science Teaching, and 1996 Washington State science teacher of the year.

Scott Holman received his B.S. in Biology from Southern Oregon University in Ashland, Oregon. Scott's original intentions after graduation were to go into teaching. Job opportunities led him and his wife to Portland, Oregon where Scott gained employment with us at Vernier Software. At Vernier he has found a rewarding opportunity to work with computers and help teach teachers how to integrate technology into the biology classroom. Since that time, he has presented numerous workshops on the use of computer and calculator interfaces in the classroom.

Scott has also been involved in the research and development of numerous sensors and probes (Exercise Heart Rate Monitor, CO_2 Gas Sensor, Biology Gas Pressure Sensor, and Respiration Monitor Belt). In addition to this book, he has co-authored the "Water Quality with Computers" and the "Water Quality with Calculators" lab books.

Scott served as the biology specialist at Vernier Software & Technology for six years. In addition, he assisted with water quality and chemistry technical support.

Proper safety precautions must be taken to protect teachers and students during experiments described herein. Neither the authors nor the publisher assumes responsibility or liability for the use of material described in this publication. It cannot be assumed that all safety warnings and precautions are included.

Contents

Experiments

Appendixes

Sensors Used in Experiments

		Temperature	Gas Pressure	CO_2	Colorimeter	Conductivity	Dissolved Oxygen	EKG	Heart Rate	O_2	pH	Respiration Belt
1	Energy in Food	1										
2	Limitations on Cell Size: Surface Area to Volume					1						
3	Acids and Bases										1	
4	Diffusion Through Membranes					1						
5	Conducting Solutions					1						
6A	Enzyme Action: Testing Catalase Activity									1		
6B	Enzyme Action		1									
7	Photosynthesis				1							
8	The Effect of Alcohol on Biological Membranes				1							
9	Biological Membranes				1							
10	Transpiration		1									
11A	Cell Respiration									1		
11B	Cell Respiration			1								
11C	Cell Respiration		1									
11D	Cell Respiration			1						1		
12A	Respiration of Sugars by Yeast			1								
12B	Sugar Fermentation		1									
13	Population Dynamics				1							
14	Interdependence of Plants and Animals						1				1	
15	Biodiversity and Ecosystems	1										
16A	Effect of Temperature on Respiration			1								
16B	Effect of Temperature on Fermentation		1									
17	Aerobic Respiration	1					1					
18	Acid Rain										1	
19	Dissolved Oxygen in Water	1					1					
20	Watershed Testing	1				1	1				1	
21	Physical Profile of a Lake	1				1	1				1	
22	Osmosis		1									
23A	Effect of Temperature Cold-Blooded Organisms									1		
23B	Effect of Temperature Cold-Blooded Organisms			1								
24A	Lactaid Action			1								
24B	Lactaid Action		1									
25	Primary Productivity						1					
26	Control of Human Respiration		1									1
27	Heart Rate and Physical Fitness								1			
28	Monitoring EKG							1				
29	Ventilation and Heart Rate								1			
30	Oxygen Gas and Human Respiration									1		
31A	Photosynthesis and Respiration									1		
31B	Photosynthesis and Respiration			1								
31C	Photosynthesis and Respiration			1						1		

Preface

This book contains thirty-one student experiments using Logger *Pro* 3 data collection software and Windows and Macintosh computers for collecting, displaying, printing, graphing, and analyzing data. Such experiments can comprise more than fifty percent of the experiments included in a quality high school biology course. We are convinced of the importance of "hands-on" experiments. Like balances, burners, and beakers, computers can be an integral and indispensable equipment component in a biology laboratory.

Vernier probes will often result in more accurate temperature, pH, voltage, colorimetric, conductivity, and pressure data measurements in the biology lab. Your students will perform many new experiments with measurements not previously obtainable in the classroom. Vernier Software, with its high quality and comparatively inexpensive hardware, supported with well-written, thorough, and easy-to-use software, has made it possible and relatively simple for high school and university instructors to completely integrate the computer into their biology classes. We hope this book helps in the task.

You will find a wide range of experiments in this book. Whether your biology class is high school or college, Advanced Placement, honors or general biology, you should find a large number of experiments in this book that match the scope and objectives of your course. Following each student experiment there is an extensive Teacher Information section with sample results, answers to questions, directions for preparing solutions, and other helpful hints regarding the planning and implementation of a particular experiment.

This book has been revised to include instructions for Logger *Pro* 3 software. If you are using an earlier version of Logger *Pro,* instructions for all experiments can be found in the Word Files for Older Book folder on the book CD.

Experiments in this book can be used unchanged or they can be modified using the word-processing files provided on the accompanying CD. Students will respond differently to the design of the experiments, depending on teaching styles of their teachers, math background, and the scope and level of the biology course. Here are some ways to use the experiments in this book.

- Unchanged. You can photocopy the student sheets, distribute them, and have the experiments done following the procedures as they are. Many students will be more comfortable if most of the computer steps used in data collection and analysis are included in each experiment.
- Slightly modified. The CD accompanying the book are for this purpose. Before producing student copies, you can change the directions to make them better fit your teaching circumstances. See *Appendix A* for instructions on using the CD.
- Extensively modified. This, too, can be accomplished using the accompanying CD. Some teachers will want to decrease the degree of detail in student instructions.

We hope and expect that experienced biology teachers will significantly modify the procedures given in this manual.

We feel it is **VERY IMPORTANT** to have your students perform Experiment 1, "Energy in Food." This experiment has introductory details that are not included in other experiments in the book. These include specific directions for methods of examining data on graphs, viewing data tables, analyzing data, and printing. The experiment can be completed quickly, leaving students plenty of time to explore other important capabilities of the Logger*Pro* software.

We also feel it is **IMPORTANT** for teachers to read over information presented in the appendices in the back of the book. There is valuable information here that can help make you more comfortable with your initial use of Logger*Pro* software, a Vernier interface, and Vernier sensors. Here is a short summary of the information available in each appendix:

- *AppendixA* tells you how to use the word-processing files found on the CD.
- *Appendix B* provides instructions on using LabPro as a remote device.
- *Appendix C* provides information on Vernier products for biology.
- *Appendix D* provides safety information associated with the experiments in this book.
- *Appendix E* provides a list of equipment and supplies used in these experiments.

We are thankful to Alex Plank, David Vernier, Christine Vernier, and Dan Holmquist of Vernier Software & Technology for the advice and editing help they provided.

Above all, we are grateful to Kristine Holman and Marcia Masterman for their patience and understanding throughout this project.

Scott Holman David Masterman

Experiment

1

Energy in Food

Food supplies energy for all animals—without it we could not live. The quantity of energy stored in food is of great interest to humans. The energy your body needs for running, talking, and thinking comes from the foods you eat. Not all foods contain the same amount of energy, nor are all foods equally nutritious for you. An average person should consume a minimum of 2,000 kilocalories per day. That is equivalent to 8,360 kilojoules. Calories and joules are both units of energy. We will use joules in this lab since it is the accepted SI metric standard.

You can determine energy content of food by burning a portion of it and capturing the heat released to a known amount of water. This technique is called *calorimetry*. The energy content of the food is the amount of heat produced by the combustion of 1 gram of a substance. It is measured in kilojoules per gram (kJ/g).

OBJECTIVES

In this experiment, you will

- Use a computer to measure temperature changes.
- Monitor the energy given off by food as it burns.
- Determine and compare the energy content of different foods.

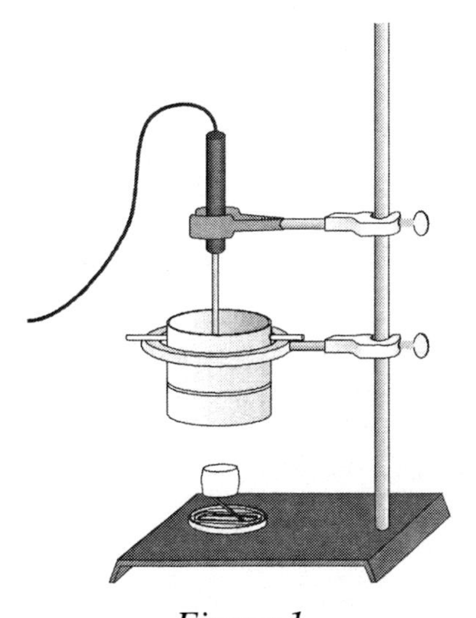

Figure 1

MATERIALS

computer
Vernier computer interface
Logger*Pro*
Temperature Probe
100 mL graduated cylinder
balance
food holder
two food samples (nut, popcorn, or marshmallow)
matches
ring stand and 10 cm ring
small can
split 1-hole stopper
two stirring rods
utility clamp
warm and cool water
wooden splint
two 1-hole rubber stoppers

PROCEDURE

1. Obtain and wear goggles.
2. Obtain a sample of food and a food holder similar to the one shown in Figure 1. Mount the food onto the food holder so that it can burn without damaging the holder. Find and record the initial mass of the food sample and food holder. **CAUTION:** *Do not eat or drink in the laboratory.*
3. Connect the Temperature Probe to the computer interface. Prepare the computer for data collection by opening the file "01 Energy in Food" from the *Biology with Computers* folder of Logger*Pro*.
4. Set up the apparatus shown in Figure 1.
   a. Determine the mass of an empty can. Record the value in Table 1.
   b. Place about 50 mL of cold water into the can.
   c. Determine and record the mass of the can plus the water.
   d. Insert a stirring rod through the holes in the top of the can and hold it in place with two one-hole stoppers. Position the can 2.5 cm (~1 inch) above the food sample.
   e. Use a utility clamp and split stopper to suspend the temperature probe in the water. The probe should not touch the bottom or side of the can.
5. Click [Collect] to begin data collection. Record the initial (minimum) temperature of the water in Table 1.
6. Remove the food sample from under the can and use a wooden splint to light it. Quickly place the burning food sample directly under the center of the can. Stir the water constantly. **CAUTION:** *Keep hair and clothing away from an open flame.*
7. If the temperature of the water exceeds 60°C, blow the flame out. Do not stop the computer yet.
8. After 4 minutes, if the food is still burning, blow the flame out. Record the maximum temperature of the water in Table 1.
9. Once the water temperature begins to decrease, end data collection by clicking [Stop].
10. Determine the final mass of the food sample and food holder.
11. Place burned food, matches, and wooden splints in the container supplied by your instructor.
12. You can confirm your data by clicking the Statistics button, [STAT]. The minimum temperature (t_2) and maximum temperature (t_1) are listed in the floating box.
13. Repeat Steps 4 – 12 for a second food sample. Be sure to use a new 50 mL portion of cold water.

DATA

Table 1		
Measurements	Sample 1	Sample 2
Food used		
Mass of empty can (g)		
Mass of can plus water (g)		
Minimum temperature of water (°C)		
Maximum temperature of water (°C)		
Initial mass of food (g)		
Final mass of food (g)		

Table 2		
Calculations	Sample 1	Sample 2
Mass of water (g)		
Δt of water (°C)		
Δmass of food (g)		
Energy gained by water (J)		
Energy content of food (J/g)		

PROCESSING THE DATA

Record the following calculations in Table 2. Show your work in Table 3.

1. Calculate the change in mass of water. Show your calculations.

2. Calculate the change in mass of each food sample. Show your calculations.

3. Calculate the changes in the temperature of the water, Δt. Record this in Table 2. Show your calculations.

4. Calculate the energy gained by the heated water. Show your calculations. To do this, use the following equation:

$$\text{Energy gained by water} = (\text{mass of water}) \times (\Delta t \text{ of water}) \times (4.18 \text{ J/g°C})$$

5. Convert the energy you calculated in Step 3 to kilojoules (1 kJ = 1000 J).

6. Use your answer in Step 4 to calculate the energy content of each food sample (in kJ/g):

$$\text{Energy content of food} = \text{Energy gained by water} / \Delta\text{mass of food}$$

Table 3		
Calculation	Sample 1	Sample 2
Δm		
Δt		
Energy gained		
Energy content		

7. Record your results and the results of other groups below.

Table 4			
Class Results			
Food Type	Food Type	Food Type	Food Type
Energy content (kJ/g)			
Average			

QUESTIONS

1. Which of the foods has the greatest energy content?
2. Which of the tested foods is the best energy source? Why?
3. What was the original energy source of the foods tested?
4. Why might some foods with a lower energy content be better energy sources than other foods with a higher energy content?
5. Would you expect the energy content values that you measured to be close to the value listed in dietary books? Why?

EXTENSION

1. Determine the energy content of other combustible foods.

TEACHER INFORMATION

Energy in Food

1. This experiment serves as an introduction to the use of Logger*Pro* and temperature probes. In the procedure, students are encouraged to explore many of the menu options available in Logger*Pro*.

2. Any four nuts could be used for this experiment. Walnuts, pecans, peanuts, and almonds are easy to obtain and give excellent results.

3. The water can should be approximately 1-1/2 to 2 inches in diameter and about 3 inches long. A small juice can will do. Drill two holes in the can just under the metal rim, large enough so that a solid glass rod can easy fit in. The can will be suspended by the glass rod with a one-hole rubber stopper at each end. The rubber stoppers will rest on a metal 4 inch ring.

4. The food stand is made using a cork stopper, size 7 or larger, and a paper clip. Straighten one end of the paper clip and push it into the bottom of the cork stopper. Bend the other end of the paper clip into a ring so it will cup the food sample.

5. The two rubber stoppers on the very end of the stirring rod holding the can will prevent the can from slipping off the ring stand.

6. Heat is lost to the environment during this experiment as the fuel is burning. Therefore, the energy content students measure will not be similar to the published values. This lab is still valid, however, since the heat lost to the environment is nearly proportional in every experiment. If the physical conditions in every experiment are the same, the energy contents will be proportional. Several key factors include:
   - The distance from the bottom of the can to the flame (or table top) should be equal.
   - The cans should be of equal dimensions.
   - The flame should not be in a breeze.

7. Provide each lab group with a container to discard their burnt foods. The charred pieces will make a mess otherwise. Soot will accumulate on the outside of the calorimeter can. Provide a paper towel for students to set the can onto between experiments. Soap may be needed after this experiment!

8. The stored calibration for the Stainless Steel Temperature Probe or Direct-Connect Temperature Probe works well for this experiment.

SAMPLE RESULTS

Table 1			
Measurements	Sample 1	Sample 2	Sample 3
Food used	walnut	almond	pine nut
Mass of empty can (g)	28.51	28.51	28.51
Mass of can plus water (g)	77.35	78.27	76.99
Initial temperature of water (°C)	21.6	22.5	22.3
Final temperature of water (°C)	47.7	49.8	31.4
Initial mass of food (g)	12.85	12.92	12.30
Final mass of food (g)	12.25	12.18	12.10

Table 2			
Calculations	Sample 1	Sample 2	Sample 3
Mass of water (g)	49.18	49.76	48.48
Δt of water (°C)	26.1	27.3	9.1
Δmass of food (g)	0.60	0.74	0.20
Energy gained by water (J)	5330	5680	1840
Energy content of food (J/g)	8880	7670	9220

Since this experiment is designed for 9th and 10th grade students, the mathematics has been simplified. Here is a more complete description of some of the mathematical reasoning.

The law of conservation of energy states that the energy lost by the food should equal the energy gained by the water.

$$\Delta E_{\text{food}} = \Delta E_{\text{water}}$$

The energy lost to the environment is nearly proportional in every experiment so it can be ignored. The energy gained by the water can be calculated using the equation below where m_{water} is the mass of the water in grams, C_p is the heat capacity of water which is equal to 4.186 J/g°C, and Δt is the change in temperature in °C.

$$\Delta E_{\text{water}} = m_{\text{water}} \bullet C_p \bullet \Delta t$$

Since the energy gained by the water is equal to the energy lost by the food, then the energy lost by the food can be found by using this equation.

$$\Delta E_{\text{food}} = \Delta E_{\text{water}} = m_{\text{water}} \bullet C_p \bullet \Delta t$$

To calculate how much energy would be lost by one gram of the same food, divide the energy lost by the food by the mass of food that did burn.

$$\text{Energy content} = \frac{\Delta E_{\text{food}}}{1\ \text{gram}} = \frac{m_{\text{water}} \bullet C_p \bullet \Delta t}{\Delta m_{\text{food}}}$$

After converting joules to kilojoules, this gives us an answer similar to that in the student handout. We define the term *energy content* to be that amount of energy that can be obtained by the combustion of one gram of food.

The energy contents of a few sample foods are listed below. Note that these should be proportional to the measured energy contents, not equal to them, since heat was lost to the environment, and combustion was not complete.

Food	Energy content (kJ/g)
almond	26.8
brazil nuts	29.0
cashews	25.5
coconut (dry)	23.8
kidney beans	25.9

Food	Energy content (kJ/g)
lard	37.6
lima bean	14.3
peanuts	25.9
walnut	29.3

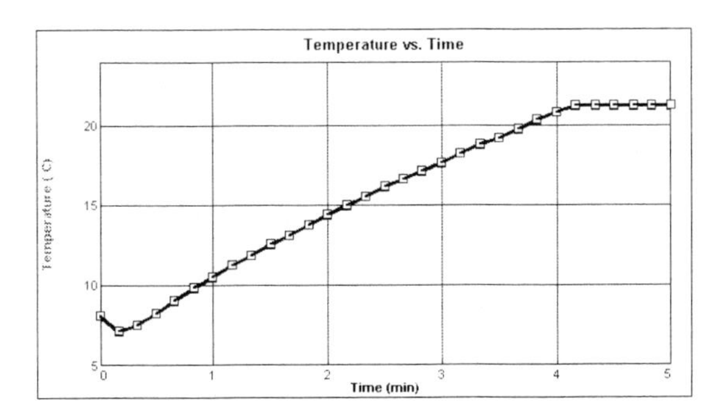

Walnut burning and warming water

ANSWERS TO QUESTIONS

1. Answers may vary, depending upon the type of food used. In the above experiment, pine nuts had the highest heat content, followed by walnuts, then almonds.
2. Answers may vary. The food with the highest energy content is the best energy source.
3. The sun is the source of energy for land plants. Plants can transform this radiant energy into chemical energy. This energy is used to manufacture many different substances, possibly in the form of fats, carbohydrates, or other high energy chemicals. When these chemicals are broken down, they release energy.
4. Some high energy foods might be indigestible. Food with a high energy content might be high in cholesterol or saturated fats. These may be harmful to some people in large amounts.
5. No, the measured values should be lower than those listed in a reference book.
   - There would be a certain amount of heat lost to the air and is not used to heat the water.
   - Soot that collects indicates incomplete combustion of the food. The published values assume complete combustion.

Experiment

2

Limitations on Cell Size: Surface Area to Volume

In order for cells to survive, they must constantly exchange ions, gases, nutrients, and wastes with their environment. These exchanges take place at the cell's surface. To perform this function efficiently, there must be an adequate ratio between the cell's volume and its surface area. As a cell's volume increases, its surface area increases, but at a decreased rate. If you continued to increase the cell's volume, it would soon be unable to efficiently exchange materials and the cell would die. This is the reason that the kidney cell of an elephant is the same general size as a mouse kidney cell.

In this lab activity, you will use agar cubes, which have a high salt content, as cell models. You will investigate how increasing a cell's surface area while maintaining an equal volume affects the rate of material exchange with the environment. When the agar cubes are placed in distilled water, they will begin to dissolve, releasing sodium and chloride ions. The solution's conductivity, measured by a Conductivity Probe, is proportional to the ion concentration in the solution.

OBJECTIVES

In this experiment, you will

- Use agar cubes cut into various size blocks to simulate cells.
- Use a Conductivity Probe to measure the quantity of ions in a solution.
- Determine the relationship between the surface area and volume of a cell.

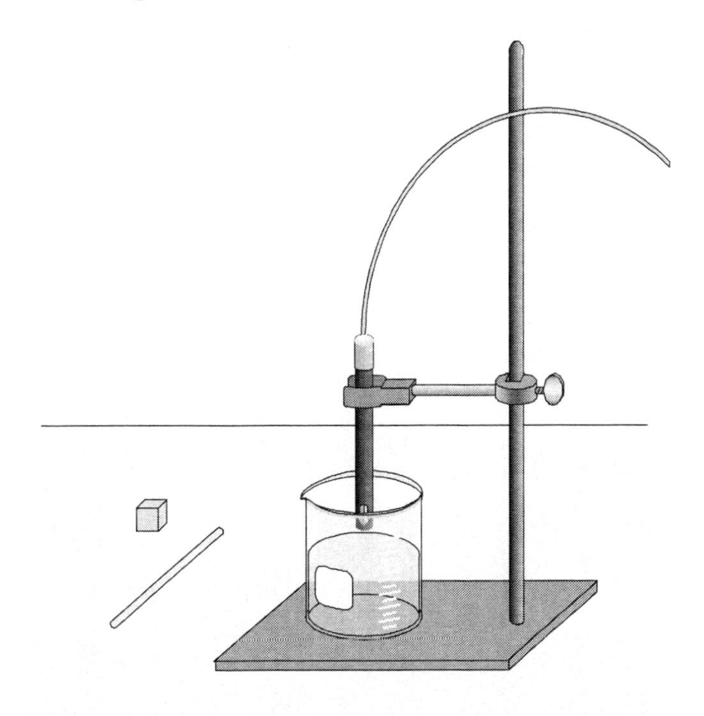

Figure 1

MATERIALS

computer	agar cubes
Vernier computer interface	distilled water
Logger*Pro*	ring stand
Vernier Conductivity Probe	utility clamp
600 mL beaker	metric ruler
glass stirring rod	lined graph paper
scalpel or razor blade (optional)	

PRE-LAB ACTIVITY

1. Obtain three agar cubes.

2. Cut one agar cube into 8 equal cubes. Cut a second cube into 64 equal cubes. Leave the third cube whole.

3. Using a metric ruler, determine the total surface area and the total volume of each agar cube. Note: If the cube is cut into eight pieces, the surface area for that cube is the sum of the surface areas for each of the eight pieces.

4. Record the calculated surface areas and volumes in Table 1.

5. Use your answers from Step 4 to calculate the surface-to-volume ratio. Divide the surface area of each cube by its volume. Enter the results in Table 1. Use the example below to help you with your calculations.

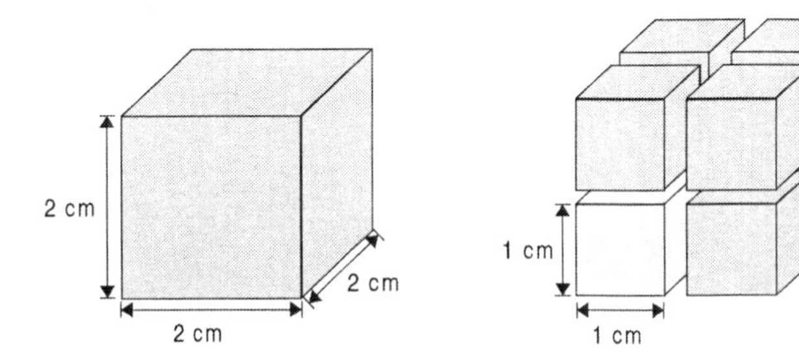

Surface Area:	6 sides (2 cm X 2 cm) = 24 cm^2	6 sides (1 cm X 1 cm) X 8 cubes = 48 cm^2
Volume:	2 cm X 2 cm X 2 cm = 8 cm^3	(1 cm X 1 cm X 1 cm) X 8 cubes = 8 cm^3
Surface-to-volume ratio:	3 to 1	6 to 1

PROCEDURE

1. Set up the utility clamp, Conductivity Probe, and ring stand as shown in Figure 1.

2. Connect the Conductivity Probe to the computer interface. Prepare the computer for data collection by opening the file "02 Limits on Cell Size" from the *Biology with Computers* folder of Logger*Pro*.

3. Set the selector switch on the side of the Conductivity Probe to the 0 – 20000 μS/cm range.

4. Pour 300 mL of distilled water into the 600 mL beaker. Position the Conductivity Probe in the water so the tip is about 2 cm from the bottom of the beaker.

5. Place the solid agar cube sample in the beaker and begin data collection by clicking Collect. Stir the water using the stirring rod. Data collection will automatically end after two minutes.

6. Determine the rate of ion exchange by performing a linear regression of the data:
   a. Click the Linear Fit button, , to perform a linear regression. A floating box will appear with the formula for a best fit line.
   b. Record the slope of the line, *m*, as the rate in Table 1.

7. Empty the water from the beaker and rinse it thoroughly. Rinse the probe with clean, distilled water. Blot the outside of the probe tip dry with a tissue or paper towel. It is *not* necessary to dry the *inside* of the hole near the probe end.

8. Move your data to a stored run. To do this, choose Store Latest Run from the Experiment menu.

9. Repeat Steps 4 – 8 for the sample of agar cut into 8 pieces.

10. Repeat Steps 4 – 8 for the sample of agar cut into 64 pieces.

DATA AND CALCULATIONS

Table 1				
Pieces	Surface area (cm^2)	Volume (cm^3)	Surface-to-volume ratio	Rate (μS/cm s)
1				
8				
64				

PROCESSING THE DATA

1. On Page 2 of this experiment file, plot the surface-to-volume ratio on the x-axis and rate on the y-axis.

QUESTIONS

1. What is the relationship between rate of ion exchange and surface-to-volume ratio?
2. Why is it important for a cell to have a large surface-to-volume ratio?
3. Which is more efficient at exchanging materials, a small or a large cell? Explain.
4. Some cells in the body have adapted to the task of absorption and excretion of large amounts of materials. In what ways have these cells adapted to this task?

5. How does cell growth affect the cell's surface-to-volume ratio?

6. In order for a cell to continue being efficient at exchanging materials, what must it do to maintain its surface-to-volume ratio as it grows larger?

TEACHER INFORMATION

Experiment
2

Limitations on Cell Size: Surface Area to Volume

1. Bouillon cubes are a convenient substitute for agar cubes. There are several brands of bouillon cubes that can be used. The beef flavor was found to be easier to cut than the chicken. Because of the high fat content in the bouillon cubes, you should soak the Conductivity Probes in warm, soapy water at the end of the day. This will remove any deposits left on the surface of the graphite electrodes. Note: Do not use an abrasive to clean the surfaces of the electrodes.
2. Preparation of Agar cubes:
   - Add 12 gm of powdered agar per Liter of distilled water.
   - Heat while stirring continuously.
   - Add 100 gm of NaCl and mix.
   - Heat to boiling.
   - Remove from heat and continue stirring.
   - After 5 minutes of cooling, pour into glass dish or tray and let cool.
   - When cool, remove from dish and cut into 2 cm cubes.
3. Agar can be purchased from most science supply companies.
4. It is important for students to make the surface area, volume, and surface-to-volume ratio calculations prior to doing the experiment.
5. The prepared agar can be stored overnight by wrapping it in plastic wrap and placing it in a refrigerator.
6. The stored conductivity calibration (for 0-20,000μS/cm) works well for this experiment.

ANSWERS TO QUESTIONS

1. As the surface area and the surface-to-volume ratio of a cell increases, so does its ability to exchange materials. The rate increased almost linearly with increasing surface-to-volume ratio.
2. As cells perform their normal functions, they need to exchange materials with the world outside of the cellular membrane. Cells with larger surface-to-volume ratios are able to exchange materials more efficiently.
3. The smaller a cell, the larger its surface-to-volume ratio. As a cell grows larger, the volume increases faster than the surface area. Small cells are more efficient at exchanging materials than large cells because they have more surface area to work with in relation to their size.
4. In the small intestine, the surface is covered by epithelial cells that have increased their surface areas by forming tiny folds of their cellular membrane called *microvilli.* The microvilli dramatically increase the surface area of these cells and their ability to absorb nutrients. In general, cells adapt to the task of absorption and excretion by increasing the size of their cellular membrane while keeping their volume low.

5. As cells grow larger, their volume increases faster than their surface area. This leads to a decrease in the efficiency of the cell at exchanging materials across the cell membrane.
6. To be able to efficiently exchange materials, cells must divide by mitosis and cytokinesis when they become too large.

SAMPLE DATA

Table 1

Pieces	Surface area (cm^2)	Volume (cm^3)	Surface-to-volume ratio	Rate (μS/cm s)
1	24	8	3	3.22
8	48	8	6	6.93
64	96	8	12	13.23

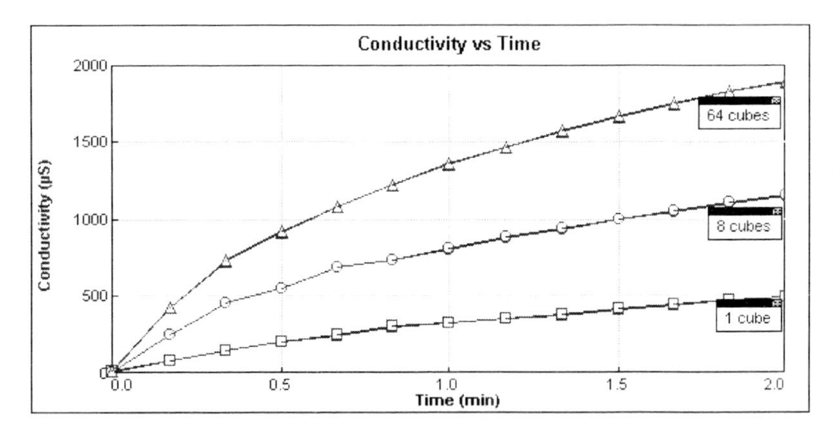

Conductivity resulting from cubes in one, eight, and sixty-four pieces.

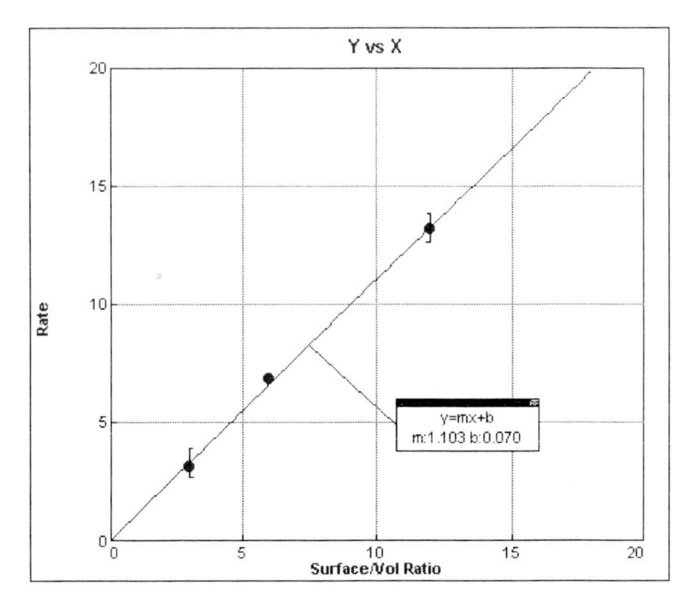

Rate of ion exchange vs. surface-to-volume ratio

Experiment

3

Acids and Bases

Organisms are often very sensitive to the effect of acids and bases in their environment. They need to maintain a stable internal pH in order to survive—even in the event of environmental changes. Many naturally occurring biological, geological, and man-made chemicals are capable of stabilizing the environment's pH. This may allow organisms to better survive in diverse environments found throughout the earth. Teams will work in pairs, using one computer and two pH systems. One team will measure the effect of acid on biological materials, while the other team will measure the effect of base on biological materials. Each group will test the biological materials assigned to them, and all groups will share their data at the end of the class.

OBJECTIVES

In this experiment, you will

- Add an acid to a material and note the extent that it resists changes in pH.
- Add a base to a material and note the extent that it resists changes in pH.
- Work with classmates to compare the ability of different materials to resist pH changes.

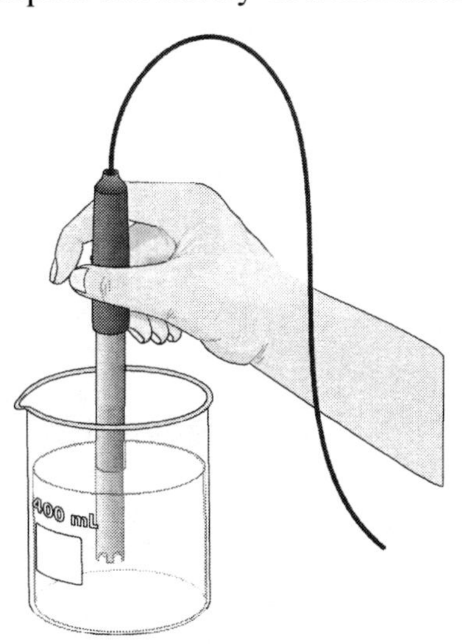

Figure 1

MATERIALS

computer
Vernier computer interface
Logger*Pro*
Vernier pH Sensor (one per team)
Various biological organisms (or parts of an organism), such as yeast, potato, orange juice, or a plant leaf solution.
Various non-biological materials, such as an antacid, buffer, carbonated water or soda, salt, or Alka-Seltzer solution.
two 250 mL beakers
one rinse bottle with distilled water
0.10 M HCl (acid) with dropper
0.10 M NaOH (base) with dropper
50 mL graduated cylinder
goggles
lab apron
two 50 mL beakers
Various simple biological materials, such as egg white, vitamin C, or gelatin solution.

PROCEDURE

1. Obtain and wear goggles.

2. Team A will use the pH probe in CH 1, while Team 2 will use the pH probe in CH 2. Before each use of the pH probe, you need to rinse the tip of the electrode thoroughly with distilled water. To do this, hold the pH electrode above a rinse beaker and use the rinse bottle to thoroughly rinse the electrode tip.

   Important: Do not let the pH electrode dry out. Keep it in a 250 mL beaker with about 100 mL of tap water when not in use. The tip of the probe is made of glass—it is fragile. Handle with care!

3. Connect the probes to the computer interface. Prepare the computer for data collection by opening the file "03 Acids and Bases" from the *Biology with Computers* folder of Logger*Pro*.

Testing the effect of acid and base on water

4. Label one of the 50 mL beakers *acidic* and label the other *basic*. Place 20 mL of distilled water in each beaker.

5. Rinse the pH probe thoroughly with distilled water, then place it into the beaker to be tested:
   - Team A: Place your probe in the beaker labeled *acidic*.
   - Team B: Place your probe in the beaker labeled *basic*.

6. Click [Collect] to begin making pH measurements.

7. The group will be entering the number of drops of acid or base added to the beaker. Before you begin, determine the initial pH of the solution. Click [Keep], then type "0" in the text box and press ENTER.

8. Add acid or base to the solution. Stir each solution thoroughly after addition. **CAUTION:** Handle the hydrochloric acid with care. It can cause painful burns if it comes in contact with the skin. Sodium hydroxide solution is caustic. Avoid spilling it on your skin or clothing.
   - Team A: Add 5 drops of acid to the beaker labeled *acidic*.
   - Team B: Add 5 drops of base to the beaker labeled *basic*.

9. When the pH readings are stable click [Keep]. Enter the total number of drops of acid or base you have added to the water in the beaker. Type "5" in the text box and press ENTER.

10. Repeat Steps 8 through 9, adding 5 drops at a time until each team has added a total of 30 drops.

11. Click [Stop] when you have added a total of 30 drops.

12. Rinse the pH probe thoroughly and place the probe into the beaker of tap water. Clean the two 50 mL beakers.

13. Move your data to a stored data run. To do this, choose Store Latest Run from the Experiment menu. This will allow the data you obtained for water to be included in every future graph.

Testing the effect of acid and base on other materials

14. Test the effect of acid and base on a material assigned to you by your instructor:
    a. Obtain 20 mL of a solution to test from your instructor.
    b. Repeat Steps 5 – 12.
    c. Record the volume and pH values from the table in Table 1. Run 1 data will be the data collected using water. The data labeled Latest will be the data for your tested material.
    d. (optional) Print a copy of your graph. Enter your name(s) and the number of copies of the graph. The graph should have four lines on it—water with acid, water with base, your material with acid, and your material with base.
15. If time permits, repeat Step 14 for as many materials as you can. Before starting the next experiment, delete the latest run by choosing Delete Data Set ▶ Latest from the Data menu.
16. Obtain the pH values of any materials you did not test from your classmates. These values should be listed on the board. Record these values in Table 1.
17. Subtract the ΔpH of the acid from the ΔpH of the base to determine the Total Buffer Range. Record these values in Table 1.

DATA TABLE

Table 1										
Material Tested	Add	pH, after adding this many drops								
		0	5	10	15	20	25	30	ΔpH	Total Buffer Range
	acid									
	base									
	acid									
	base									
	acid									
	base									
	acid									
	base									
	acid									
	base									
	acid									
	base									
	acid									
	base									
	acid									
	base									
	acid									
	base									

PROCESSING THE DATA

1. Make a series of graphs of the data obtained from other students. Alternatively, if instructed by your teacher, obtain a printout of each plot from other student teams. Construct the graphs so they appear similar to the plot your team made:
   - The horizontal axis has Volume scaled from 0 to 30 drops.
   - The vertical axis has pH scaled from 0 to 12.
   - The data you obtained for water should be included in every graph.
   - Construct one graph from the data in each row of Table 1.

2. Make a list of each material that was tested by the teams in your class. Place the most acidic material at the top of the list and the most basic material at the bottom of the list. Use the value corresponding to 0 drops of acid or base, as this value represents the natural acidity of the material.

Table 2		
Material	Initial pH	Rank
		most acidic
		2
		3
		4
		5
		6
		7
		least acidic

3. Put the materials tested into the following three categories:

Biological Organisms	Biological Chemicals	Non-Biological Chemicals

4. Calculate the pH change for each material. Record this in Table 1.

5. Make a second list of each material in Table 1. Place the material that had the largest Total Buffer Range at the top of the list in Table 3 and the smallest range at the bottom of the list.

Table 3		
Material	Total Buffer Range	Rank
		greatest change
		2
		3
		4
		5
		6
		7
		8
		least change

QUESTIONS

1. How should the pH of a material to test in the *Acidic* beaker compare to that in the *Basic* beaker before any acid or base is added? Why?
2. Referring to Question 1, does your data support your hypothesis? If not, what might cause the differences?
3. Generally, what was the effect of adding HCl to each solution? Was this true for every solution? Why do you think this happened the way it did?
4. Generally, what was the effect of adding NaOH to each solution? Was this true for every solution? Why do you think this happened the way it did?
5. Compare the various graphs of each substance. Why was it of value to include the plot of water in acid and water in base with every experiment?
6. Which class of materials, biological organisms, biological chemicals, or non-biological chemicals reacted most dramatically to the addition of acid or base? How does this relate to their complexity?
7. Which of the materials in Table 3 is the best buffer? The poorest buffer?

EXTENSION

1. Bring in common materials from home to test. How do you think they will respond? How did their response compare to your predictions?

Experiment

3

TEACHER INFORMATION

Acids and Bases

1. To prepare the 0.1 M NaOH solution, use 4.0 g of solid NaOH pellets per 1 L of solution. **HAZARD ALERT:** Corrosive solid; skin burns are possible; much heat evolves when added to water; very dangerous to eyes; wear face and eye protection when using this substance. Wear gloves. Hazard Code: B—Hazardous.

2. To prepare the 0.1 M HCl solution, use 8.6 mL of concentrated acid per 1 L of solution. **HAZARD ALERT:** Highly toxic by ingestion or inhalation; severely corrosive to skin and eyes. Hazard Code: A—Extremely hazardous.

   The hazard information reference is: Flinn Scientific, Inc., *Chemical & Biological Catalog Reference Manual, 2000*, (800) 452-1261, www.flinnsci.com. See *Appendix D* of this book, *Biology with Computers*, for more information.

3. Try to make a 1% solution of the materials to test. It is not too critical to be exact. Add ~10 grams of material for each liter of solution.

4. Have the students help design the list of materials to use. Try to keep the three classes of materials balanced—biological organisms or tissues, biological chemicals, and non-biological chemicals.

   Good organisms or tissues to use might include blended liver, plant leaves, potato roots, yeast, fruit juices (from real fruit—not those <10% varieties!) or Euglena (if you culture them). Try to avoid oily materials—they will be difficult to clean off the probe.

   Good chemicals include starch, enzymes, gelatin, vitamin Bs or C, casein, egg white, or other simple, non-oily biochemicals.

   Good non-biological materials include a mix of buffers with non-buffers. Buffers might include soda water, Alka-Seltzer, phosphate buffer, Tums, etc. An interesting combination is aspirin and Bufferin. Good non-buffers include table salt and nitrogen fertilizer. It is fun to include rocks—try marble (calcium carbonate—a buffer in acid) and quartz.

5. The stored calibration will work fine for this experiment. However, you may want to calibrate the pH probes before students use them. You can save this calibration as part of the experiment file. Refer to the teacher's section of experiment 18 for instructions on calibrating a pH Sensor.

6. Vernier Software sells a pH buffer package for preparing buffer solutions with pH values of 4, 7, and 10 (order code PHB). Simply add the capsule contents to 100 mL of distilled water. You can also prepare pH buffers using the following recipes:

   - pH 4.00: Add 2.0 mL of 0.1 M HCl to 1 L of 0.1 M potassium hydrogen phthalate.
   - pH 7.00: Add 582 mL of 0.1 M NaOH to 1 L of 0.1 M potassium dihydrogen phosphate.
   - pH 10.00: Add 214 mL of 0.1 M NaOH to 1 L of 0.05 M sodium bicarbonate.

7. Teams of students need to work together using the same computer and one interface. One team will be adding acid while a cooperating team adds base. They will need to keep synchronized! For example, when the acid group has added 20 drops, the base group must

also have added 20 drops, or the plot will not be meaningful. You may also modify the experiment and use one pH Sensor at each computer.

SAMPLE DATA

Table 1										
Material Tested	Add	pH, after adding this many drops								
		0	5	10	15	20	25	30	ΔpH	Total Buffer Range
tap water	acid	6.81	2.91	2.53	2.35	2.19	2.08	1.99	–4.8	9.3
	base	6.95	10.83	11.08	11.24	11.34	11.42	11.48	4.5	
aspirin	acid	3.1	2.8	2.7	2.6	2.5	2.5	2.4	–0.7	1.3
	base	3.0	3.1	3.2	3.3	3.5	3.5	3.6	0.6	
Bufferin	acid	3.2	3.2	3.2	3.2	3.2	3.3	3.3	0.1	0.8
	base	3.2	3.4	3.6	3.7	3.8	3.9	4.1	0.9	
liver	acid	4.52	4.42	4.38	4.33	4.28	4.22	4.18	–0.3	1.6
	base	4.99	5.07	5.33	5.51	5.71	5.91	6.3	1.3	
egg white	acid	9.52	8.97	8.3	7.62	7.25	6.98	6.8	–2.7	3.6
	base	9.45	9.82	9.88	9.96	10.12	10.25	10.36	0.9	
gelatin	acid	5.75	5.11	4.84	4.68	4.45	4.2	4.06	–1.7	6.1
	base	5.75	7.62	9.18	9.56	9.78	9.99	10.13	4.4	
soda water	acid	4.65	3.36	2.6	2.5	2.4	2.3	2.3	–2.4	3.3
	base	4.45	4.7	4.9	5.0	5.1	5.2	5.3	0.9	
potato	acid	6.14	6.03	5.96	5.87	5.81	5.75	5.69	–0.4	3.1
	base	6.14	6.32	6.7	7.33	7.76	8.46	8.84	2.7	

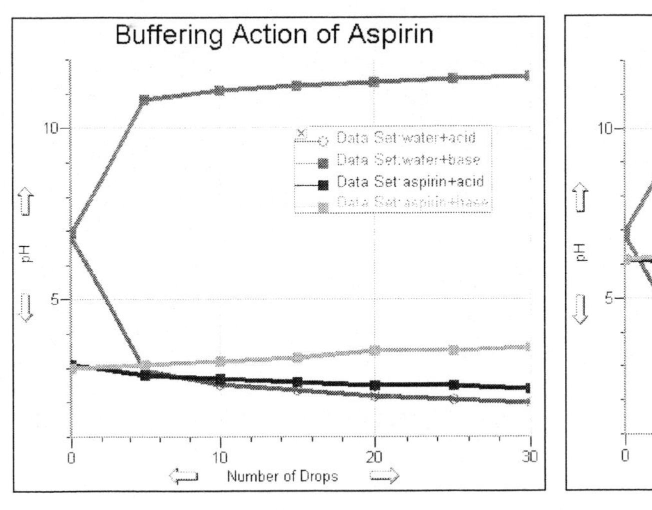

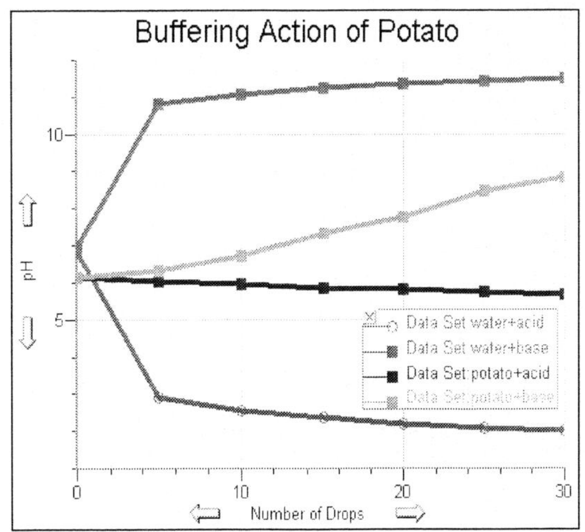

Buffering action of aspirin and potato following the addition of acid and base

Classification of Materials		
Organisms or Tissues	Biological Chemicals	Non-Biological Chemicals
liver	aspirin	Bufferin
potato	gelatin	soda water
egg		water

Table 2		
Material	Initial pH	Rank
aspirin	3.0	most acidic
Bufferin	3.2	2
soda water	4.5	3
liver	4.7	4
gelatin	5.8	5
potato	6.1	6
tap water	6.9	7
egg	9.5	least acidic

Table 3		
Material	Total Buffer Range	Rank
tap water	9.3	worst
gelatin	6.1	2
egg	3.6	3
soda	3.3	4
potato	3.1	5
liver	1.6	6
aspirin	1.3	7
Bufferin	0.8	best

ANSWERS TO QUESTIONS

1. The values should be the same, since the same solution is in each beaker.

2. The actual results may vary. Possible reasons include:
   - The beakers were not cleaned equally well by the cooperative groups.
   - The probes were not calibrated equally, or differed slightly in their response.

3. The effect HCl had on each solution was to decrease its pH. Not all materials responded equally, and several did not respond much at all. These were better buffers.

4. The effect NaOH had on each solution was to increase its pH. Not all materials responded equally, and several did not respond much at all. These were better buffers.

5. Water acted as a control. The similarities and differences among the graphs can be noted more easily when each is compared to a single substance, such as water.

6. Non-biological chemicals, such as water and salt, reacted most dramatically to the addition of acid or base. Biologically complex materials and non-biological buffers resisted pH changes most. The non-biological materials were most divergent in their behavior. This is especially true if any of the graphs have a different scaling than the others.

   The order in which material reacted most dramatically to the addition of acid or base is: water, gelatin, egg white, soda water, potato, liver, aspirin, and Bufferin. The ranking by complexity is similar—water is the simplest material, followed by the two proteins. The soda water has a natural buffer, the bicarbonate ion. Soda water is very simple. Potato roots and livers are cellular material, thus more complex than any of the above. Finally, aspirin and Bufferin are simple chemicals with great buffering capacity.

   As a general rule, simple chemicals may or may not be good buffers, depending upon their make-up. Complex biological materials are almost always better buffers than simple ones, since there are usually a greater number of chemicals in cellular matter that serve as buffers.

7. Answers may vary. See Table 1 for sample values. Of these data, Bufferin is the best buffer and water is the poorest buffer.

Experiment

4

Diffusion through Membranes

Diffusion is a process that allows ions or molecules to move from where they are more concentrated to where they are less concentrated. This process accounts for the movement of many small molecules across a cell membrane. Diffusion is the process by which cells acquire food and exchange waste products. Oxygen, for instance, might diffuse in pond water for use by fish and other aquatic animals. When animals use oxygen, more oxygen will diffuse to replace it from the neighboring environment. Waste products released by aquatic animals are diluted by diffusion and dispersed throughout the pond.

It is important to consider how the rate of diffusion of particles might be affected or altered.

- Diffusion may be affected by the steepness of the concentration gradient (the difference between the number of ions or molecules in one region of a substance and that in an adjoining region). The direction that a diffusing molecule or ion might travel in any particular direction is random. While the particles are diffusing, is there a net movement from where they are concentrated to where they are less concentrated?
- Diffusion might be affected by other different, neighboring particles. For instance, if oxygen diffuses towards a single-celled pond organism at a certain rate, will that rate be altered if some other molecule suddenly surrounded the organism? Would the presence of other molecules block or enhance the diffusion of a molecule, or would the molecule's rate be independent of particles that do not alter the concentration gradient?

One way to measure the rate of diffusion of ions is to monitor their concentration in solution over a period of time. Since ions are electrically charged, water solutions containing ions will conduct electricity. A Conductivity Probe is capable of monitoring ions in solution. This probe however, will not measure the amount of electrically neutral molecules dissolved in water. Salts, such as sodium chloride, produce ions when they dissolve in water. If you place a salt solution in a container such as dialysis tubing, the salt can travel through the very small holes in the tubing. When dialysis tubing containing a solution of salt ions is placed into a beaker of water, the ions can diffuse out of the tubing and into the surrounding water. In this way, you will be able to measure the diffusion of salts in a solution of water and determine how concentration gradients and the presence of other particles affect the diffusion of the salt across a membrane.

OBJECTIVES

In this experiment, you will

- Use a computer and Conductivity Probe to measure the ionic concentration of various solutions.
- Study the effect of concentration gradients on the rate of diffusion.
- Determine if the diffusion rate for a molecule is affected by the presence of a second molecule.

MATERIALS

computer
Vernier computer interface
Logger*Pro*
Vernier Conductivity Probe
three 18 × 150 mm test tubes with rack
1%, 5%, and 10% salt water
400 mL beaker
dialysis tubing, 2.5 cm × 12 cm
dropper pipet or Beral pipet
scissors
stirring rod
5% sucrose (table sugar) solution
dental floss or clamp
ring stand and utility clamp

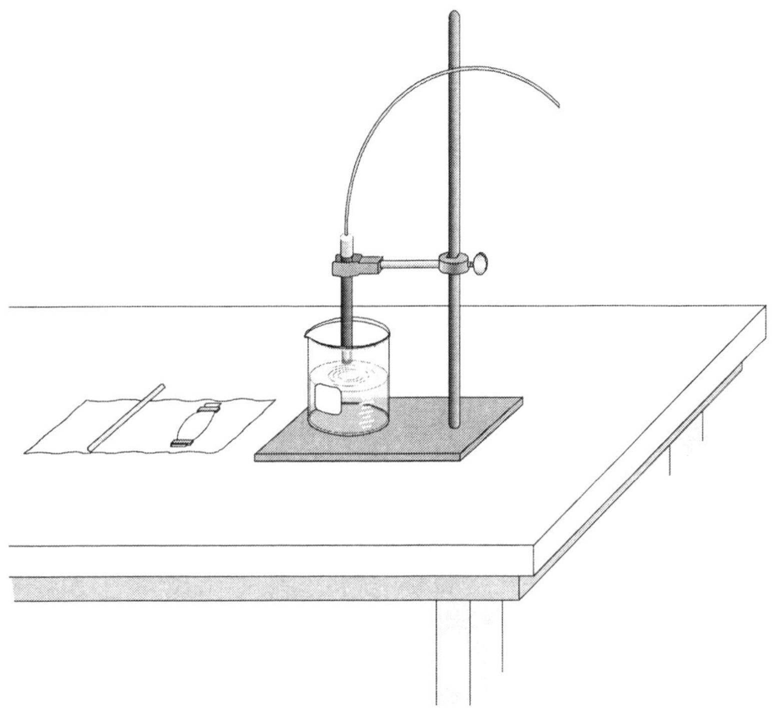

Figure 1

PROCEDURE

1. Connect the Conductivity Probe to the computer interface. Check to be sure the Conductivity Probe is set to the intermediate setting, 2000 μS/cm (equivalent to a concentration of 1000 mg/L).

2. Prepare the computer for data collection by opening the file "04 Membrane Diffusion" from the *Biology with Computers* folder of Logger*Pro*.

Part I Concentration gradients

3. Test whether different concentration gradients affect the rate of diffusion. You will place three solutions of differing salt concentrations (1%, 5%, and 10%) in distilled water. Each salt solution will be placed in a dialysis tube and allowed to diffuse into the surrounding water. When salt diffuses, the conductivity of water in the beaker will increase.

4. In Table 1, predict what you believe will happen in this set of experiments. How will the rate of diffusion change when a 10% salt solution is placed in contact with pure water compared to when a 1% salt solution is placed in contact with pure water?

5. Prepare the dialysis tubing. Obtain a piece of wet dialysis tube and a dialysis tubing clamp or a short (approximately 10 cm) length of dental floss. Using the clamp or floss, tie one end of the tube closed about 1 cm from the end, as shown in Figure 2.

6. Place a 1% salt solution into a section of dialysis tubing. To do this,
   a. Obtain about 15 mL of a 1% salt water solution in a test tube.
   b. Using a funnel or Beral pipet, transfer about 10 mL of the 1% salt water into the dialysis tube, as in Figure 2. Note: To open the tube, you may need to rub the tubing between your fingers a bit.
   c. Tie off the top of the dialysis tube with a clamp or a new length of dental floss. Try not to allow any air into the dialysis tube. The tube should be very firm after it is tied or clamped. Trim off any excess dental floss extending more than 1 cm from either knot.
   d. Wash the outside of the tubing with tap water thoroughly, so that there is no salt water adhering to the tubing.

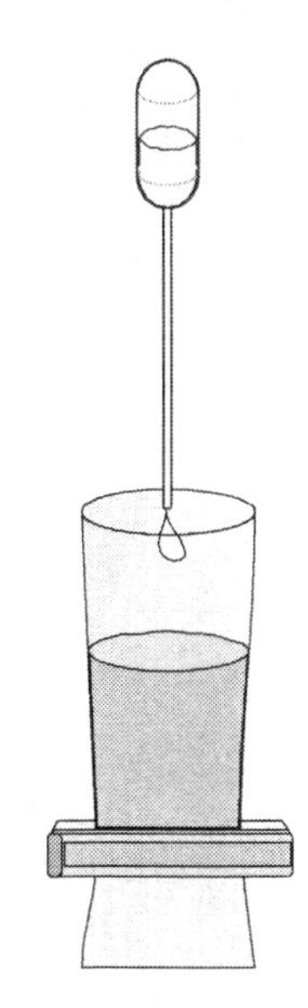

Figure 2

7. Place 300 mL of water into a 400 mL beaker. If the conductivity of the tap water is low (50 mg/L or less), use tap water to fill the beaker. Otherwise, use distilled water.

8. Position the Conductivity Probe into the water as shown in Figure 1. Place the dialysis tube into the water. Be sure the tubing is submerged completely under the water. **Important:** Be sure to position the Conductivity Probe and dialysis tubing the same distance apart in each trial.

9. After stirring the solution for 30 seconds, begin data collection by clicking Collect. Stir the solution slowly and continuously throughout the two-minute data collection period. Data collection will automatically end after two minutes have passed.

10. Determine the rate of diffusion. To do this:
    a. Move the mouse pointer to the point where the data values begin to increase. Hold down the mouse button. Drag the pointer to the end of the data and release the mouse button.
    b. Click the Linear Fit button, , to perform a linear regression. A floating box will appear with the formula for a best fit line.
    c. Record the slope of the line, *m*, as the rate of diffusion in Table 2.
    d. Close the linear regression floating box.

11. Remove one of the clamps. If the dialysis tubing is tied off with floss, use a pair of scissors and carefully cut one of the dental floss knots and discard the floss. If you accidentally make a cut in the tubing, replace it.

12. Empty all of the liquid out of the dialysis tube. Squeeze the excess liquid out with your fingers.

13. Obtain 15 mL of a 5% salt solution in a test tube. Repeat Steps 6 – 12, substituting this 5% salt solution for the 1% solution.

14. Obtain 15 mL of a 10% salt solution in a test tube. Repeat Steps 6 – 12, substituting this 10% salt solution for the 1% solution.

15. Examine your data closely and make a conclusion. Record your conclusion in Table 1.

Part II Effect of other molecules

16. Measure the rate of diffusion of salt while it is in the presence of another, non-conducting solution. Since sugar does not form ions in solution, it should not conduct electricity. Sugar will be added to the water to determine whether it interferes with the diffusion of salt.

17. In Table 1, predict what you believe will happen in this set of experiments. Will the non-conducting sugar in the water block or reduce the diffusion rate of salt? Why?

Test to determine if water or a sugar solution conducts electricity.

18. Place some water in a clean 400 mL beaker.

19. Test the total dissolved solids concentration of the water by placing a clean Conductivity Probe into it. Record the total dissolved solid concentration value in Table 3. The total dissolved solids value should be displayed in the meter at the right of the screen.

20. Obtain 300 mL of a 5% sugar solution in a clean 400 mL beaker.

21. Test the total dissolved solids concentration of the 5% sugar solution by placing a clean Conductivity Probe into it. Record the total dissolved solids value in Table 3. The total dissolved solids value should be displayed in the meter on the right of the screen.

Test if 5% sugar interferes with the diffusion of a 5% salt solution.

22. Repeat Steps 6 – 12, with the following changes:
    - Use 300 mL of sugar water in place of the water in Step 7.
    - Substitute a 5% salt solution for the 1% solution.
    - Record the rate of diffusion in Table 4.

23. Examine your data closely and make a conclusion. Record your conclusion in Table 1.

DATA

Table 1

	Prediction	Conclusion
Part I		
Part II		

Part I

Table 2	
Salt concentration (%)	Rate of diffusion (mg/L/s)
1	
5	
10	

Part II

Table 3	
Solution	Concentration (mg/L)
Distilled water	
Sugar water	

Table 4: Summary of Data	
Solution	Rate of diffusion (mg/L/s)
5% salt	
5% salt / 5% sugar	

QUESTIONS

1. How did you arrive at your conclusion for Part I?
2. How did your conclusion compare to your prediction for Part I? Can you account for any differences?
3. If the rates in any of the three experiments varied in Part I, calculate how much faster each rate was compared to the other. For instance, if the rate of the 1% solution was 50 mg/L/s and the rate of the 10% solution was 250 mg/L/min, then the rate of diffusion for the 10% solution would be (250/50) five times the rate of the 1% salt solution.
4. Compare the ionic concentration of pure water with a sugar water solution. How do you account for this?
5. How did your conclusion compare to your prediction for Part II? Can you account for any differences?

EXTENSIONS

1. Make a plot of the rate of diffusion *vs.* the salt concentration in the dialysis bag. Using your plot, estimate the rate of diffusion of a 3% salt solution.
2. If the results of the experiments in Part I can be extrapolated to diffusion in living systems, how would a single-celled organism respond in an oxygen rich pond compared to an oxygen poor pond? Explain.
3. Design an experiment to determine the effect of temperature on the diffusion of salt. Perform the experiment you designed.

4. Ectotherms are organisms whose body temperature varies with the surrounding environment. On the basis of on your data from Extension Question 3, how do you expect the oxygen consumption of ectotherms to vary as the temperature varies? Explain.

5. If waste products of a single celled organism were released by the organism into the pond, how would that affect the organism's ability to obtain oxygen as readily?

TEACHER INFORMATION

Diffusion through Membranes

1. If the water in your area is very soft, you may want to use tap water instead of distilled water. Test to see if the conductivity of the tap water is less than about 50 mg/L salt.
2. Provide each group with pre-cut, hydrated dialysis tubing. The tubing must be soaked in water for at least ten minutes prior to use. The tubing should be soft and flexible.
3. Use dialysis tubing clamps if at all possible, as this will speed things up greatly. If desired, use dental floss or string to tie off the dialysis tubing. The floss works exceptionally well. You may want to show students how to tie off the dialysis tubes.
4. Have students check their dialysis tubes for leakage. This should be done before each experiment. Leaky tubes should be replaced.
5. Any sugar may be used in Part II. Table sugar is inexpensive and readily available.
6. To prepare 5% sugar solution, add 50 grams of sugar to make one liter of solution (300 mL per group is needed).
7. To prepare 1% salt solution, add 10 grams of NaCl to make one liter of solution (15 mL per group is needed).
8. To prepare 5% salt solution, add 50 grams of NaCl to make one liter of solution (30 mL per group is needed).
9. To prepare 10% salt solution, add 100 grams of NaCl to make one liter of solution (15 mL per group is needed).
10. The stored conductivity calibration (for 0-2000μS/cm) works well for this experiment. The ionic concentration is approximately proportional to the conductivity of the solution.

SAMPLE RESULTS

The following data may be different from students' results.

Part I

Table 2

Salt concentration (%)	Rate of diffusion (mg/L/s)
1	0.25
5	1.23
10	2.19

Part II

Table 3	
Solution	Conductivity (mg/L)
Distilled water	40.3
Sugar water	41.2

Table 4: Summary of Data	
Solution	Rate of diffusion (mg/L/s)
5% salt	1.23
5% salt / 5% sugar	1.21

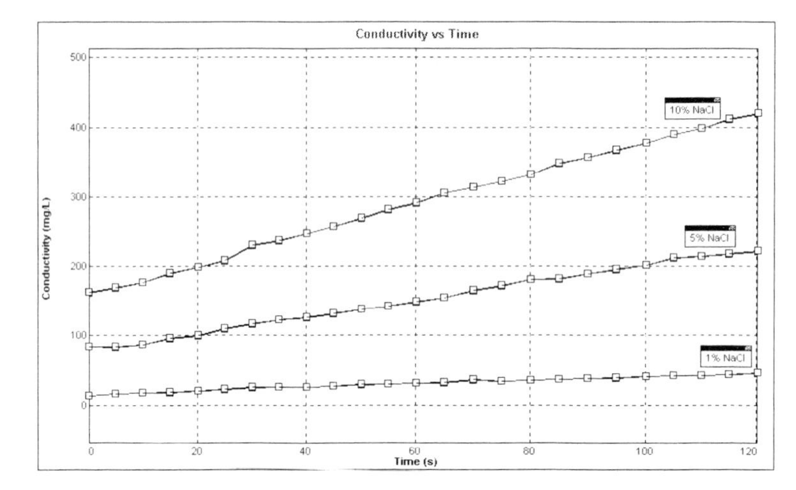

Diffusion through dialysis tubing of differing salt concentrations

ANSWERS TO QUESTIONS

1. Rate of diffusion should increase with increasing salt concentration.
2. The rate of diffusion should increase as the concentration gradient becomes steeper. The rate of the 10% salt solution should be the greatest and the rate of the 1% salt solution should be the lowest of the three.
3. The rate of the 10% salt solution should be approximately ten times that of the 1% solution, while the rate of the 5% salt solution should be five times that of the 1% solution.
4. The conductivity should be the same, as neither will conduct appreciably. Neither molecule is electrically charged.
5. Student answers will vary. The rates of diffusion should be the same.

Experiment

5

Conducting Solutions

In this experiment, you will study the electrical conductivity of water and various water solutions. A solution can contain molecules, ions, or both. Some substances, such as sucrose ($C_{12}H_{22}O_{11}$) and glucose ($C_6H_{12}O_6$), dissolve to give a solution containing mostly molecules. An equation representing the dissolving of sucrose (table sugar) in water is:

$$C_{12}H_{22}O_{11}(s) \longleftrightarrow C_{12}H_{22}O_{11}(aq)$$

where (s) refers to a solid substance and (aq) refers to a substance dissolved in water. Other substances, such as calcium chloride ($CaCl_2$), dissolve in water to produce a solution containing mostly ions. An equation is:

$$CaCl_2(s) \longleftrightarrow Ca^{2+}(aq) + 2\ Cl^-(aq)$$

Calcium ions are necessary for muscle contraction, mitochondrial activity, bone formation, and many other metabolic processes. Organisms may obtain minerals such as calcium from their water supply, since ions dissolve in water.

You will determine conductivity of the solutions using a computer-interfaced Conductivity Probe. The unit of conductivity in this experiment is microsiemens per centimeter, or μS/cm.

OBJECTIVES

In this experiment, you will

- Write equations for the dissolving of substances in water.
- Use a Conductivity Probe to test the electrical conductivity of solutions.
- Determine whether molecules or ions are responsible for electrical conductivity of solutions.

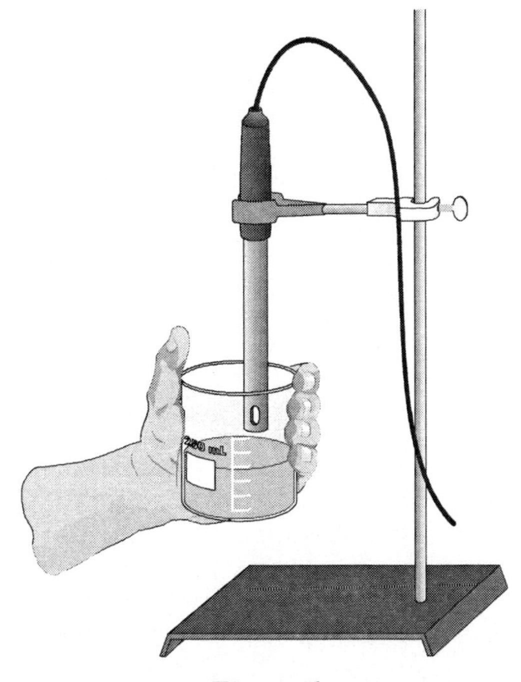

Figure 1

MATERIALS

computer
Vernier computer interface
Vernier Conductivity Probe
Logger*Pro*
sodium chloride, NaCl, solution
calcium chloride, $CaCl_2$, solution
aluminum chloride, $AlCl_3$, solution
ring stand
ethanol, C_2H_6O, solution
sucrose, $C_{12}H_{22}O_{11}$, solution
glucose, $C_6H_{12}O_6$, solution
stream or lake water
ocean water (optional)
various foods in solution
distilled water
utility clamp

PRE-LAB EXERCISES

Many of the materials you will be using today are found in common household items. A list of common names or uses can be found below:

Sodium chloride, NaCl	Common household salt
Calcium chloride, $CaCl_2$	Used to pickle cucumbers, or to help concrete cure in cold weather
Acetic acid, CH_3COOH	Vinegar
Ethanol, C_2H_6O	Found in gasoline or in alcoholic beverages. Usually obtained from yeast fermentation
Fructose, $C_6H_{12}O_6$	Fruit sugar
Sucrose, $C_{12}H_{22}O_{11}$	Table sugar, beet or cane sugar
Glucose, $C_6H_{12}O_6$	Corn or blood sugar

1. An equation representing the dissolving of sucrose in water is:

$$C_{12}H_{22}O_{11}(s) \longleftrightarrow C_{12}H_{22}O_{11}(aq)$$

Like solid sucrose, the substances glucose, $C_6H_{12}O_6(s)$, and ethanol, $C_2H_6O(l)$, dissolve in water to yield solutions containing mostly molecules. Write equations showing the dissolving of these two liquids in water in Table 1.

2. An equation showing the dissolving of $CaCl_2$ in water is:

$$CaCl_2(s) \longleftrightarrow Ca^{2+}(aq) + 2\ Cl^-(aq)$$

Like $CaCl_2$, the substances NaCl and $AlCl_3$ dissolve in water to give solutions containing mostly ions. Write equations showing the dissolving of these two solids in water in Table 2.

PROCEDURE

1. Obtain and wear goggles.
2. Secure the Conductivity Probe with the ring stand and utility clamp as shown in Figure 1.
3. Connect the Conductivity Probe to the computer interface. Check to be sure the Conductivity Probe is set to 0 – 20,000 µS/cm.
4. Prepare the computer for data collection by opening the file “05 Conducting Solutions” from the *Biology with Computers* folder of Logger*Pro*.

5. Test the conductivity of each solution listed in the data table. You can do the tests in any sequence.
   a. Carefully raise each vial and its contents up around the Conductivity Probe until the hole near the probe end is completely submerged in the solution being tested. **Important:** Since the two electrodes are positioned on either side of the hole, this part of the probe must be completely submerged.
   b. Briefly swirl the beaker contents. Once the conductivity reading in the meter has stabilized, record the value in Table 3.
   c. Before testing the next solution, clean the electrodes by surrounding them with a 250 mL beaker and rinse them with distilled water from a wash bottle. Blot the outside of the probe end dry using a tissue. It is *not* necessary to dry the *inside* of the hole near the probe end.

DATA

Table 1	
$C_6H_{12}O_6$(s)	C_2H_6O(l)

Table 2	
NaCl(s)	$AlCl_3$(s)

Table 3		
Solution	Material	Conductivity (μS/cm)
1	Distilled water	
2	Sodium chloride, NaCl	
3	Calcium chloride, $CaCl_2$	
4	Aluminum chloride, $AlCl_3$	
5	Ethanol, C_2H_6O	
6	Sucrose, $C_{12}H_{22}O_{11}$	
7	Glucose, $C_6H_{12}O_6$	
8	Tap water	
9	Stream water	
10	Ocean water	
11		
12		

QUESTIONS

1. Which solutions conduct electricity best, those containing mostly ions or those containing mostly molecules?
2. Does distilled water conduct electricity well? Explain.
3. Does tap water conduct electricity? Account for this observation.
4. Consider the conductivity readings for the NaCl, $CaCl_2$, and $AlCl_3$ solutions. What trend do you observe? Account for this trend.
5. How does the conductivity of ocean water compare to pond or stream water? How can you account for this?
6. Which foods in solution conducted electricity well? How can you account for this?
7. Suggest three other substances whose water solutions would conduct electricity well. Explain how you decided on your choices.

EXTENSIONS

1. Test your predictions for Question 7 above.

TEACHER INFORMATION

Conducting Solutions

1. Two or more sets of the solutions can be made available in small beakers or jars.
2. The solutions can be prepared (using distilled water) as follows:

   0.05 M $NaCl$ (2.93 g/liter)
   0.05 M $AlC1_3$ (6.7 g/liter)
   0.05 M ethanol (2.3 g or 2.9 mL/liter)
   0.05 M sucrose (17.1 g/liter)
   0.05 M $CaCl_2$ (5.55 g/liter)
   0.05 M glucose (9.0 g/liter)

3. A variety of food suspensions may be used. Both plant and animal foods might be considered.
4. To prepare food suspensions, cut the food into small pieces and blend for 5 to 10 seconds, or until finely chopped. Strain the food through cheesecloth and collect the resulting filtrate for testing. This way, students will be testing the resulting dilute solution that will contain varying amounts of ions and molecules. Avoid foods that are high in oil or fat content, as they may leave residues on the electrodes of the Conductivity Probe (see the probe user's guide that was shipped with the probe for further information).
5. Several sources of water can be tested, including stream, tap, ocean, and lake water. Students may want to bring samples in from home to test.
6. The calibration that is stored within the experiment file will work fine for a comparison of different solutions. For more accurate conductivity readings, you (or your students) can do a 2-point calibration for each Conductivity Probe using air (0 conductivity value) and the calibration solution that came with the Conductivity Probe (1000 μS/cm value).
7. If you make measurements of ocean water, you will need to dilute samples to 1/4 of their original concentration by adding 100 mL of the salt-water sample to 300 mL of distilled water. This diluted sample can then be measured using the Conductivity Probe at the high-range setting. Multiply the conductivity reading by 4 to obtain the actual conductivity.

SAMPLE RESULTS

Table 1		
Solution	Material	Conductivity (μS/cm)
1	Distilled water	0
2	Sodium chloride, $NaCl$	5214
3	Calcium chloride, $CaCl_2$	9362
4	Aluminum chloride, $AlCl_3$	11707
5	Ethanol, C_2H_6O	0
6	Sucrose, $C_{12}H_{22}O_{11}$	0
7	Glucose, $C_6H_{12}O_6$	0
8	Tap water	varies (20 – 1000)

ANSWERS TO QUESTIONS

1. The solutions containing mostly ions conduct best.
2. Distilled water does not conduct well because it contains few ions.
3. Tap water does conduct electricity. It contains Ca^{2+} Mg^{2+}, Fe^{3+} CO_3^{-2}, HCO_3^- and other ions that dissolve into water as it flows through and over soil and rocks.
4. The conductivity increases from NaCl through $AlC1_3$ because of the increasing number of ions. A formula unit of NaCl contributes two ions, $CaCl_2$ three ions, and $AlCl_3$ four total ions.
5. Ocean water conducts much more than pond water. It must have many more ions in it than pond water.
6. Answers may vary.
7. Any soluble ionic solid, and some soluble molecular substances, will give a conducting solution. Some common ionic solids that give conducting solutions include
   - The "no-salt" substitute, potassium chloride (KCl).
   - Salt peter, sodium nitrate ($NaNO_3$).
   - Ammonium chloride (NH_4Cl).
   - Epsom salts, magnesium sulfate ($MgSO_4$).
   - Drano®, sodium hydroxide (NaOH).
   - Muratic acid, hydrochloric acid (HCl), is an example of a conducting solution made by dissolving a molecular substance.

ACKNOWLEDGMENT

We would like to thank Dan Holmquist for helping to develop, test, and write this laboratory experiment.

Experiment
6A

Enzyme Action: Testing Catalase Activity

Many organisms can decompose hydrogen peroxide (H_2O_2) enzymatically. Enzymes are globular proteins, responsible for most of the chemical activities of living organisms. They act as *catalysts,* substances that speed up chemical reactions without being destroyed or altered during the process. Enzymes are extremely efficient and may be used over and over again. One enzyme may catalyze thousands of reactions every second. Both the temperature and the pH at which enzymes function are extremely important. Most organisms have a preferred temperature range in which they survive, and their enzymes most likely function best within that temperature range. If the environment of the enzyme is too acidic or too basic, the enzyme may irreversibly *denature,* or unravel, until it no longer has the shape necessary for proper functioning.

H_2O_2 is toxic to most living organisms. Many organisms are capable of enzymatically destroying the H_2O_2 before it can do much damage. H_2O_2 can be converted to oxygen and water, as follows:

$$2\ H_2O_2 \rightarrow 2\ H_2O + O_2$$

Although this reaction occurs spontaneously, enzymes increase the rate considerably. At least two different enzymes are known to catalyze this reaction: *catalase,* found in animals and protists, and *peroxidase,* found in plants. A great deal can be learned about enzymes by studying the rates of enzyme-catalyzed reactions. The rate of a chemical reaction may be studied in a number of ways including:

- measuring the rate of appearance of a product (in this case, O_2, which is given off as a gas)
- measuring the rate of disappearance of substrate (in this case, H_2O_2)
- measuring the pressure of the product as it appears (in this case, O_2).

In this experiment, you will measure the rate of enzyme activity under various conditions, such as different enzyme concentrations, pH values, and temperatures. It is possible to measure the concentration of oxygen gas formed as H_2O_2 is destroyed using an O_2 Gas Sensor. If a plot is made, it may appear similar to the graph shown.

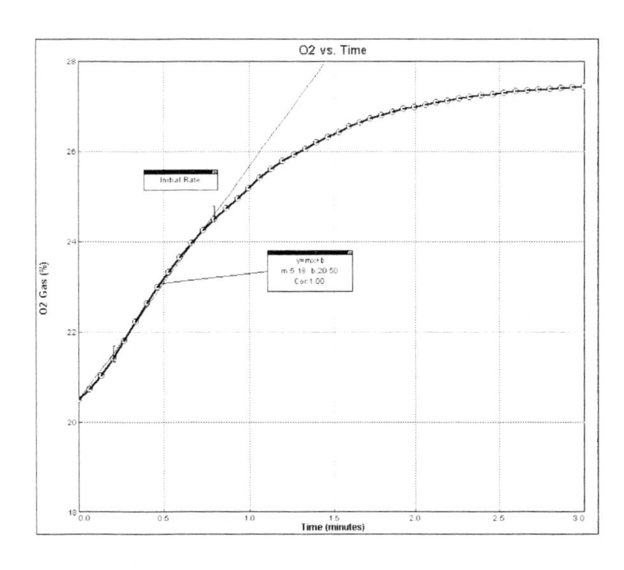

At the start of the reaction, there is no product, and the concentration is the same as the atmosphere. After a short time, oxygen accumulates at a rather constant rate. The slope of the curve at this initial time is constant and is called the *initial rate.* As the peroxide is destroyed, less of it is available to react and the O_2 is produced at lower rates. When no more peroxide is left, O_2 is no longer produced.

OBJECTIVES

In this experiment, you will

- Use a computer and an Oxygen Gas Sensor to measure the production of oxygen gas as hydrogen peroxide is destroyed by the enzyme catalase or peroxidase at various enzyme concentrations.
- Measure and compare the initial rates of reaction for this enzyme when different concentrations of enzyme react with H_2O_2.
- Measure the production of oxygen gas as hydrogen peroxide is destroyed by the enzyme catalase or peroxidase at various temperatures.
- Measure and compare the initial rates of reaction for the enzyme at each temperature.
- Measure the production of oxygen gas as hydrogen peroxide is destroyed by the enzyme catalase or peroxidase at various pH values.
- Measure and compare the initial rates of reaction for the enzyme at each pH value.

Figure 1

MATERIALS

computer
Vernier computer interface
Logger*Pro*
Vernier O_2 Gas Sensor
400 mL beaker
10 mL graduated cylinder
250 mL Nalgene bottle
three dropper pipettes
3.0% H_2O_2
enzyme suspension
three 18 × 150 mm test tubes
ice
pH buffers
test tube rack
thermometer

PROCEDURE

1. Obtain and wear goggles.

2. Connect the Oxygen Gas Sensor to the computer interface. Prepare the computer for data collection by opening the file "06A Enzyme (O2)" from the *Biology with Computers* folder of Logger*Pro*.

Part I Testing the Effect of Enzyme Concentration

3. Place three test tubes in a rack and label them 1, 2, and 3. Fill each test tube with 3 mL of 3.0% H_2O_2 and 3 mL of water.

4. Initiate the enzyme catalyzed reaction.

   a. Using a clean dropper pipette, add 5 drops of enzyme suspension to test tube 1.
   b. Begin timing with a stopwatch or clock.
   c. Cover the opening of the test tube with a finger and gently invert the test tube two times.
   d. Pour the contents of the test tube into a clean 250 mL Nalgene bottle.
   e. Place the O_2 Gas Sensor into the bottle as shown in Figure 1. Gently push the sensor down into the bottle until it stops. The sensor is designed to seal the bottle without the need for unnecessary force.
   f. When 30 seconds has passed, Click Collect to begin data collection.

5. When data collection has finished, remove the O_2 gas sensor from the Nalgene bottle. Rinse the bottle with water and dry with a paper towel.

6. Move your data to a stored run. To do this, choose Store Latest Run from the Experiment menu.

7. Collect data for test tubes 2 and 3:

   - Add 10 drops of the enzyme solution to test tube 2. Repeat Steps 4 – 6.
   - Add 20 drops of the enzyme solution to test tube 3. Repeat Steps 4 – 5.

8. Using the mouse, select the initial linear region of your data on the graph. Click on the Linear Fit button, . Click OK and a best-fit linear regression line will be shown for each run selected. In your data table, record the value of the slope, *m*, for each of the three solutions. (The linear regression statistics are displayed in a floating box for each of the data sets.)

9. To print a graph of concentration *vs.* volume showing all three data runs:

   a. Label all three curves by choosing Text Annotation from the Insert menu, and typing "5 Drops" (or "10 Drops", or "20 Drops") in the edit box. Then drag each box to a position near its respective curve. Adjust the position of the arrow head.
   b. Print a copy of the graph, with all three data sets and the regression lines displayed. Enter your name(s) and the number of copies of the graph you want.

10. Determine the rate of reaction for each of the time intervals listed in Table 3 using the procedure outlined in Step 8. Record the rates for all three data runs in the Table 3.

Part II Testing the Effect of Temperature

Your teacher will assign a temperature range for your lab group to test. Depending on your assigned temperature range, set up your water bath as described below. Place a thermometer in your water bath to assist in maintaining the proper temperature.

- 0 – 5°C: 400 mL beaker filled with ice and water.
- 20 – 25°C: No water bath needed to maintain room temperature.
- 30 – 35°C: 400 mL beaker filled with very warm water.
- 50 – 55°C: 400 mL beaker filled with hot water.

11. Rinse the three numbered test tubes used for Part I. Fill each test tube with 3 mL of 3.0% H_2O_2 and 3 mL of water. Place the test tubes in the water bath. The test tubes should be in the water bath for 5 minutes before proceeding to Step 12. Record the temperature of the water bath, as indicated on the thermometer, in the space provided in Table 4.
12. Find the rate of enzyme activity for test tubes 1, 2, and 3:
    - Add 10 Drops of the enzyme solution to test tube 1. Repeat Steps 4 – 6.
    - Add 10 drops of the enzyme solution to test tube 2. Repeat Steps 4 – 6.
    - Add 10 drops of the enzyme solution to test tube 3. Repeat Steps 4 – 5.
13. Repeat Step 8 and record the reaction rate for each data set in Table 4. Calculate and record the average rate in Table 4.
14. Record the average rate and the temperature of your water bath from Table 4 on the class data table. When the entire class has reported their data, record the class data in Table 5.

Part III Testing the Effect of pH

15. Place three clean test tubes in a rack and label them pH 4, pH 7, and pH 10.
16. Add 3 mL of 3% H_2O_2 and 3 mL of a pH buffer to each test tube, as in Table 1.

Table 1		
pH of buffer	Volume of 3% H_2O_2 (mL)	Volume of buffer (mL)
pH 4	5	5
pH 7	5	5
pH 10	5	5

17. Using the test tube labeled pH 4, add 10 drops of enzyme solution and repeat Steps 4 – 6.
18. Using the test tube labeled pH 7, add 10 drops of enzyme solution and repeat Steps 4 – 6.
19. Using the test tube labeled pH 10, add 10 drops of enzyme solution and repeat Steps 4 – 5.
20. Repeat Steps 8 and 9 to calculate the rate of reaction and print your graph. Record the reaction rate for each pH value in Table 6.

DATA

Part I Effect of Enzyme Concentration

Table 2	
Test tube label	Slope, or rate (%/min)
5 Drops	
10 Drops	
20 Drops	

Table 3 Time intervals (Minutes)					
Rates	0-0.5 min	0.5-1.0 min	1.0-1.5 min	1.5-2.0 min	2.0-3.0 min
5 Drops					
10 Drops					
20 Drops					

Part II Effect of Temperature

Table 4	
Test tube label	Slope, or rate (%/min)
Trial 1	
Trial 2	
Trial 3	
Average	
Temperature range:____°C	

Table 5 (Class Data)	
Temperature tested	Average rate

Part III Effect of pH

Table 6	
Test tube label	Slope, or rate (%/min)
pH 4	
pH 7	
pH 10	

PROCESSING THE DATA

1. On Page 2 of this experiment file, create a graph of the rate of enzyme activity *vs.* temperature. Plot the rate values for the class data in Table 5 on the y-axis, and the temperature on the x-axis. Use this graph to answer the questions for Part II.

QUESTIONS

Part I Effect of Enzyme Concentration

1. How does changing the concentration of enzyme affect the rate of decomposition of H_2O_2?
2. What do you think will happen to the rate of reaction if one increases the concentration of enzyme to twenty-five drops? Predict what the rate would be for 25 drops.
3. When is the reaction rate highest? Explain why.
4. When is the reaction rate lowest? Why?

Part II Effect of Temperature

5. At what temperature is the rate of enzyme activity the highest? Lowest? Explain.
6. How does changing the temperature affect the rate of enzyme activity? Does this follow a pattern you anticipated?
7. Why might the enzyme activity decrease at very high temperatures?

Part III Effect of pH

8. At what pH is the rate of enzyme activity the highest? Lowest?
9. How does changing the pH affect the rate of enzyme activity?

EXTENSIONS

1. Different organisms often live in very different habitats. Design a series of experiments to investigate how different types of organisms might affect the rate of enzyme activity. Consider testing a plant, an animal, and a protist.
2. Presumably, at higher concentrations of H_2O_2, there is a greater chance that an enzyme molecule might collide with H_2O_2. If so, the concentration of H_2O_2 might alter the rate of oxygen production. Design a series of experiments to investigate how differing concentrations of the substrate hydrogen peroxide might affect the rate of enzyme activity.
3. Design an experiment to determine the effect of boiling the catalase on the rate of reaction.
4. Explain how environmental factors affect the rate of enzyme-catalyzed reactions.

Enzyme Action: Testing Catalase Activity

1. This experiment may take a single group several lab periods to complete. A good breaking point is after the completion of Step 10, when students have tested the effect of different enzyme concentrations. Alternatively, if time is limited, different groups can be assigned one of the three tests and the data can be shared.
2. Your hot tap water may be in the range of 50-55°C for the hot-water bath. If not, you may want to supply pre-warmed temperature baths for Part II, where students need to maintain very warm water. Warn students not to touch the hot water.
3. Many different organisms may be used as a source of catalase in this experiment. If enzymes from an animal, a protist, and a plant are used by different teams in the same class, it will be possible to compare the similarities and differences among those organisms. Often, either beef liver, beef blood, or living yeast are used.
4. To prepare the yeast solution, dissolve 7 g (1 package) of dried yeast per 100 mL of 2% glucose solution. A 2% glucose is made by adding 20 g of glucose to enough distilled water to make 1 L of solution. Incubate the suspension in 37 – 40°C water for at least 10 minutes to activate the yeast. Test the experiment before the students begin. The yeast may need to be diluted if the reaction occurs too rapidly. The reaction in Step 4, with 6 mL of 1.5% hydrogen peroxide, and 5 drops of suspension produces enough oxygen to exceed a measured concentration of 22% in 40 to 60 seconds.
5. To prepare a liver suspension, homogenize 0.5 to 1.5 g of beef liver in 100 mL of cold water. You will need to test the suspension before use, as its activity varies greatly depending on its freshness. Dilute the suspension until the reaction in Step 4, with 6 mL of 1.5% hydrogen peroxide, and 5 drops of suspension produces enough oxygen to exceed a measured concentration of 22 % in 40 to 60 seconds. The color of the suspension will be a faint pink. Keep the suspension on ice until used in an experiment.
6. 3% H_2O_2 may be purchased from any supermarket. If refrigerated, bring it to room temperature before starting the experiment.
7. To extend the life of the O_2 Gas Sensor, always store the sensor upright in the box in which it was shipped.
8. Vernier Software sells a pH buffer package for preparing buffer solutions with pH values of 4, 7, and 10 (order code PHB). Simply add the capsule contents to 100 mL of distilled water.
9. You can also prepare pH buffers using the following recipes:
   - pH 4: Add 2.0 mL of 0.1 M HCl to 1000 mL of 0.1 M potassium hydrogen phthalate.
   - pH 7: Add 582 mL of 0.1 M NaOH to 1000 mL of 0.1 M potassium dihydrogen phosphate.
   - pH 10: Add 214 mL of 0.1 M NaOH to 1000 mL of 0.05 M sodium bicarbonate.
10. You may need to let students know that at pH values above 10, enzymes will become denatured and the rate of activity will drop. If you have pH buffers higher than 10, have students perform an experimental run using them.

SAMPLE RESULTS

Sample class data	
Test tube label	Slope, or rate (%/min)
5 Drops	0.27
10 Drops	0.73
20 Drops	1.59
0 – 5 °C range: 4°C	0.58
20 – 25 °C range: 21 °C	0.82
30 – 35 °C range: 34°C	1.43
50 – 55 °C range: 51°C	0.36
pH 4	0.36
pH 7	0.89
pH 10	0.97

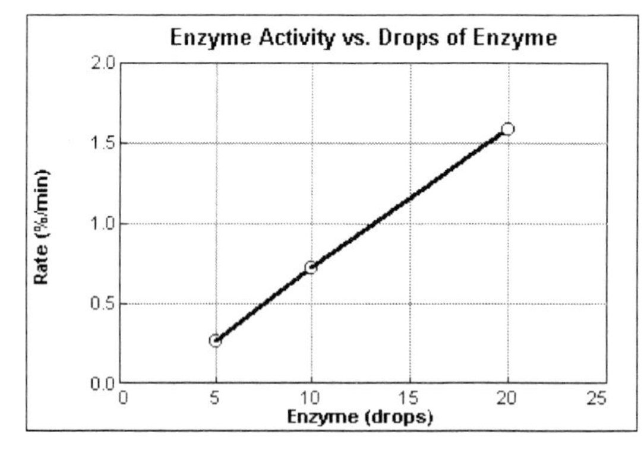

The effect of H_2O_2 concentration on the rate of enzyme activity

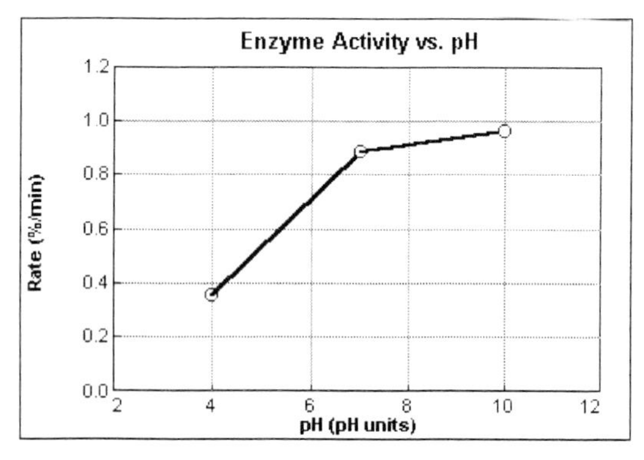

The effect of pH on the rate of enzyme activity

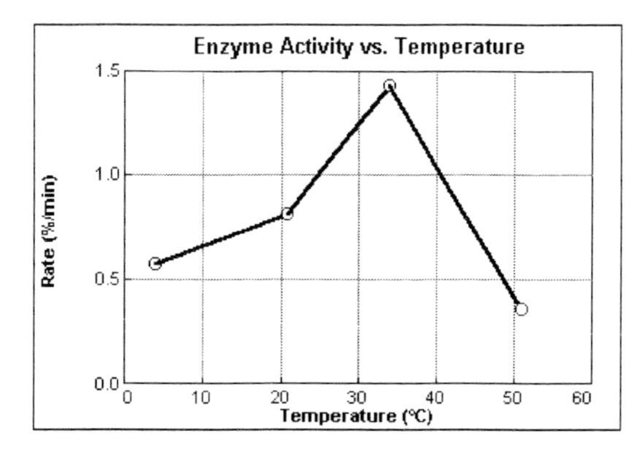

The effect of temperature on the rate of enzyme activity

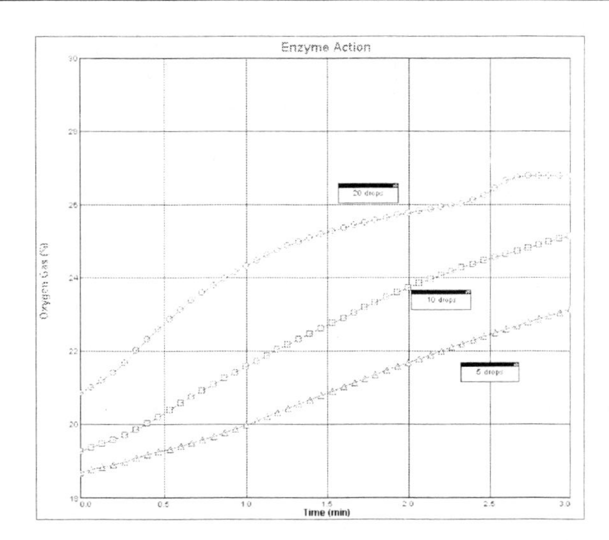

Sample Data: Effect of H_2O_2 concentration on the rate of enzyme activity.

ANSWERS TO QUESTIONS

1. The rate should be highest when the concentration of enzyme is highest. With higher concentration of enzyme, there is a greater chance of an effective collision between the enzyme and H_2O_2 molecule.

2. Roughly, the rate doubles when the concentration of enzyme doubles. Since the data are somewhat linear, the rate is proportional to the concentration. At a concentration of 25 drops, the rate would be about 2.42 %/min.

3. Student answers vary. Ideal data would have the rate being the highest during the first (and maybe second) interval. This is because there are a large number of substrate molecules in comparison to the number of enzyme molecules and there will be a maximum number of collisions between the enzyme and the substrate

4. Student answers vary. Ideal data would have the rate being lowest (the rate would be zero or would approach zero) during the last intervals. As the number of substrate molecules decreases and the number of product molecules increases, the number of collisions between the enzyme and the substrate decreases.

5. The temperature at which the rate of enzyme activity is the highest should be close to 30°C. The lowest rate of enzyme activity should be at 60°C.

6. The rate increases as the temperature increases, until the temperature reaches about 50°C. Above this temperature, the rate decreases.

7. At high temperatures, enzymes lose activity as they are denatured.

8. Student answers may vary. Activity is usually highest at pH 10 and lowest at pH 4.

9. Student answers may vary. Usually, the enzyme activity increases from pH 4 to 10. At low pH values, the protein may denature or change its structure. This may affect the enzyme's ability to recognize a substrate or it may alter its polarity within a cell.

Experiment

6B

Enzyme Action: Testing Catalase Activity

Many organisms can decompose hydrogen peroxide (H_2O_2) enzymatically. Enzymes are globular proteins, responsible for most of the chemical activities of living organisms. They act as *catalysts,* as substances that speed up chemical reactions without being destroyed or altered during the process. Enzymes are extremely efficient and may be used over and over again. One enzyme may catalyze thousands of reactions every second. Both the temperature and the pH at which enzymes function are extremely important. Most organisms have a preferred temperature range in which they survive, and their enzymes most likely function best within that temperature range. If the environment of the enzyme is too acidic or too basic, the enzyme may irreversibly *denature*, or unravel, until it no longer has the shape necessary for proper functioning.

H_2O_2 is toxic to most living organisms. Many organisms are capable of enzymatically destroying the H_2O_2 before it can do much damage. H_2O_2 can be converted to oxygen and water, as follows:

$$2\ H_2O_2 \rightarrow 2\ H_2O + O_2$$

Although this reaction occurs spontaneously, enzymes increase the rate considerably. At least two different enzymes are known to catalyze this reaction: *catalase,* found in animals and protists, and *peroxidase*, found in plants. A great deal can be learned about enzymes by studying the rates of enzyme-catalyzed reactions. The rate of a chemical reaction may be studied in a number of ways including:

- measuring the pressure of the product as it appears (in this case, O_2)
- measuring the rate of disappearance of substrate (in this case, H_2O_2)
- measuring the rate of appearance of a product (in this case, O_2 which is given off as a gas)

In this experiment, you will measure the rate of enzyme activity under various conditions, such as different enzyme concentrations, pH values, and temperatures. It is possible to measure the pressure of oxygen gas formed as H_2O_2 is destroyed. If a plot is made, it may appear similar to the graph shown.

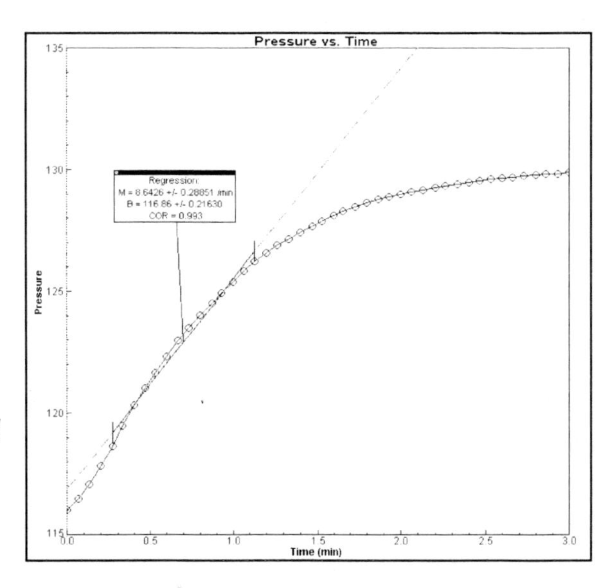

At the start of the reaction, there is no product, and the pressure is the same as the atmospheric pressure. After a short time, oxygen accumulates at a rather constant rate. The slope of the curve at this initial time is constant and is called the *initial rate*. As the peroxide is destroyed, less of it is available to react and the O_2 is produced at lower rates. When no more peroxide is left, O_2 is no longer produced.

OBJECTIVES

In this experiment, you will

- Use a computer and Gas Pressure Sensor to measure the production of oxygen gas as hydrogen peroxide is destroyed by the enzyme catalase or peroxidase at various enzyme concentrations.
- Measure and compare the initial rates of reaction for this enzyme when different concentrations of enzyme react with H_2O_2.
- Measure the production of oxygen gas as hydrogen peroxide is destroyed by the enzyme catalase or peroxidase at various temperatures.
- Measure and compare the initial rates of reaction for the enzyme at each temperature.
- Measure the production of oxygen gas as hydrogen peroxide is destroyed by the enzyme catalase or peroxidase at various pH values.
- Measure and compare the initial rates of reaction for the enzyme at each pH value.

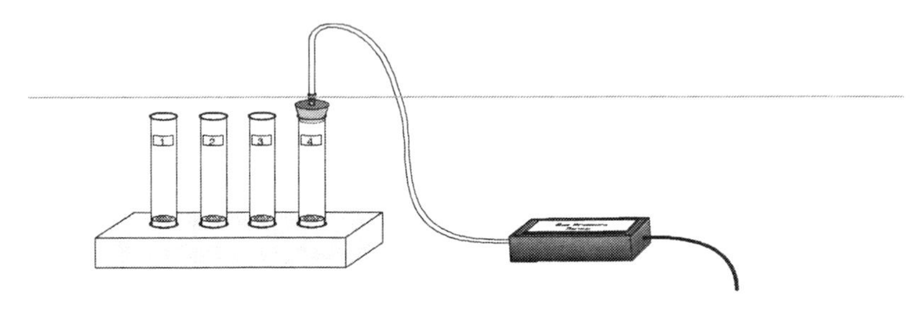

Figure 1

MATERIALS

computer
Vernier computer interface
Logger*Pro*
Vernier Gas Pressure Sensor
1-hole rubber stopper assembly
10 mL graduated cylinder
250 mL beaker of water
3% H_2O_2
600 mL beaker
enzyme suspension
four 18 x 150 mm test tubes
ice
pH buffers
test tube rack
thermometer
three dropper pipettes

PROCEDURE

1. Obtain and wear goggles.
2. Connect the Gas Pressure Sensor to the computer interface. Prepare the computer for data collection by opening the file "06B Enzyme (Pressure)" from the *Biology with Computers* folder of Logger*Pro*.
3. Connect the plastic tubing to the valve on the Gas Pressure Sensor.

Part I Testing the Effect of Enzyme Concentration

4. Place four test tubes in a rack and label them 1, 2, 3, and 4. Partially fill a beaker with tap water for use in Step 5.
5. Add 3 mL of water and 3 mL of 3% H_2O_2 to each test tube.

6. Using a clean dropper pipette, add 1 drop of enzyme suspension to Test Tube 1. **Note**: Be sure not to let the enzyme fall against the side of the test tube.

Table 1		
Test tube label	Volume of 3% H_2O_2 (mL)	Volume of water (mL)
1	3	3
2	3	3
3	3	3
4	3	3

7. Stopper the test tube and gently swirl to thoroughly mix the contents. The reaction should begin. The next step should be completed as rapidly as possible.

8. Connect the free-end of the plastic tubing to the connector in the rubber stopper as shown in Figure 3. Click Collect to begin data collection. Data collection will end after 3 minutes.

9. If the pressure exceeds 130 kPa, the pressure inside the tube will be too great and the rubber stopper is likely to pop off. Disconnect the plastic tubing from the Gas Pressure Sensor if the pressure exceeds 130 kPa.

10. When data collection has finished, disconnect the plastic tubing connector from the rubber stopper. Remove the rubber stopper from the test tube and discard the contents in a waste beaker.

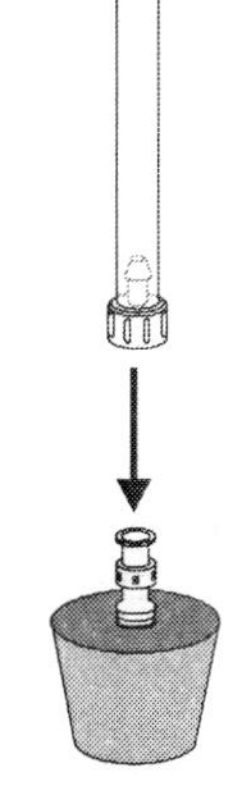

Figure 2

11. Find the rate of enzyme activity:
    a. Move the mouse pointer to the point where the data values begin to increase. Hold down the mouse button. Drag the mouse pointer to the point where the pressure values no longer increase and release the mouse button.
    b. Click the Linear Fit button, , to perform a linear regression. A floating box will appear with the formula for a best-fit line.
    c. Record the slope of the line, *m*, as the rate of enzyme activity in Table 4.
    d. Close the linear regression floating box.

12. Find the rate of enzyme activity for test tubes 2 – 4:
    a. Add 2 drops of the enzyme solution to test tube 2. Repeat Steps 7 – 11.
    b. Add 3 drops of the enzyme solution to test tube 3. Repeat Steps 7 – 11.
    c. Add 4 drops of the enzyme solution to test tube 4. Repeat Steps 7 – 11.

Part II Testing the Effect of Temperature

13. Place four clean test tubes in a rack and label them T 0–5, T 20–25, T 30–35, and T 50–55.

14. Add 3 mL of 3% H_2O_2 and 3 mL of water to each test tube, as shown in Table 2.

Table 2		
Test tube label	Volume of 3% H_2O_2 (mL)	Volume of water
T 0 – 5	3	3
T 20 – 25 (room temp)	3	3
T 30 – 35	3	3
T 50 – 55	3	3

15. Measure the enzyme activity at 0 – 5°C:
    a. Prepare a water bath at a temperature in the range of 0 – 5°C by placing ice and water in a 600 mL beaker. Check that the temperature remains in this range throughout this test.
    b. Place Test Tube T 0 – 5 in the cold water bath until the temperature of the mixture reaches a temperature in the 0 – 5°C range. Record the actual temperature of the test-tube contents in the blank in Table 4.
    c. Add 2 drops of the enzyme solution to Test Tube T 0 – 5. Repeat Steps 7 – 11.
16. Measure the enzyme activity at 30 – 35°C:
    a. Prepare a water bath at a temperature in the range of 30 – 35°C by placing warm water in a 600 mL beaker. Check that the temperature remains in this range throughout this test.
    b. Place Test Tube T 30 – 35 in the warm water bath until the temperature of the mixture reaches a temperature in the 30 – 35°C range. Record the actual temperature of the test-tube contents in the blank in Table 4.
    c. Add 2 drops of the enzyme solution to Test Tube T 30 – 35. Repeat Steps 7 – 11.
17. Measure the enzyme activity at 50 – 55°C:
    a. Prepare a water bath at a temperature in the range of 50 – 55°C by placing hot water in a 600 mL beaker (hot tap water will probably work fine). Check that the temperature remains in this range throughout this test.
    b. Place Test Tube T 50 – 55 in the warm water bath until the temperature of the mixture reaches a temperature in the 50 – 55°C range. Record the actual temperature of the test-tube contents in the blank in Table 4.
    c. Add 2 drops of the enzyme solution to Test Tube T 50 – 55. Repeat Steps 7 – 11.
18. Measure the enzyme activity at 20 – 25°C (room temperature):
    a. Record the temperature of Test Tube T 20 – 25 in Table 4.
    b. In the tube labeled T 20 – 25, add 2 drops of the enzyme solution. Repeat Steps 7 – 11.

Part III Testing the Effect of pH

19. Place three clean test tubes in a rack and label them pH 4, pH 7, and pH 10.

20. Add 3 mL of 3% H_2O_2 and 3 mL of each pH buffer to each test tube, as in Table 3.

Table 3		
pH of buffer	Volume of 3% H_2O_2 (mL)	Volume of buffer (mL)
pH 4	3	3
pH 7	3	3
pH 10	3	3

21. In the tube labeled pH 4, add 2 drops of the enzyme solution. Repeat Steps 7 – 11.

22. In the tube labeled pH 7, add 2 drops of the enzyme solution. Repeat Steps 7 – 11.

23. In the tube labeled pH 10, add 2 drops of the enzyme solution. Repeat Steps 7 – 11.

DATA

Table 4	
Test tube label	Slope, or rate (kPa/min)
1 Drop	
2 Drops	
3 Drops	
4 Drops	
0 – 5°C range: _____ °C	
20 – 25°C range: _____ °C	
30 – 35°C range: _____ °C	
50 – 55°C range: _____ °C	
pH 4	
pH 7	
pH 10	

PROCESSING THE DATA

Enzyme concentration plot

1. On Page 2 of this experiment file, create a graph of the rate of enzyme activity *vs.* enzyme concentration. The rate values should be plotted on the y-axis, and the number of drops of enzyme on the x-axis. The rate values are the same as the slope values in Table 4.

Temperature plot

2. On Page 3 of this experiment file, create a graph of the rate of enzyme activity *vs.* temperature. The rate values should be plotted on the y-axis, and the temperature on the x-axis. The rate values are the same as the slope values in Table 4.

pH plot

3. On Page 4 of this experiment file, create a graph of rate of enzyme activity *vs.* pH. The rate values should be plotted on the y-axis, and the pH on the x-axis. The rate values are the same as the slope values in Table 4.

QUESTIONS

Part I Effect of Enzyme Concentration

1. How does changing the concentration of enzyme affect the rate of decomposition of H_2O_2?

2. What do you think will happen to the rate of reaction if the concentration of enzyme is increased to five drops? Predict what the rate would be for 5 drops.

Part II Effect of Temperature

3. At what temperature is the rate of enzyme activity the highest? Lowest? Explain.

4. How does changing the temperature affect the rate of enzyme activity? Does this follow a pattern you anticipated?

5. Why might the enzyme activity decrease at very high temperatures?

Part III Effect of pH

6. At what pH is the rate of enzyme activity the highest? Lowest?

7. How does changing the pH affect the rate of enzyme activity? Does this follow a pattern you anticipated?

EXTENSIONS

1. Different organisms often live in very different habitats. Design a series of experiments to investigate how different types of organisms might affect the rate of enzyme activity. Consider testing a plant, an animal, and a protist.

2. Presumably, at higher concentrations of H_2O_2, there is a greater chance that an enzyme molecule might collide with H_2O_2. If so, the concentration of H_2O_2 might alter the rate of oxygen production. Design a series of experiments to investigate how differing concentrations of the substrate hydrogen peroxide might affect the rate of enzyme activity.

3. Design an experiment to determine the effect of boiling the catalase on the reaction rate.

4. Explain how environmental factors affect the rate of enzyme-catalyzed reactions.

TEACHER INFORMATION

Enzyme Action: Testing Catalase Activity

1. This experiment may take a single group several lab periods to complete. A good breaking point is after the completion of Step 12, when students have tested the effect of different enzyme concentrations. Alternatively, if time is limited, different groups can be assigned one of the three tests and the data can be shared.

2. Your hot tap water may be in the range of 50-55°C for the hot-water bath. If not, you may want to supply pre-warmed temperature baths for Step 17, where students need to maintain very warm water. Warn students not to touch the hot water.

3. Many different organisms may be used as a source of catalase in this experiment. If enzymes from an animal, a protist, and a plant are used by different teams in the same class, it will be possible to compare the similarities and differences among those organisms. Often, either beef liver, beef blood, or living yeast are used.

4. To prepare the yeast solution, dissolve 7 g (1 package) of dried yeast per 100 mL of 2% glucose solution. Incubate the suspension in 37 – 40°C water for at least 10 minutes to activate the yeast. Test the experiment before the students begin. The yeast may need to be diluted if the reaction occurs too rapidly. The reaction in Step 12, with 3 mL of 3% hydrogen peroxide, 3 mL of water, and 2 drops of suspension should produce a pressure of 1.3 atmospheres in 40 to 60 seconds.

   To prepare a 2% sugar solution, add 20 grams of sugar to make one liter of solution (100 mL per group is needed).

5. To prepare a liver suspension, homogenize 0.5 to 1.5 g of beef liver in 100 mL of cold water. You will need to test the suspension before use, as its activity varies greatly depending on its freshness. Dilute the suspension until the reaction in Step 12, with 3 mL of 3% hydrogen peroxide, 3 mL of water, and 2 drops of suspension produces a pressure of 130 kPa in 40 to 60 seconds. The color of the suspension will be a faint pink. Keep the suspension on ice until used in an experiment.

6. 3% H_2O_2 may be purchased from any supermarket. If refrigerated, bring it to room temperature before starting the experiment.

7. Emphasize to your students the importance of providing an airtight fit with all plastic-tubing connections and when closing valves or twisting the stopper into a test tube.

8. All of the pressure valves, tubing, and connectors used in this experiment are included with Vernier Gas Pressure Sensors shipped after February 15, 1998. These accessories are also helpful when performing respiration/fermentation experiments such as Experiments 11C, 12B, and 16B in this manual.

   If you purchased your Gas Pressure Sensor at an earlier date, Vernier has a Pressure Sensor Accessories Kit (PS-ACC) that includes all of the parts shown here for doing pressure-related experiments. Using this kit allows for easy assembly of a completely airtight system. The kit includes the following parts:

   - two ribbed, tapered valve connectors inserted into a No. 5 rubber stopper
   - one ribbed, tapered valve connectors inserted into a No. 1 rubber stopper
   - two Luer-lock connectors connected to either end of a piece of plastic tubing
   - one two-way valve
   - one 20-mL syringe
   - two tubing clamps for transpiration experiments

9. The accessory items used in this experiment are the #1 single hole stopper fitted with a tapered valve connector and the section of plastic tubing fitted with Luer-lock connectors.

10. The length of plastic tubing connecting the rubber stopper assemblies to each gas pressure sensor must be the same for all groups. It is best to keep the length of tubing reasonably small to keep the volume of gas in the test tube low. **Note:** If pressure changes during data collection are too small, you may need to decrease the total gas volume in the system. Shortening the length of tubing used will help to decrease the volume.

11. If the Vernier Gas Pressure Sensor or Biology Gas Pressure Sensor is unavailable, the Vernier Pressure Sensor may be used as an alternative.

12. Vernier Software sells a pH buffer package for preparing buffer solutions with pH values of 4, 7, and 10 (order code PHB). Simply add the capsule contents to 100 mL of distilled water.

13. You can also prepare pH buffers using the following recipes:

   - pH 4: Add 2.0 mL of 0.1 M HCl to 1000 mL of 0.1 M potassium hydrogen phthalate.
   - pH 7: Add 582 mL of 0.1 M NaOH to 1000 mL of 0.1 M potassium dihydrogen phosphate.
   - pH 10: Add 214 mL of 0.1 M NaOH to 1000 mL of 0.05 M sodium bicarbonate.

14. You may need to let students know that at pH values above 10 enzymes will become denatured and the rate of activity will drop. If you have pH buffers higher than 10, have students perform an experimental run using them.

SAMPLE RESULTS

Table 5	
Test tube label	Slope, or rate (kPa/min)
1 Drop	10.23
2 Drops	44.98
3 Drops	59.36
4 Drops	98.26
0 – 5 °C range: 4°C	41.43
20 – 25 °C range: 21°C	48.02
30 – 35 °C range: 34°C	73.85
50 – 55 °C range: 51°C	27.55
pH 4	36.57
pH 7	66.86
pH 10	75.27

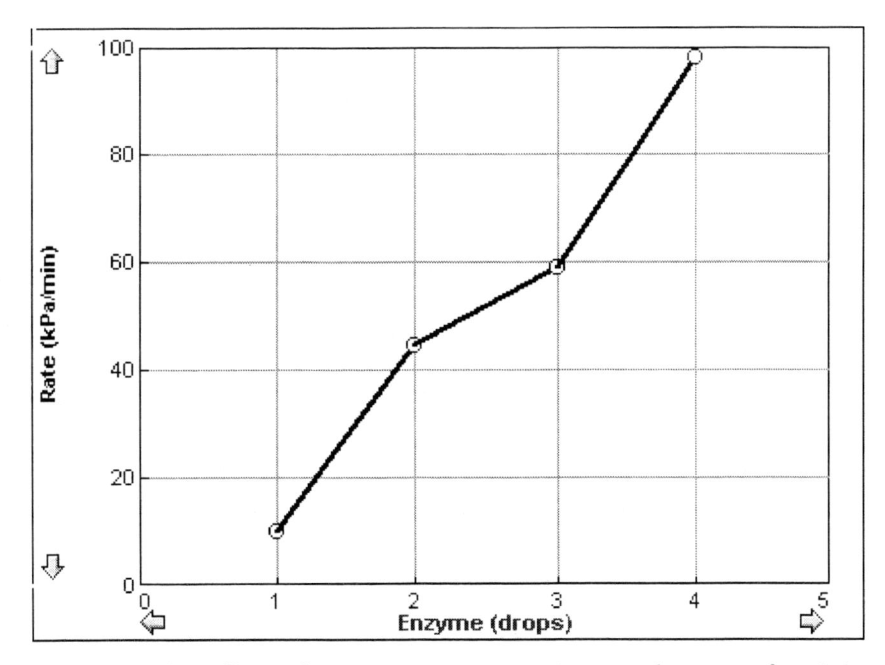

Figure 1: The effect of enzyme concentration on the rate of activity.

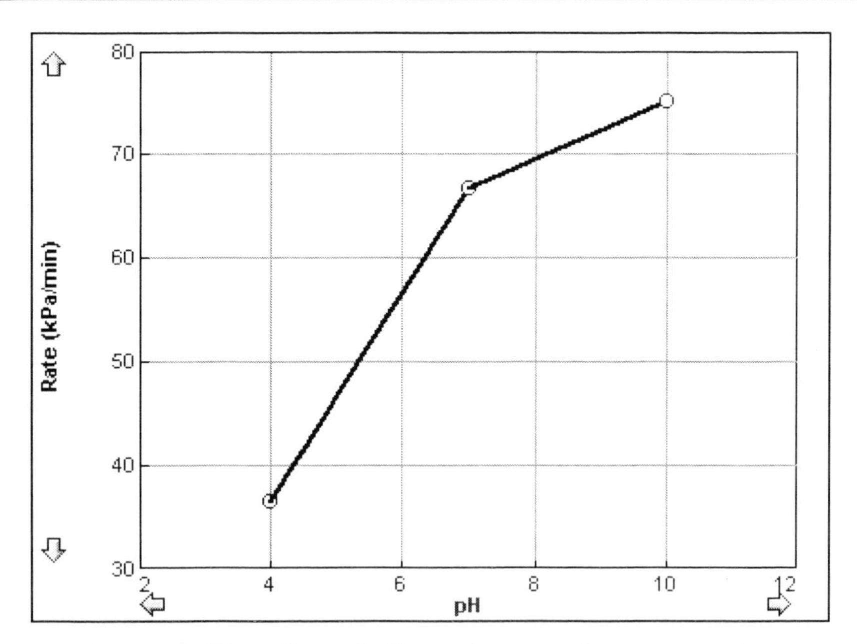

Figure 2: The effect of pH on the rate of enzyme activity

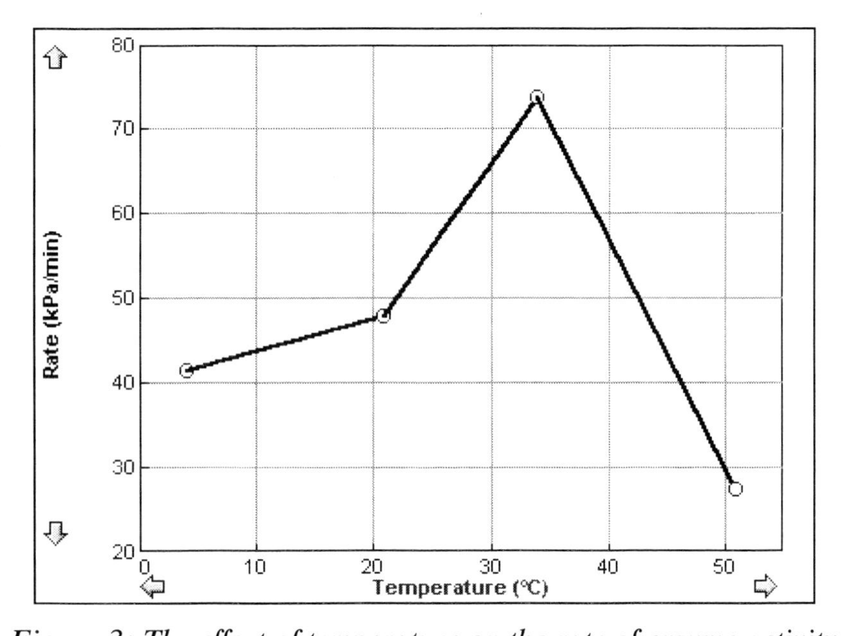

Figure 3: The effect of temperature on the rate of enzyme activity

ANSWERS TO QUESTIONS

1. The rate should be highest when the concentration of enzyme is highest. With higher concentration of enzyme, there is a greater chance of an effective collision between the enzyme and H_2O_2 molecule.

2. Roughly, the rate doubles when the concentration of enzyme doubles. Since the data are somewhat linear, the rate is proportional to the concentration. At a concentration of 5 drops, the rate in the above experiment should be about 111 kPa/min.

3. The temperature at which the rate of enzyme activity is the highest should be close to 30°C. The lowest rate of enzyme activity should be at 60°C.

4. The rate increases as the temperature increases, until the temperature reaches about 50°C. Above this temperature, the rate decreases.

5. At high temperatures, enzymes lose activity as they are denatured.

6. Student answers may vary. Activity is usually highest at pH 10 and lowest at pH 4.

7. Student answers may vary. Usually, the enzyme activity increases from pH 4 to 10. At low pH values, the protein may denature or change its structure. This may affect the enzyme's ability to recognize a substrate or it may alter its polarity within a cell.

Experiment

7

Photosynthesis

The process of photosynthesis involves the use of light energy to convert carbon dioxide and water into sugar, oxygen, and other organic compounds. This process is often summarized by the following reaction:

$$6\ H_2O + 6\ CO_2 + \text{light energy} \rightarrow C_6H_{12}O_6 + 6\ O_2$$

This process is an extremely complex one, occurring in two stages. The first stage, called the *light reactions of photosynthesis*, requires light energy. The products of the light reactions are then used to produce glucose from carbon dioxide and water. Because the reactions in the second stage do not require the direct use of light energy, they are called the *dark reactions of photosynthesis.*

In the light reactions, electrons derived from water are "excited" (raised to higher energy levels) in several steps, called photosystems I and II. In both steps, chlorophyll absorbs light energy that is used to excite the electrons. Normally, these electrons are passed to a cytochrome-containing electron transport chain. In the first photosystem, these electrons are used to generate ATP. In the second photosystem, excited electrons are used to produce the reduced coenzyme nicotinamide adenine dinucleotide phosphate (NADPH). Both ATP and NADPH are then used in the dark reactions to produce glucose.

In this experiment, a blue dye (2,6-dichlorophenol-indophenol, or DPIP) will be used to replace NADPH in the light reactions. When the dye is oxidized, it is blue. When reduced, however, it turns colorless. Since DPIP replaces NADPH in the light reactions, it will turn from blue to colorless when reduced during photosynthesis.

OBJECTIVES

In this experiment, you will

- Use a Colorimeter to measure color changes due to photosynthesis.
- Study the effect of light on photosynthesis.
- Study the effect that the boiling of plant cells has on photosynthesis.
- Compare the rates of photosynthesis for plants in different light conditions.

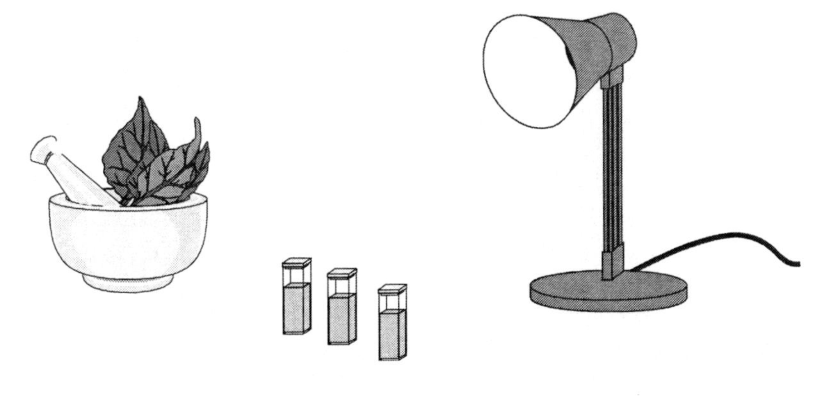

Figure 1

MATERIALS

computer
Vernier computer interface
Logger*Pro*
Vernier Colorimeter
two cuvettes with lids
aluminum foil covered cuvette with lid
100 watt floodlight
stopwatch
600 mL beaker
250 mL beaker
two small test tubes
5 mL pipet
pipet pump or bulb
two eyedroppers or Beral pipets
10 mL DPIP/phosphate buffer solution
unboiled chloroplast suspension
boiled chloroplast suspension
ice

PROCEDURE

1. Obtain and wear goggles.
2. Obtain two plastic Beral pipets, two cuvettes with lids, and one aluminum foil covered cuvette with a lid. Mark one Beral pipet with a U (unboiled) and one with a B (boiled). Mark the lid for the cuvette with aluminum foil with a D (dark). For the remaining two cuvettes, mark one lid with a U (unboiled) and one with a B (boiled).
3. Connect the Colorimeter to the computer interface. Prepare the computer for data collection by opening the file "07 Photosynthesis" from the *Biology with Computers* folder of Logger*Pro*.
4. You are now ready to calibrate the Colorimeter. Prepare a blank by filling a cuvette 3/4 full with distilled water. To correctly use a Colorimeter cuvette, remember:
   - All cuvettes should be wiped clean and dry on the outside with a tissue.
   - Handle cuvettes only by the top edge of the ribbed sides.
   - All solutions should be free of bubbles.
   - Always position the cuvette with its reference mark facing toward the white reference mark at the top of the cuvette slot on the Colorimeter.
5. Calibrate the Colorimeter.
   a. Open the Colorimeter lid.
   b. Holding the cuvette by the upper edges, place it in the cuvette slot of the Colorimeter. Close the lid.
   c. If your Colorimeter has a CAL button, press the < or > button on the Colorimeter to select a wavelength of 635 nm (Red) for this experiment. Press the CAL button until the red LED begins to flash. Then release the CAL button. When the LED stops flashing, the calibration is complete. Proceed directly to Step 6. If your Colorimeter does not have a CAL button, continue with this step to calibrate your Colorimeter.

First Calibration Point

d. Choose Calibrate ▸ CH1: Colorimeter (%T) from the Experiment menu and then click Calibrate Now.

e. Turn the wavelength knob on the Colorimeter to the "0% T" position.

f. Type "0" in the edit box.

g. When the displayed voltage reading for Reading 1 stabilizes, click Keep.

Second Calibration Point

h. Turn the knob of the Colorimeter to the Red LED position (635 nm).

i. Type "100" in the edit box.

j. When the displayed voltage reading for Reading 2 stabilizes, click Keep, then click Done.

6. Obtain a 600 mL beaker filled with water and a flood lamp. Arrange the lamp and beaker as shown in Figure 2. The beaker will act as a heat shield, protecting the chloroplasts from warming by the flood lamp. Do not turn the lamp on until Step 10.

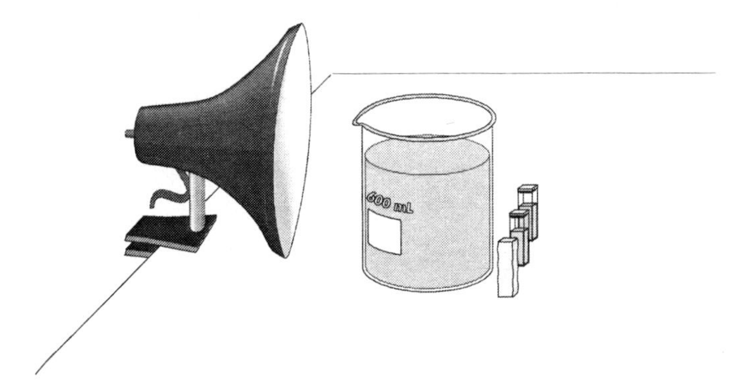

Figure 2

7. Locate the unboiled and boiled chloroplast suspension prepared by your instructor. Before removing any of the chloroplast suspension, gently swirl to resuspend any chloroplast which may have settled out. Using the pipet marked U, draw up ~1 mL of unboiled chloroplast suspension. Using the pipet marked B, draw up ~1 mL of boiled chloroplast suspension. Set both pipettes in the small beaker filled with ice at your lab station to keep the chloroplasts cooled.

8. Add 2.5 mL of DPIP/phosphate buffer solution to each of the cuvettes. **Important**: perform the following steps as quickly as possible and proceed directly to Step 9.

   a. Cuvette U: Add 3 drops of *unboiled* chloroplasts. Place the lid on the cuvette and gently mix; try not to introduce bubbles in the solution. Place the cuvette in front of the lamp as shown in Figure 2. Mark the cuvette's position so that it can always be placed back in the same spot.

   b. Cuvette D: Add 3 drops of *unboiled* chloroplasts. Place the lid on the cuvette and gently mix; try not to introduce bubbles in the solution. Place the foil-covered cuvette in front of the lamp as shown in Figure 2 and mark its position. Make sure that no light can penetrate the cuvette.

   c. Cuvette B: Add 3 drops of *boiled* chloroplasts. Place the lid on the cuvette and gently mix; try not to introduce bubbles in the solution. Place the cuvette in front of the lamp as shown in Figure 2. Mark the cuvette's position so that it can be placed back in the same spot.

9. Take absorbance readings for each cuvette. Invert each cuvette two times to resuspend the chloroplast before taking a reading. If any air bubbles form, gently tap on the cuvette lid to knock them loose.
   a. Cuvette U: Place the cuvette in the cuvette slot of the Colorimeter and close the lid. Allow 10 seconds for the readings displayed in the meter to stabilize, then record the absorbance value in Table 1. Remove the cuvette and place it in its original position in front of the lamp.
   b. Cuvette D: Remove the cuvette from the foil sleeve and place it in the cuvette slot of the Colorimeter. Close the Colorimeter lid and wait 10 seconds. Record the absorbance value displayed in the meter in Table 1. Remove the cuvette and place it back into the foil sleeve. Place the cuvette in its original position in front of the lamp.
   c. Cuvette B: Place the cuvette in the cuvette slot of the Colorimeter and close the lid. Allow 10 seconds for the readings displayed in the meter to stabilize, then record the absorbance value in Table 1. Remove the cuvette and place it in its original position in front of the lamp.
10. Turn on the lamp.
11. Repeat Step 9 when 5 minutes have elapsed.
12. Repeat Step 9 when 10 minutes have elapsed.
13. Repeat Step 9 when 15 minutes have elapsed.
14. Repeat Step 9 when 20 minutes have elapsed.

PROCESSING THE DATA

1. Go to Page 2 of the experiment file and manually the data recorded in Table 1 into the appropriate column in the table. To type click on the table cell with the mouse pointer. You will see a blinking cursor in the cell. Type your data point and press ENTER. The cursor will move down to the next cell. The graph will update after each data point is entered.
2. Calculate the rate of photosynthesis for each of the three cuvettes tested.
   a. Click the Linear Fit button, ,to perform a linear regression. A dialog box will appear. Select the three data sets you wish to perform a linear regression on and click OK. A floating box will appear with the formula for a best fit line for each data set selected.
   b. In Table 2, record the slope of the line, *m*, as the rate of photosynthesis for each data set.
   c. Close the linear regression floating boxes.

DATA

Table 1			
Time (min)	Absorbance unboiled	Absorbance in dark	Absorbance boiled
0			
5			
10			
15			
20			

Table 2	
Chloroplast	Rate of photosynthesis
Unboiled	
Dark	
Boiled	

QUESTIONS

1. Is there evidence that chloroplasts were able to reduce DPIP in this experiment? Explain.
2. Were chloroplasts able to reduce DPIP when kept in the dark? Explain.
3. Were boiled chloroplasts able to reduce DPIP ? Explain.
4. What conclusions can you make about the photosynthetic activity of spinach?

EXTENSION - PLANT PIGMENT CHROMATOGRAPHY

Paper chromatography is a technique used to separate substances in a mixture based on the movement of the different substances up a piece of paper by capillary action. Pigments extracted from plant cells contain a variety of molecules, such as chlorophylls, beta carotene, and xanthophyll, that can be separated using paper chromatography. A small sample of plant pigment placed on chromatography paper travels up the paper due to capillary action. Beta carotene is carried the furthest because it is highly soluble in the solvent and because it forms no hydrogen bonds with the chromatography paper fibers. Xanthophyll contains oxygen and does not travel quite as far with the solvent because it is less soluble than beta carotene and forms some hydrogen bonds with the paper. Chlorophylls are bound more tightly to the paper than the other two, so they travel the shortest distance.

The ratio of the distance moved by a pigment to the distance moved by the solvent is a constant, R_f. Each type of molecule has its own R_f value.

$$R_f = \frac{\text{distance traveled by pigment}}{\text{distance traveled by solvent}}$$

OBJECTIVES

In this experiment, you will

- Separate plant pigments.
- Calculate the R_f values of the pigments.

MATERIALS

50 mL graduated cylinder
chromatography paper
spinach leaves
coin
goggles
cork stopper
pencil
scissors
solvent
ruler

PROCEDURE

Obtain and wear goggles! ***Caution:*** *The solvent in this experiment is flammable and poisonous. Be sure there are no open flames in the lab during this experiment. Avoid inhaling fumes. Wear goggles at all times. Notify your teacher immediately if an accident occurs.*

1. Obtain a 50 mL graduated cylinder with 5 mL of solvent in the bottom.
2. Cut the chromatography paper so that it is long enough to reach the solvent. Cut one end of the paper into a point.
3. Draw a pencil line 2.0 cm above the pointed end of the paper.
4. Use the coin to extract the pigments from the spinach leaf. Place a small section of the leaf on top of the pencil line. Use the ribbed edge of the coin to push the plant cells into the chromatography paper. Repeat the procedure 10 times making sure to use a different part of the leaf each time.
5. Place the chromatography paper in the cylinder so the pointed end just touches the solvent. Make sure the pigment is not in the solvent.
6. Stopper the cylinder and wait until the solvent is approximately 1 cm from the top of the paper. Remove the chromatography paper and mark the solvent front before it evaporates.
7. Allow the paper to dry. Mark the bottom of each pigment band. Measure the distance each pigment moved from the starting line to the bottom of the pigment band. Record the distance that each of the pigments and the solvent moved, in millimeters.
8. Identify each of the bands and label them on the chromatography paper.
   - beta carotene: yellow to yellow orange
   - xanthophyll: yellow
   - chlorophyll *a*: bright green to blue green
   - chlorophyll *b*: yellow green to olive green
9. Staple the chromatogram to the front of your lab sheet.
10. Discard the solvent as directed by your teacher.

DATA

Table 1			
Band number	Distance traveled (mm)	Band color	Identity
1			
2			
3			
4			
5*			
Distance solvent front moved = ________ mm			

* The fifth band may not appear.

PROCESSING THE DATA

Calculate the R_f values and record in Table 2.

Table 2	
Molecule	R_f
beta carotene	
xanthophyll	
chlorophyll *a*	
chlorophyll *b*	

QUESTIONS

1. What factors are involved in the separation of the pigments?
2. Would you expect the R_f value to be different with a different solvent?
3. Why do the pigments become separated during the development of the chromatogram?

ADDITIONAL EXTENSIONS

1. Repeat the paper chromatography with various species of plants. What similarities do you see? What differences are there?

2. Use colored filters around the cuvettes to test the effect of red, blue, and green light on the photosynthetic activity of spinach.

3. Vary the distance of the floodlight source to determine the effect of light intensity on photosynthesis.

4. Compare the photosynthetic activity of spinach with that of chloroplasts from other plants.

5. Investigate the effect of temperature on the photosynthetic activity of spinach.

6. Explain why the rate of photosynthesis varies under different environmental conditions.

Experiment
7

TEACHER INFORMATION

Photosynthesis

1. It is necessary to suspend the spinach chloroplasts in a 0.5 M sucrose solution. To prepare the 0.5 M sucrose solution, add 171 g of sucrose to distilled water to make a total volume of 1 liter. Store this solution in the refrigerator overnight before preparing the chloroplast solution. Place the blender and beaker in the freezer overnight also.

2. To prepare the phosphate buffer solution:

   a. Add 174 g of K_2HPO_4 (dibasic) to distilled water to make a total volume of 1 liter.

   b. Add 136 g of KH_2PO_4 (monobasic) to distilled water to make a total volume of 1 liter.

   c. Combine the two solutions until the pH is 6.5. About 685 mL of monobasic should be added to 315 mL of dibasic to obtain a solution with a pH of 6.5.

3. To prepare the DPIP (2,6-dichlorophenol-indophenol) solution, add 0.072 g of DPIP to distilled water to make a total volume of 1 liter. Store this solution in a dark bottle and refrigerate.

4. Prepared DPIP and phosphate buffer solution are available from *Carolina Biological Supply Company*. Prepared solutions should be mixed in a 1:1:3 ratio of DPIP solution to phosphate buffer solution to distilled water.

5. Just prior to doing the experiment, prepare the DPIP solution students will use in Step 9 of the student procedure. Combine the DPIP solution with phosphate buffer and distilled water in the following ratio:

   - 100 mL of 0.1% DPIP solution
   - 100 mL phosphate buffer
   - 300 mL distilled water

6. To prepare a chloroplast suspension,

   a. Place fresh spinach leaves in the light for a few hours. Be sure they remain cool and hydrated.

   b. Cover the blades of a blender with cold 0.5 M sucrose.

   c. Pack the spinach leaves into the blender until they are about three centimeters above the blades.

   d. Blend the spinach with three 10 second bursts. Wait 30 seconds between bursts.

   e. Filter the mixture through cheesecloth into a cold beaker. Keep the beaker on ice. You will need to squeeze the cheesecloth to release as much liquid as possible.

   f. Split the chloroplast suspension into two equal parts. Set one part aside for Item 7 below.

   g. Distribute 2 mL portions into covered, cooled vials. The vials should be light-tight. Black electrical tape works well to cover the vials.

   h. Keep the chloroplast suspensions on ice.

7. Obtain the spinach solution that was set aside in Item 6 and boil it for 5 minutes. Distribute 2 mL portions of boiled chloroplasts into darkened, taped vials. Keep these suspensions on ice.

8. Test the chloroplast suspension prior to use. If the DPIP is reduced too rapidly, further dilute the suspension before distributing it into the vials.

9. The success of this lab is greatly dependent on the ability of the students to synchronize taking absorbance readings. Encourage the students to read over the procedure carefully before beginning.

10. The procedure used to calibrate at 0% and 100% transmittance is similar to that used with most spectrophotometers. The blank cuvette can be in the colorimeter for the 0% calibration (as well as the 100% calibration).

11. Prepare the solvent for chromatography by mixing 9 parts (by volume) petroleum ether with 1 part acetone. Distribute 5 mL of solvent to each of the cylinders **Warning:** *Be very careful with the petroleum ether as it is highly flammable.* Extinguish all flames in the room.

    HAZARD ALERT: Petroleum Ether: Flammable liquid; flammable; dangerous fire risk. Hazard code: C—Somewhat hazardous.

    HAZARD ALERT: Acetone: Dangerous fire risk; flammable; slightly toxic by ingestion and inhalation. Hazard code: C—Somewhat hazardous

    The hazard information reference is: Flinn Scientific, Inc., *Chemical & Biological Catalog Reference Manual, 2000*, (800) 452-1261, www.flinnsci.com. See *Appendix D* of this book, *Biology with Computers*, for more information.

12. Remember to handle paper strips by the edges only

13. Remind students to wear goggles and to keep the tubes stoppered

14. Refer to the MSDS for proper disposal techniques.

SAMPLE RESULTS

Table 1

Time (min)	Unboiled Absorbance	Dark Absorbance	Boiled Absorbance
0	0.480	0.456	0.480
5	0.394	0.466	0.470
10	0.326	0.466	0.490
15	0.250	0.451	0.466
20	0.139	0.456	0.466

Table 2	
Chloroplast	Rate of photosynthesis
Unboiled	0.0165
Dark	0.0003
Boiled	0.0007

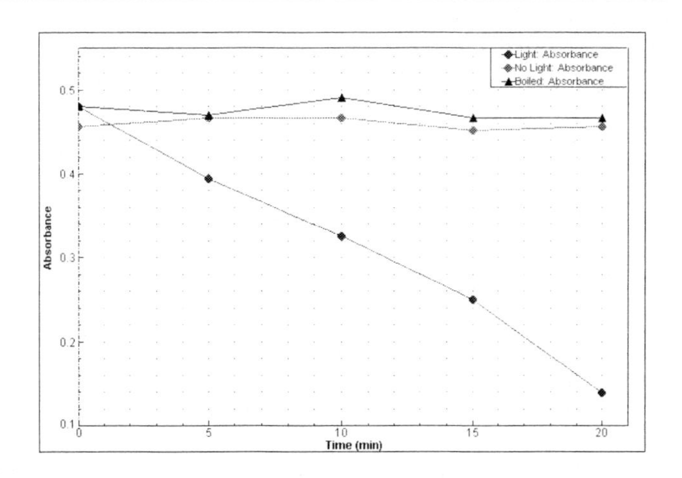

ANSWERS TO QUESTIONS

1. Yes, there is evidence that spinach chloroplasts were able to reduce DPIP in this experiment, as the color of the solution changed when the light intensity changed.

2. No, chloroplasts were not able to reduce DPIP while kept in the dark, as evidenced by the lack of color change of DPIP and the low rate of photosynthesis. The reduced form of DPIP is colorless, and since the solution remained blue, the DPIP did not participate in the light reactions of photosynthesis.

3. Boiled chloroplasts were not able to reduce DPIP, as the heat destroyed the photosynthetic machinery of chloroplasts. The chloroplasts were inactive, even in the presence of light.

4. Based upon student data, chloroplasts should reduce DPIP when exposed to light. Chloroplasts should not be able to reduce DPIP when placed in the dark, nor when destroyed. The amount of reduced DPIP is proportional to the amount of light the chloroplasts are exposed to. The implication is that the amount of sugar produced in photosynthesis depends upon the duration of exposure to light.

EXTENSION SAMPLE RESULTS

Table 1			
Band number	Distance traveled (mm)	Band color	Identity
1	141	faint yellow	carotene
2	60	yellow	xanthrophyll
3	30	bright green	chlorophyll *a*
4	15	olive green	chlorophyll *b*
5*			
Distance solvent front moved = 150 mm			

* band 5 may not appear

Table 2	
Molecule	R_f
beta carotene	.94
xanthophyll	.40
chlorophyll *a*	.20
chlorophyll *b*	.10

ANSWERS TO EXTENSION QUESTIONS

1. The distance that each pigment migrates up the paper is determined by its solubility and its attraction to the fibers in the paper. Pigments with the greatest solubility and least attraction to the fibers in the paper travel the greatest distance up the paper.

2. Since the R_f factor depends on a molecule's ability to dissolve in the solvent, the pigment may not travel as far up the paper.

3. The pigments become separated because they are not equally soluble in the solvent. They are attracted, to different degrees, to the fibers in the paper through the formation of intermolecular bonds, such as hydrogen bonds.

Experiment

8

The Effect of Alcohol on Biological Membranes

The primary objective of this experiment is to determine the stress that various alcohols have on biological membranes. Membranes within cells are composed mainly of lipids and proteins and often serve to help maintain order within a cell by containing cellular materials. Different membranes have a variety of specific functions.

One type of membrane-bound vacuole found in plant cells, the *tonoplast*, is quite large and usually contains water. In beet plants, this membrane-bound vacuole also contains a water-soluble red pigment, *betacyanin*, that gives the beet its characteristic color. Since the pigment is water soluble and not lipid soluble, it remains in the vacuole when the cells are healthy. If the integrity of a membrane is disrupted, however, the contents of the vacuole will spill out into the surrounding environment. This usually means the cell is dead.

In this experiment, you will test the effect of three different alcohols (methanol, ethanol, and 1-propanol) on membranes. Ethanol is found in alcoholic beverages. Methanol, sometimes referred to as wood alcohol, can cause blindness and death. Propanol is fatal if consumed. One possible reason why they are so dangerous to living organisms is that they might damage cellular membranes. Methanol, ethanol, and 1-propanol are very similar alcohols, differing by the number of carbon and hydrogen atoms within the molecule. Methanol, CH_3OH, is the smallest, ethanol, CH_3CH_2OH, is intermediate in size, and 1-propanol, $CH_3CH_2CH_2OH$, is the largest of the three molecules.

If beet membranes are damaged, the red pigment will leak out into the surrounding environment. The intensity of color in the environment should be proportional to the amount of cellular damage sustained by the beet.

To measure the color intensity, you will be using a Colorimeter. In this device, blue light from the LED light source will pass through the solution and strike a photocell. The alcohol solutions used in this experiment are clear. If the beet pigment leaks into the solution, it will color the solution red. A higher concentration of colored solution absorbs more light and transmits less light than a solution of lower concentration. The computer-interfaced Colorimeter monitors the light received by the photocell as either an *absorbance* or a *percent transmittance* value.

You are to prepare five solutions of differing alcohol concentrations (0%, 10%, 20%, 30%, and 40%) for each of the three alcohols. A small piece of beet is placed in each solution. After ten minutes, each alcohol solution is transferred to a small, rectangular cuvette that is placed into the Colorimeter. The amount of light that penetrates the solution and strikes the photocell is used to compute the absorbance of each solution. The absorbance is directly related to the amount of red pigment in the solution. By plotting the percent alcohol vs. the amount of pigment (that is, the absorbance), you can assess the amount of damage various alcohols cause to cell membranes.

OBJECTIVES

In this experiment, you will

- Use a Colorimeter to measure the color intensity of beet pigment in alcohol solutions.
- Test the effect of three different alcohols on membranes.
- Test the effect of different alcohol concentrations on membranes.

MATERIALS

computer
Vernier computer interface
Logger*Pro*
Vernier Colorimeter
four graduated Beral pipets
10 mL 1-propanol
10 mL ethanol
10 mL methanol
three 18 × 150 mm test tubes with rack
100 mL beaker
beet root
cotton swabs
forceps
knife
lab apron
microplate, 24-well
one pair gloves
ruler (cm)
tap water
timer or stopwatch
tissues (preferably lint-free)
toothpick

PROCEDURE

1. Obtain and wear goggles, an apron, and gloves. **CAUTION:** The compounds used in this experiment are flammable and poisonous. Avoid inhaling vapors. Avoid contacting them with your skin or clothing. Be sure there are no open flames in the lab during this experiment. Notify your teacher immediately if an accident occurs.

2. Obtain the following materials:

   a. Place about 10 mL of methanol in a medium sized test tube. Label this tube M.
   b. Place about 10 mL of ethanol in a medium sized test tube. Label this tube E.
   c. Place about 10 mL of 1-propanol in a medium sized test tube. Label this tube P.
   d. Place about 30 mL of tap water in a small beaker.

3. Prepare five methanol solutions (0%, 10%, 20%, 30% and 40%). Using Beral pipets, add the number of *drops* of water specified in Table 1 to each of five wells.

   Use a different Beral pipet to add alcohol to each of five wells in the microwell plate. See Table 1 to determine the number of drops of alcohol to add to each well.

Table 1

Well number	H_2O	Alcohol	Concentration of alcohol (%)
	drops	drops	
1	64	0	0
2	57	7	10
3	51	13	20
4	44	20	30
5	38	26	40

4. Clean the pipet used to transfer alcohol. To do this, wipe the outside clean and empty it of liquid. Draw up a little ethanol into the pipette and use the liquid to rinse the inside of the pipette. Discard the ethanol.

5. Prepare five ethanol solutions. To do so, repeat Step 3, substituting ethanol for methanol. Place each solution in the second row of wells. See Figure 1.

6. Prepare five 1-propanol solutions. To do so, clean your pipette and repeat Step 3, substituting 1-propanol for methanol. Place each solution in the third row of wells. See Figure 1.

7. Now, obtain a piece of beet from your instructor. Cut 15 squares, each 0.5 cm x 0.5 cm x 0.5 cm in size. They should easily fit into a microwell without being wedged in. While cutting the beet, be sure:
   - There are no ragged edges.
   - No piece has any of the outer skin on it.
   - All of the pieces are the same size.
   - The pieces do not dry out.

8. Rinse the beet pieces several times using a small amount of water. Immediately drain off the water. This will wash off any pigment released during the cutting process.

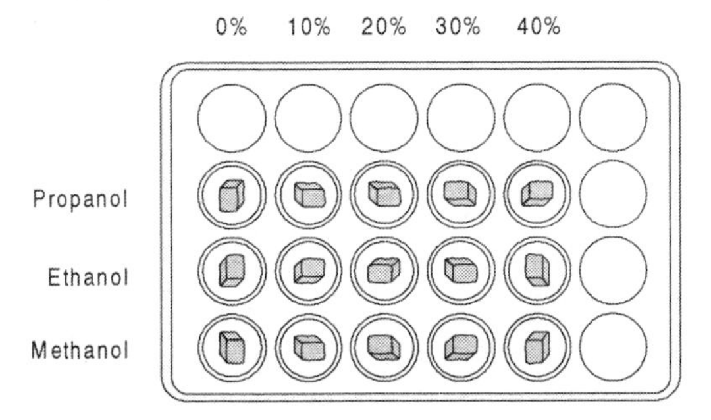

Figure 1

9. Set the timer to 10 minutes and begin timing. Use forceps to place a piece of beet into each of 15 wells, as shown in Figure 1. Stir the beet in the alcohol solution once every minute with a toothpick. Be careful not to puncture or damage the beet.

10. Connect the Colorimeter to the computer interface. While one team member is performing Step 9, another team member should prepare the computer for data collection by opening the file "08 Alcohol and Membranes" from the *Biology with Computers* folder of Logger*Pro*.

11. You are now ready to calibrate the Colorimeter. Prepare a *blank* by filling a cuvette 3/4 full with distilled water. To correctly use a Colorimeter cuvette, remember:
    - All cuvettes should be wiped clean and dry on the outside with a tissue.
    - Handle cuvettes only by the top edge of the ribbed sides.
    - All solutions should be free of bubbles.
    - Always position the cuvette with its reference mark facing toward the white reference mark at the top of the cuvette slot on the Colorimeter.

12. Calibrate the Colorimeter.

    a. Open the Colorimeter lid.

    b. Holding the cuvette by the upper edges, place it in the cuvette slot of the Colorimeter. Close the lid.

    c. If your Colorimeter has a CAL button, Press the < or > button on the Colorimeter to select a wavelength of 470 nm (Blue) for this experiment. Press the CAL button until the red LED begins to flash. Then release the CAL button. When the LED stops flashing, the calibration is complete. Proceed directly to Step 13. If your Colorimeter does not have a CAL button, continue with this step to calibrate your Colorimeter.

    First Calibration Point

    d. Choose Calibrate ▸ CH1: Colorimeter (%T) from the Experiment menu and then click Calibrate Now.

    e. Turn the wavelength knob on the Colorimeter to the "0% T" position.

    f. Type "0" in the edit box.

    g. When the displayed voltage reading for Reading 1 stabilizes, click Keep.

    Second Calibration Point

    h. Turn the knob of the Colorimeter to the Green LED position (565 nm).

    i. Type "100" in the edit box.

    j. When the displayed voltage reading for Reading 2 stabilizes, click Keep, then click Done.

13. After 10 minutes, remove the beet pieces from the wells. Remove them in the same order that they were placed into the wells. Discard the beet pieces and retain the colored solutions.

14. You are now ready to collect absorbance data for the alcohol solutions.

    a. Click Collect.

    b. Empty the water from the cuvette. Use a cotton swab to dry the cuvette after the water has been emptied from it.

    c. Transfer all of the 0% methanol solution from Well 1 into the cuvette using a Beral pipet. Wipe the outside with a tissue and place it in the colorimeter. After closing the lid, wait for the absorbance value displayed in the meter to stabilize.

    d. Click Keep, enter "0" in the edit box and then press ENTER. The data pair you just collected should now be plotted on the graph.

15. Discard the cuvette contents into your waste beaker. Remove all of the solution from the cuvette. Use a cotton swab to dry the cuvette. Fill the cuvette with the 10% methanol solution from Well 2 using a Beral pipet. Wipe the outside with a tissue and place it in the colorimeter. After closing the lid, wait for the absorbance value displayed in the meter to stabilize. Click Keep, enter "10" in the edit box and then press ENTER.

16. Repeat Step 15, using the solutions in Wells 3, 4, and 5. When you have finished with all of the methanol solutions click Stop.

17. In the data table, record the absorbance and concentration data pairs listed in the data table.

18. Store the data for the methanol solutions by choosing Store Latest Run from the Experiment menu.

19. Repeat Steps 14 through 18, measuring the five ethanol solutions.

20. Repeat Steps 14 through 17, measuring the five propanol solutions.

21. To print a graph of concentration *vs.* absorbance showing the data for all three alcohols:
    a. Label all three curves by choosing Text Annotation from the Insert menu, and typing "Ethanol" (or "Methanol", or "1-Propanol") in the edit box. Then drag each box to a position near its respective curve. Adjust the placement of the arrow head.
    b. Print a copy of the graph, with all three data sets displayed. Enter your name(s) and the number of copies of the graph you want.
    c. Use your graph to answer the discussion questions at the end of this experiment.

DATA

Trial	Concentration (%)	Absorbance		
		Methanol	Ethanol	1-Propanol
1	0			
2	10			
3	20			
4	30			
5	40			

QUESTIONS

1. Which alcohol damaged the beet at the lowest concentrations? How did you determine this?

2. Which of the three alcohols seems to affect membranes the most? How did you come to this conclusion?

3. At what percentage of alcohol is the cellular damage highest for methanol? ethanol? 1-propanol?

CHALLENGE QUESTION

1. What is the relationship between the size of the alcohol molecule and the extent of membrane damage? Hypothesize why this might be so.

TEACHER INFORMATION

Experiment **8**

The Effect of Alcohol on Biological Membranes

1. The light source for the red-pigmented solution is the blue LED (470 nm).
2. Isopropyl alcohol may be used in place of n-propanol. If rubbing alcohol is used, remember that it is already diluted to 70% or so (see the container label). You will need to recalculate the volumes used for this alcohol. Use 95% ethanol and absolute methanol.
3. Beets stain hands and clothing. Lab aprons and gloves should be worn to prevent staining.
4. Obtain large, red beets from a local store. Students will cut these into small 0.5 cm × 0.5 cm × 0.5 cm pieces. To save class time, you may want to cut a 0.5 cm × 0.5 cm × 12 cm section for each group. They can slice 0.5 cm sections from this. The end pieces should be discarded, since they will dry out and behave differently than freshly cut surfaces.
5. The cuvette must have at least 2 mL of solution in order to get a valid absorbance reading. If students fill the cuvette as described in the procedure, they will have sufficient volume.
6. Since there is some variation in the amount of light absorbed by the cuvette if it is rotated 180°, you should use a water-proof marker to make a reference mark on the top edge of one of the clear sides of all cuvettes. Students are reminded in the procedure to align this mark with the white reference mark to the right of the cuvette slot on the colorimeter.
7. The use of a single cuvette in the procedure is to eliminate errors introduced by slight variations in the absorbance of different plastic cuvettes. If one cuvette is used throughout the experiment by a student group, this variable is eliminated.
8. We recommend the use of cotton swabs to dry a cuvette after water or a solution has been emptied from it. The cotton swab will remove any remaining droplets in the cuvette.
9. The procedure used to calibrate at 0% and 100% transmittance is similar to that used with most spectrophotometers. The blank cuvette can be in the colorimeter for the 0% calibration (as well as the 100% calibration).
10. This experiment gives you a good opportunity to discuss the relationship between percent transmittance and absorbance. You can also discuss the mathematical relationship between absorbance and percent transmittance, as represented by the formula:

$$A = \log(100/\%T)$$

SAMPLE RESULTS

Trial	Concentration (%)	Absorbance		
		Methanol	Ethanol	1-Propanol
1	0	0.011	0.016	0.018
2	10	0.014	0.017	0.077
3	20	0.026	0.038	0.494
4	30	0.051	0.141	0.683
5	40	0.144	0.662	0.805

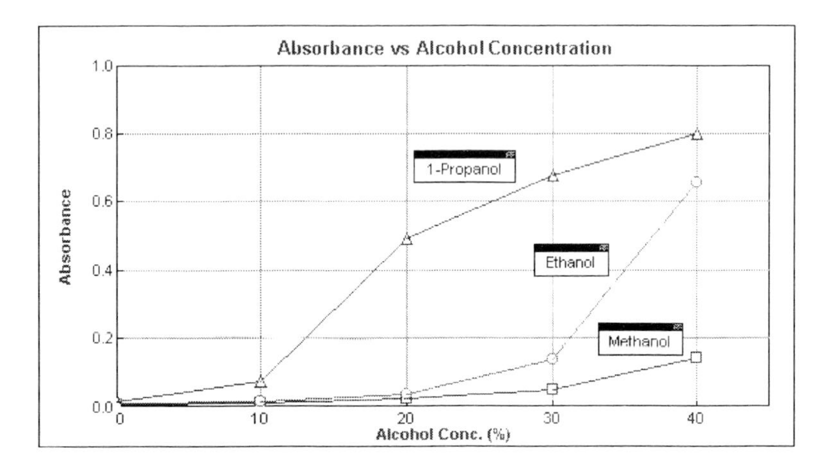

Absorbance readings in different concentrations of alcohols

ANSWERS TO QUESTIONS

1. Propanol damaged membranes more at low concentrations. At 20% alcohol, propanol caused three times the pigment to be released into the solution, compared to the other two alcohols.
2. Propanol seems to affect membranes the most, as there is more pigment (cellular damage) at any concentration of alcohol.
3. Student answers may vary. In most experiments, the percentage of methanol, ethanol, and 1-propanol that is most damaging is 60%.
4. Answers may vary.

CHALLENGE QUESTION

1. The larger the molecule, the greater the extent of membrane damage. Propanol (a three-carbon molecule) is larger than ethanol (a two-carbon molecule) and methanol (a one-carbon molecule). Lipids can be composed of hydrocarbon chains, which are non-polar. As the alcohol increases in length, it becomes less polar than smaller alcohols, and are able to mix with lipids to a greater extent. The –OH branch of the alcohol can mix with water in all three molecules. The longer alcohols can mix very effectively with both lipids and water. This tends to disrupt the membrane, much as detergents and soaps do. For this reason, larger alcohols might cause more extensive membrane damage.

Experiment

9

Biological Membranes

The primary objective of this experiment is to determine the stress that various factors, such as osmotic balance, detergents, and pH, have on biological membranes. Membranes within cells are composed mainly of lipids and proteins. They often serve to help maintain order within a cell by containing cellular materials.

One type of vacuole in the cells of plants, the *tonoplast*, is quite large and usually contains water. In beet plants, this membrane-bound vacuole also contains a water soluble red pigment, *betacyanin*, that gives the beet its characteristic color. Since the pigment is water soluble and not lipid soluble, it is contained in the vacuole when the cells are healthy. If the integrity of a membrane is disrupted, however, the contents of the vacuole will spill out into the surrounding environment and color it red. This usually means the cell is dead. If beet membranes are damaged, the red pigment will leak out into the surrounding environment. The intensity of color in the environment should be proportional to the amount of cellular damage.

You will test the effect of osmotic balance, detergents, and pH changes on biological membranes. The presence of certain salts is essential for most plant growth, but too much salt can kill plants. Even salts that are not transported across cell membranes can affect plants—by altering the osmotic balance. Osmosis is the movement of water across a semipermeable membrane from a region of low solute concentration to a region of higher solute concentration. It can greatly affect a cell's water content when the amount of water inside the cell is different than the amount outside the cell. You will test to see how this osmotic stress affects the cellular membrane integrity.

Detergents are designed to make lipids soluble in water. Since biological membranes are made of both lipids and water soluble materials, they are disrupted by detergents. You will design tests to determine the effect of this detergent on biological membranes.

The pH of an environment is critical for living things. If the environment is too acidic or too basic, organisms cannot survive. You will design tests to determine the effect of pH on biological membranes.

You will be using the Colorimeter shown in Figure 1. In this device, blue light from the LED light source will pass through the solution and strike a photocell. The alcohol solutions used in this experiment are clear. If the beet pigment leaks into the solution, it will color the solution red. A higher concentration of colored solution absorbs more light and transmits less light than a solution of lower concentration. The computer-interfaced Colorimeter monitors the light received by the photocell as either an *absorbance* or a *percentage transmittance* value. The *absorbance* of light will be used to monitor the extent of cellular membrane damage.

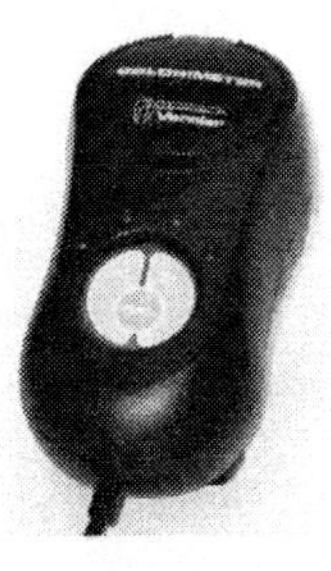

Figure 1

OBJECTIVES

In this experiment, you will

- Use a computer and a Colorimeter to measure color changes due to disrupted cell membranes.
- Determine the effect of osmotic balance on biological membranes.
- Determine the effect of detergents on biological membranes.
- Determine the effect of pH on biological membranes.

MATERIALS

computer
Vernier computer interface
Logger*Pro*
Vernier Colorimeter
100 mL beaker
three 18 × 150 mm test tubes w/ rack
microplate, 24-well
two 2 mL pipets
pipet pump or bulb
15% salt solution
beet root
cotton swabs
detergent solution
Beral pipet(s) to transfer solutions
forceps
knife
lab apron
pair of gloves
pH buffer solutions
test tube rack
ruler (cm)
tap water
timer or stopwatch
tissues (preferably lint free)
toothpicks

PROCEDURE

1. Obtain and wear goggles, an apron, and gloves.

Testing for the effect of osmotic stress

2. Obtain 10 mL of 15% salt solution and about 10 mL of tap water. Place each into a labeled test tube.
3. Prepare six salt solutions: 0%, 3%, 6%, 9%, 12%, and 15%. If your instructor has supplied graduated pipets with pipet pumps, add the *mL* of water specified in Table 1 to five of the six wells in the microwell plate. If your instructor has supplied Beral pipets, add the number of *drops* of water specified in Table 1 to each of four wells.
4. Use a different graduated pipet or Beral pipet to add 15% salt solution to five of the six wells in the microwell plate. See Table 1 to determine the mL or drops of salt solution to add to each well.
5. Clean the pipet used to transfer the salt solution.
6. Obtain a piece of beet from your instructor. Cut six squares, each 0.5 cm × 0.5 cm × 0.5 cm in size. They should fit into a microwell easily, without being wedged in. Be sure that:
   - there are no ragged edges.
   - no piece has any of the outer skin on it.
   - all of the pieces are the same size.
   - the pieces do not dry out.

7. Rinse the beet pieces twice using a small amount of water. Immediately drain off the water. This will wash off any pigment released during the cutting process.

Table 1					
Well number	H_2O		15% Salt		Concentration of salt (%)
	mL	drops	mL	drops	
1	2.4	60	0.0	0	0
2	1.9	48	0.5	12	3
3	1.4	32	1.0	24	6
4	1.0	24	1.4	32	9
5	0.5	12	1.9	48	12
6	0.0	0	2.4	60	15

8. Set the timer to 15 minutes and begin timing. Using forceps, add a piece of beet to each of the six well plates as shown in Figure 2. Stir the beet in the salt solution once every minute with a toothpick. Be careful not to puncture or damage the beet.

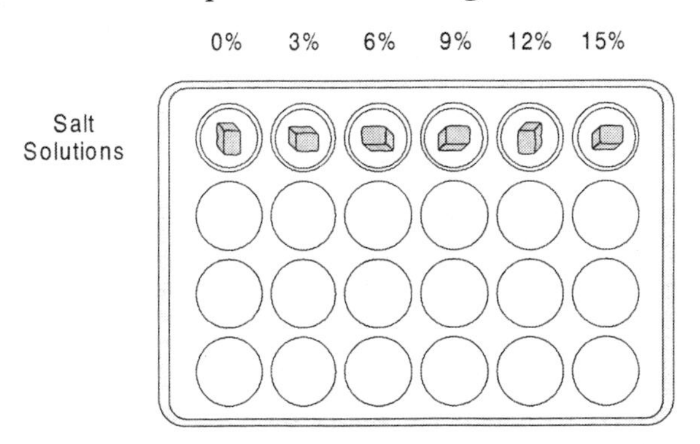

Figure 2

9. Connect the Colorimeter to the computer interface. While one team member is performing Step 8, another team member should prepare the computer for data collection by opening the file "09a Biological Membranes" from the *Biology with Computers* folder of Logger*Pro*.

10. You are now ready to calibrate the Colorimeter. Prepare a *blank* by filling a cuvette 3/4 full with distilled water. To correctly use a Colorimeter cuvette, remember:
    - All cuvettes should be wiped clean and dry on the outside with a tissue.
    - Handle cuvettes only by the top edge of the ribbed sides.
    - All solutions should be free of bubbles.
    - Always position the cuvette with its reference mark facing toward the white reference mark at the top of the cuvette slot on the Colorimeter.

11. Calibrate the Colorimeter.
    a. Open the Colorimeter lid.
    b. Holding the cuvette by the upper edges, place it in the cuvette slot of the Colorimeter. Close the lid.
    c. If your Colorimeter has a CAL button, Press the < or > button on the Colorimeter to select a wavelength of 470 nm (Blue) for this experiment. Press the CAL button until the red LED begins to flash. Then release the CAL button. When the LED stops flashing, the calibration is complete. Proceed directly to Step 12. If your Colorimeter does not have a CAL button, continue with this step to calibrate your Colorimeter.

    First Calibration Point

    d. Choose Calibrate ▸ CH1: Colorimeter (%T) from the Experiment menu and then click [Calibrate Now].
    e. Turn the wavelength knob on the Colorimeter to the "0% T" position.
    f. Type "0" in the edit box.
    g. When the displayed voltage reading for Reading 1 stabilizes, click [Keep].

    Second Calibration Point

    h. Turn the knob of the Colorimeter to the Blue LED position (470 nm).
    i. Type "100" in the edit box.
    j. When the displayed voltage reading for Reading 2 stabilizes, click [Keep], then click [Done].

12. After 15 minutes, remove the beet pieces from the wells. Remove them in the same order that they were placed in the well. Discard the beet pieces and retain the colored solutions.

13. You are now ready to collect absorbance data for the salt solutions.
    a. Click [Collect] to begin data collection.
    b. Empty the water from the cuvette. Use a cotton swab to dry the cuvette after the water has been emptied from it.
    c. Transfer all of the 0% salt solution from Well 1 into the cuvette using a Beral pipet. Wipe the outside with a tissue and place it in the colorimeter. After closing the lid, wait for the absorbance value displayed in the meter to stabilize.
    d. Click on [Keep], enter "0" in the edit box and then press ENTER. The data pair you just collected should now be plotted on the graph.

14. Collect the next data point.
    a. Discard the cuvette contents into your waste beaker. Remove all of the solution from the cuvette. Use a cotton swab to dry the cuvette.
    b. Fill the cuvette with the 3% salt solution from Well 2 using a Beral pipet. Wipe the outside with a tissue and place it in the colorimeter.
    c. After closing the lid, wait for the absorbance value displayed in the meter to stabilize. Click [Keep], enter "3" in the edit box and then press ENTER.

15. Repeat Step 14 to save and plot absorbance and concentration values of the solutions in Wells 3, 4, 5, and 6. When you have finished with all of the salt solutions, click [Stop].

16. In Table 2, record the absorbance and concentration data pairs listed in the table.

17. Print a graph with the salt solutions plotted. Enter your name(s) and number of copies of the graph. Use your graph to answer the discussion questions at the end of this experiment.

18. Discard the solutions as directed by your teacher.

Testing for the effect of detergents

19. Predict what results you expect when cells are immersed in detergent.

20. Design a set of experiments that test the effect of a detergent on biological membranes.

21. In your lab book, describe how you would test the effect of detergent on biological membranes. Think about the equipment you will need in your experiment. You might want to use some of the materials listed in the materials table. Bring the procedure and the predictions for each of the six trials to class at the next meeting.

Day 2:

1. Have your instructor check the procedure you wrote. If it is approved, carry out the experiment. Make a data table that contains the data and describe the results and conclusions of your experiment.

2. Prepare for data collection by opening the file "Exp 09b Biological Membranes" in the *Biology with Computers* folder.. Set the maximum x-axis value to the highest detergent concentration value in your test.

Testing for the effect of pH changes

3. Predict what results you expect when the pH of the cells change from acidic to basic conditions.

4. In your lab book, describe how you would test the affect of pH changes on cell membranes. Bring the procedure and the predictions for each of the six trials to class at the next meeting.

Day 3:

1. Have your instructor check the procedure you wrote. If it is approved, carry out the experiment. Make a data table that contains the data and describe the results and conclusions of your experiment.

2. Prepare for data collection by opening the file "Exp 09c Biological Membranes" in the *Biology with Computers* folder.

3. Although the Blue LED was used in the colorimeter as a light source in the above experiments, use the Green LED setting when you measure pH effects. This is because at some pH values, the betacyanin pigment turns blue. When blue, the LED's blue light is not absorbed by the pigment. **Important**: You must calibrate the colorimeter as in Step 11, using the green setting in place of the blue setting.

DATA

Table 2		
Trial	Concentration of salt (%)	Absorbance
1	0	
2	3	
3	6	
4	9	
5	12	
6	15	

QUESTIONS

1. Which concentration of salt produced the most intensely red solution? The least intensely red solution?
2. Which salt concentration(s) had the least effect on the beet membrane? How did you arrive at this conclusion?
3. Did more damage occur at high or low salt concentrations? Explain why this might be so.
4. An effective way to kill a plant is to pour salt onto the ground where it grows. How might the salt prevent the plant's growth? Is this consistent with your data?

Questions 5 – 8 refer to the experiment that tested the effect of detergents on membranes.

5. What effect did detergents have on cell membranes?
6. How did your answer in Question 5 compare to your prediction?
7. What assumptions did you make while designing your experiment that tested for the effect of detergents? How do you know they are valid assumptions to make?
8. How would you modify your experiment to either improve your results or to explore the validity of your assumptions?

Questions 9 – 12 refer to the experiment that tested the effect of pH on membranes.

9. What effect did changing the pH of the cell's environment have on cell membranes?
10. How did your answer in Question 9 compare to your prediction?
11. What assumptions did you make while designing your experiment in Day 2, testing for the effect of pH changes? How do you know they were valid assumptions to make?

TEACHER INFORMATION

Experiment
9

Biological Membranes

1. In this series of experiments, students will test the effect of salt, detergent, and pH changes on biological membranes. The first of these experiments, testing salt, is presented as a model for students. The second and third experiments will be designed by the students for homework. It often works well to have students work independently on the initial design, and collaboratively to finalize their procedures. You may want to allow sufficient class time for small group and class discussions of the various experimental designs. Emphasize that there is no one right way to do the lab.

2. To prepare a 15% salt solution, measure 15 grams of NaCl into sufficient water to make 100 mL of solution. Approximately 10 mL will be needed per group.

3. Step 3 of the student procedure has two options for measuring out volumes of 15% salt solution and water. Students can use either 2 mL graduated pipets (volumes in mL) or Beral pipets (volumes in drops). Volumes for both methods are shown in Table 1. Be sure to instruct your students as to which method they will be using.

4. To prepare a 0.5% detergent solution, measure 0.5 grams of Sodium Lauryl Sulfate (abbreviated SDS, and also called Sodium Dodecyl Sulfate) into sufficient water to make 100 mL of solution. Approximately 10 mL will be needed per group. You may want to have different groups try a variety of different detergents. SDS is the active ingredient in many shampoos.

5. Solutions of the following pH values should be available for student use: 2, 4, 6, 8, 10, and 12. To prepare the buffered solutions, make 500 mL of 0.4 M H_3PO_4 (90 mL concentrated phosphoric acid to 410 mL water) and 500 mL of 0.4 M Na_3PO_4 (76 g sodium phosphate tribasic to sufficient water to make 500 mL). Combine the two buffer solutions until the correct pH's are obtained. Each team will require about 10 mL of each buffer.

6. The light source for the red pigmented solution is the Blue LED. The Green LED should be used when the effects of pH are tested.

7. Beets will stain hands and clothing. Lab aprons and gloves should be worn to prevent staining.

8. Obtain large, red beets from a local store. Carefully cut a $0.5 \times 0.5 \times 10$ cm piece of beet just before use. Do not let the cut surfaces dry out.

9. The cuvette must have at least 2 mL of solution in order to get a valid absorbance reading. If students fill the cuvette as described in the procedure, they should be within this range.

10. Since there is some variation in the amount of light absorbed by the cuvette if it is rotated 180°, you should use a water-proof marker to make a reference mark on the top edge of one of the clear sides of all cuvettes. Students are reminded in the procedure to align this mark with the white reference mark to the right on the cuvette slot on the colorimeter.

11. The use of a single cuvette in the procedure is to eliminate errors introduced by slight variations in the absorbance of different plastic cuvettes. If one cuvette is used throughout the experiment by a student group, this variable is eliminated.

12. We recommend the use of cotton swabs to dry a cuvette after water or a solution has been emptied from it. The cotton swabs will remove any remaining droplets in the cuvette.

13. Satisfactory results can be obtained if a different cuvette is used for each solution of the experiment. For best results, the cuvettes for one student lab team should be *matched.* Each cuvette in a matched set absorbs light (when empty) at approximately the same level. To match a set of cuvettes, first calibrate the colorimeter using the method described in the section on calibration. Use a clean, dry cuvette for the 100% calibration instead of a distilled water blank. Put a reference mark on one of the clear sides of the cuvette so it is always oriented the same way in the cuvette slot. Place each cuvette in the batch into the colorimeter and record percent transmittance values for each. When you are finished, group cuvettes according to similar %T values. Each of these groups represents a set of matched cuvettes.

SAMPLE RESULTS

Testing Osmotic Stress

Table 1

Trial	Salt (%)	Absorbance
1	0	.302
2	3	.190
3	6	.201
4	9	.193
5	12	.199
6	15	.380

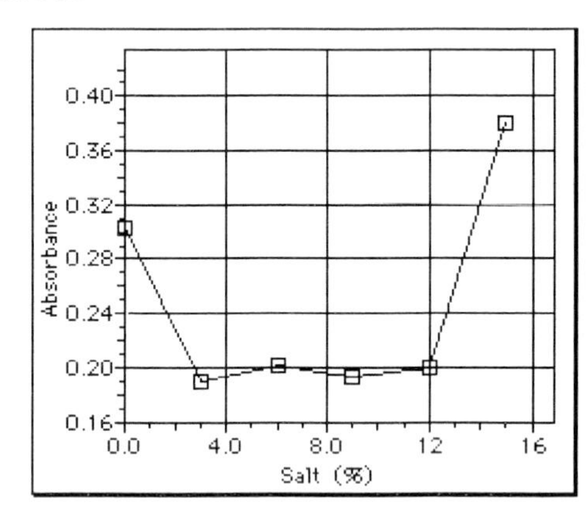

Testing the Effect of Detergents

Table 2

Trial	SDS (%)	Absorbance
1	0	.056
2	0.1	.490
3	0.2	.806
4	0.3	.833
5	0.4	.840
6	0.5	.800

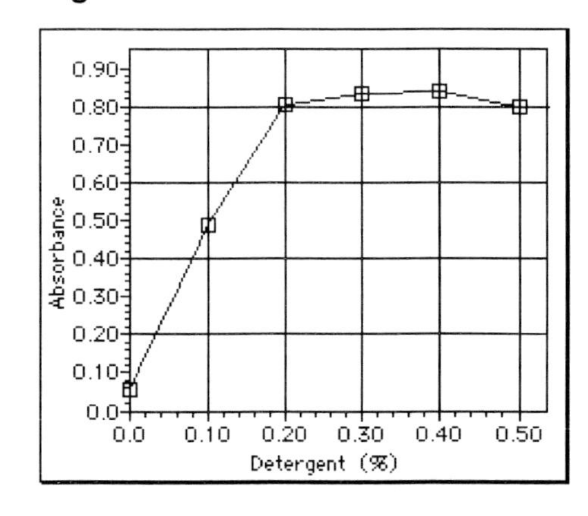

Testing the Effect of pH Changes

Table 3		
Trial	pH	Absorbance
1	2	.791
2	4	.346
3	6	.077
4	8	.065
5	10	.115
6	12	.266

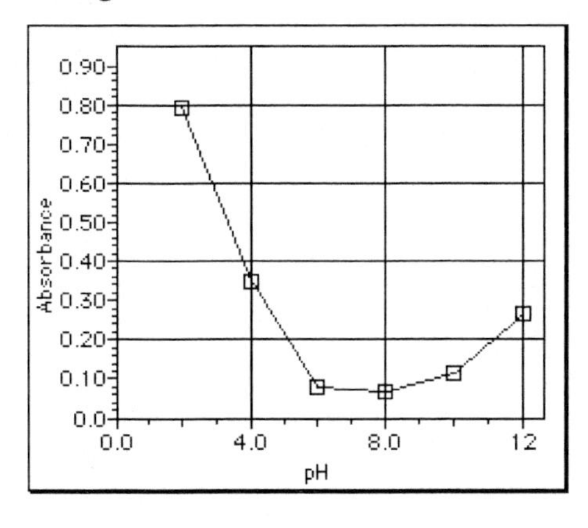

ANSWERS TO QUESTIONS

1. Answers may vary. In the above experiment, 15% salt caused the most intensely red solution, while the 3% solution was least intense.

2. Salt concentrations between 3% and 12% caused the least damage. Those solutions were only faintly red. Cellular damage was minimal.

3. More damage occurred at very high salt concentrations. This was due to osmotic stress. So much water was drawn out of the cells that they were destroyed. Plasmolysis occurred.

4. Salt on the ground might cause cellular damage by plasmolysis in plant tissues. This is supported by the evidence in Question 3.

5. Detergents had a drastic effect at even very low concentrations. The damage increased up to 0.2% SDS, and stabilized at higher concentrations. Detergents are designed to make lipids soluble in water. Since membranes have lipids and other non-polar molecules in them, membrane disruption and cellular damage are expected.

6. Answers may vary. Their responses should be consistent with their predictions.

7 - 8. Answers may vary, depending upon the design of their experiment.

9. Cell membranes were least damaged in a pH range of 6 – 10. In acidic conditions, the damage was extensive.

10. Answers may vary. Their responses should be consistent with their predictions.

11. Answers may vary, depending upon the design of their experiment.

Experiment
10

Transpiration

Water is transported in plants, from the roots to the leaves, following a decreasing water potential gradient. *Transpiration*, or loss of water from the leaves, helps to create a lower osmotic potential in the leaf. The resulting transpirational pull is responsible for the movement of water from the xylem to the mesophyll cells into the air spaces in the leaves. The rate of evaporation of water from the air spaces of the leaf to the outside air depends on the water potential gradient between the leaf and the outside air.

Various environmental factors, including those conditions which directly influence the opening and closing of the stomata, will affect a plant's transpiration rate. This experiment will measure transpiration rates under different conditions of light, humidity, temperature, and air movement. The data will be collected by measuring pressure changes as the plant takes up water into the stem.

OBJECTIVES

In this experiment, you will

- Observe how transpiration relates to the overall process of water transport in plants.
- Use a computer interface and a Gas Pressure Sensor to measure the rate of transpiration.
- Determine the effect of light intensity, humidity, wind, and temperature on the rate of transpiration of a plant cutting.

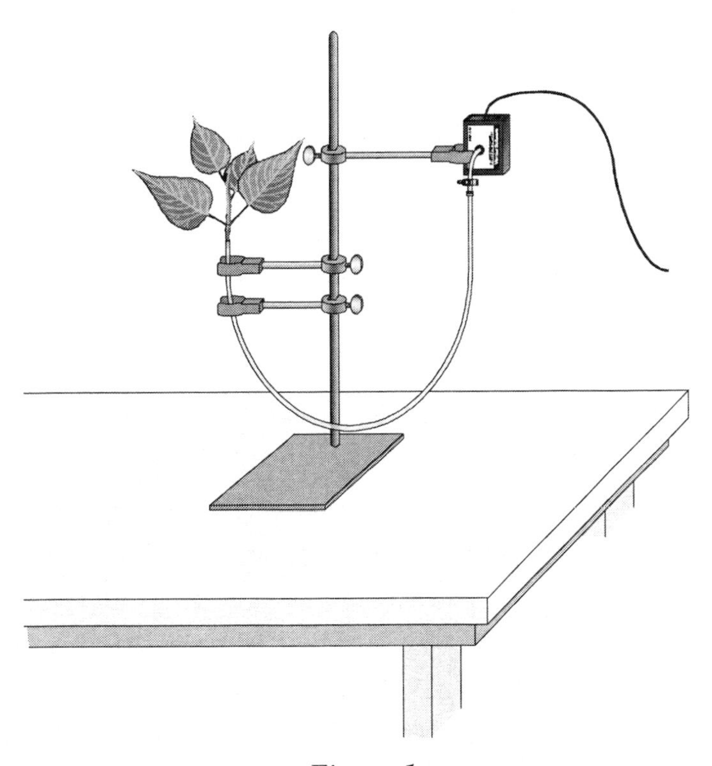

Figure 1

MATERIALS

computer
Vernier computer interface
Logger Pro
Vernier Gas Pressure Sensor
utility clamps
ring stand
plant cuttings
plastic tubing clamps
dropper or Beral pipette
razor blade or scalpel
metric ruler
masking tape
100 watt light source
plastic gallon size bag with twist tie
heater, small electric
fan with slow speed
aerosol spray container or plant mister
plastic syringe

PROCEDURE

1. Position the ring stand, utility clamps, and Gas Pressure Sensor as shown in Figure 1.

2. Prepare the plastic tubing.

   a. Connect the plastic syringe to one end of a 36-42 cm piece of plastic tubing.
   b. Place the other end of the tubing into water and use the syringe to draw water up into the tubing until it is full. Tap the tubing to expel any air bubbles that form inside the tube.
   c. Slip a plastic tubing clamp onto the tubing as shown in Figure 2.
   d. Bend the tubing into a U shape with both ends up. Remove the syringe, leaving the tubing full of water.

Figure 2

3. Select a plant which has a stem roughly the same diameter as the opening of the plastic tubing. Using a scalpel or razor blade, carefully cut the plant one inch above the soil. Place the plant under water against a hard surface and make a new cut at a 45° angle near the base of the stem.

4. Connect the plant to the tubing.

   a. The plastic tubing has a white plastic connector at one end that allows you to connect it to the valve on the Gas Pressure Sensor. Raise the end of the tubing with the connector until you see water beginning to drip out of the other end.
   b. Carefully push the cut stem of the plant down into the end of the tubing where the water is dripping out. Be careful not to allow any air bubbles to form between the cut portion of the stem and the water in the tube.
   c. Push the plant down as far as it will go without damaging the plant. At least one centimeter of the plant stem should fit into the tubing. If the stem is too large for the tubing, cut the stem at a higher point where it is smaller.
   d. Squeeze the tubing clamp shut as tight as possible as shown in Figure 3.

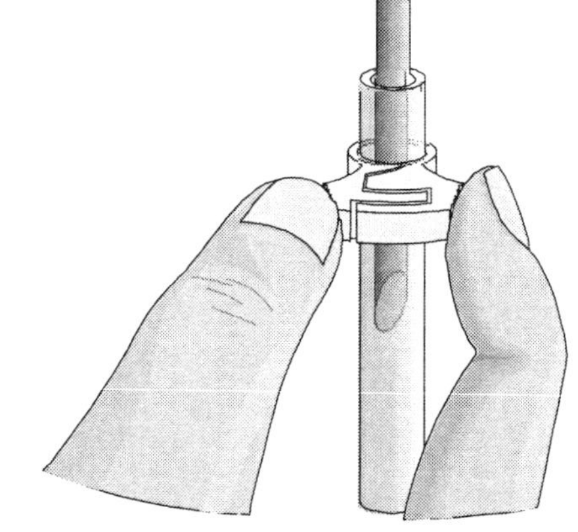

Figure 3

5. When the tubing clamp is shut tight, invert your plant cutting to check for any leaks. If water does leak out, turn the plant right-side up and try tightening the clamp further.

   Important: Be sure the tubing is filled completely with water. The water column must be flush with the stem. There should be no air visible at the base of the stem. If water moves down the tube away from the stem after it has been inserted, check for a leak in the system.

6. Connect the plastic tubing to the sensor valve. **Caution**: Do not allow water to enter the valve of the Gas Pressure Sensor.

7. Secure the plant in an upright position with the utility clamps as shown in Figure 1. It should be positioned so that the cut stem is about 8 cm below the water level at the other end of the tubing, as shown in Figure 1.

8. Place a mark on the tube at the starting water level to allow you to refill the tube to the proper level in Step 17.

9. Place your plant setup in an area where the wind, humidity, and temperature are reasonably constant. This will be your control setup.

10. Allow the system 5 minutes to adjust to the environment. While the system is adjusting, set up the computer.

11. Connect the Gas Pressure Sensor to the computer interface. Prepare the computer for data collection by opening the file "10 Transpiration" from the *Biology with Computers* folder of Logger*Pro*.

12. Check the base of the plant stem in the water tube to make sure that no air bubbles or air pockets have formed that will prevent the plant from taking up water. If an air pocket has formed, refit the plant in the tubing before initiating data collection in Step 13.

13. After the plant has equilibrated for 5 minutes, click Collect to begin data collection. Data will be collected for 15 minutes.

14. When data collection has finished, find the rate of transpiration for your plant. To do this,

    a. Move the mouse pointer to the point where the pressure values begin to decrease. Click the mouse button and drag the pointer to the end of the data, then release the mouse button.
    b. Click the Linear Fit button, , to perform a linear regression. A floating box will appear with the formula for a best fit line.
    c. Record the slope of the line, *m*, in Table 1 as the rate of transpiration for the control. Close the floating box.

15. (Optional) Double click anywhere on the graph and enter "Transpiration: Control" as the graph title. Print a copy of your graph. Enter your name(s) and the number of copies of the graph.

16. Design an experiment to simulate one of the following environmental factors, as assigned by your teacher:
    - the effect of light intensity
    - the effect of the wind blowing on the plant
    - the effect of humidity
    - the effect of temperature
    - the effect of another self-identified environmental variable

    Be sure to address the following questions in your design:
    - What is the essential question being addressed?
    - What assumptions are made about the system being measured?
    - Can those assumptions be easily verified?
    - Will the measurements provide the necessary data to answer the question under study?

17. After checking your procedure with your teacher, obtain the materials needed for the experiment and perform the tests. Refill the water level in the tube to the same level marked in Step 8. Record your values in Table 1.

PROCESSING THE DATA

1. Determine the surface area of all the leaves on your plant cutting by the following method:
    a. Cut all the leaves (not stems) off your plant and determine their mass using a balance.
    b. Estimate the total leaf surface area in cm^2 for your plant by cutting out a section of leaf 5 cm × 5 cm.
    c. Determine the mass for this leaf section and divide by 25 cm^2 to find the mass of 1 cm^2 of leaf.
    d. Divide the total mass of the leaves by the mass of 1 cm^2 to find the total leaf surface area.
    e. Record the calculated surface area in Table 1.

2. Calculate the rate of transpiration/surface area. To do this, divide the rate of transpiration by the surface area for each plant. These rate values can be expressed as $kPa/min/cm^2$. Record the rate/area in Table 1.

3. Subtract the control (rate/area) value from the experimental value. Record this adjusted rate in the last column of Table 1.

4. Record the adjusted rate for your experimental test on the board to share with the class. Record the class results in Table 2 for each of the environmental conditions tested. If a condition was tested by more than one group, take the average of the values and record in Table 2.

5. Make a bar graph that shows the effect of different environmental conditions on the transpiration of water in plant cuttings. Using the data in Table 2 plot the adjusted rate for each test on the y-axis and the test label on the x-axis.

DATA

Table 1				
Test	Slope (kPa/min)	Surface area (cm^2)	Rate/area (kPa/min/cm^2)	Adjusted rate (kPa/min/cm^2)
Experimental ______________				
Control				

Table 2	
Class Data	
Test	Adjusted rate (kPa/min/cm^2)
Light	
Humidity	
Wind	
Temperature	

QUESTIONS

1. How was the rate of transpiration affected in each of the experimental situations as compared to the control?
2. Which variable resulted in the greatest rate of water loss? Explain why this factor might increase water loss when compared to the others.
3. What adaptations enable plants to increase or decrease water loss? How might each affect transpiration?

EXTENSIONS

1. Using a compound microscope, identi:'y the vascular tissues of a plant stem. Describe the function of each tissue type identified.

   a. Obtain a section of stem from the plant you used during the transpiration experiment.
   b. Using a nut-and-bolt microtome, carefully cut 6 cross sections of the plant stem. The cross sections should be cut as thin as possible.
   c. Place each of the cross sections in a dish or cup of 50% ethanol solution for 5 minutes.
   d. Remove the cross sections from the alcohol and place them in a dish containing toludine blue O stain for 5 minutes.
   e. Rinse the cross sections with distilled water and mount them on a microscope slide with a

drop of 50% glycerin. Place a cover slip on the slide and examine the cross sections using a compound microscope.

f. On a separate sheet of paper, make a drawing of the cross sections. Identify and label the cell and tissue types described by your teacher.

2. Test cuttings from a variety of different plant species. How does each compare?

3. Count the number of stoma/cm^2 for each of the plants in Extension 1. How does this relate to the plant's ability to transpire water?

4. Design an experiment to test for the variables in Question 3.

Experiment
10

TEACHER INFORMATION

Transpiration

1. You should leave water out overnight in a beaker or cup to allow any excess dissolved air to escape. This will ensure that no air bubbles form in the tube at the cut end of the stem. If air bubbles form, it may be necessary to restart your experiment. If bubbles do form, remove the plant and tubing from the two utility clamps and allow the plant to hang towards the ground with the other end of the tubing pointing up. Carefully tap on the sides of the tubing to loosen any bubbles—they will float to the water's surface at the other end. Once all bubbles are removed, check the plant's seal at the tube. Secure your plant in the tubing and restart the data collection.
2. There is not always an immediate change in the transpiration rate. Allow the plant to spend a few extra minutes under a particular condition before initiating data collection. This will give the plant the necessary time to adjust. When the transpiration rate changes drastically the stomata will close, decreasing the transpiration rate. If the length of data collection is extended, you will be able to see on the graph when the stomata have closed and the rate slows down.
3. Many plants work well for this experiment. Plants that have been used include tomato, strawberry, bean, geranium, cyclamen, and even honeysuckle. For best results, we recommend using plants with numerous leaves. Tomato plants work very well and have been used to collect the sample data for this activity. One possible extension of this experiment would be to have the students use different plant species under similar conditions and evaluate how different plants have adapted to prevent water loss.
4. The thick-wall plastic tubing that comes with the Gas Pressure Sensor is well suited for this lab. The inner diameter of the tubing is 3 mm and may be too small for some plant specimens. Science supply companies carry thick-wall plastic tubing, with a larger inner diameter, that will work well on larger plant stems. They also sell tubing connectors that will allow you to connect the larger tubing to the tubing provided with the Gas Pressure Sensor.
5. Emphasize to your students the importance of providing an airtight fit with all plastic-tubing connections.
6. All of the tubing, and connectors used in this experiment are included with Vernier Gas Pressure Sensors shipped after February 15, 1998. These accessories are also helpful when performing respiration/fermentation experiments such as Experiments 11C, 12B, and 16B in this manual.

   If you purchased your Gas Pressure Sensor at an earlier date, Vernier has a Pressure Sensor Accessories Kit (PS-ACC) that includes all of the parts shown here for doing pressure-related experiments. Using this kit allows for easy assembly of a completely airtight system. The kit includes the following parts:

   - two ribbed, tapered valve connectors inserted into a No. 5 rubber stopper
   - one ribbed, tapered valve connectors inserted into a No. 1 rubber stopper
   - two Luer-lock connectors connected to either end of a piece of plastic tubing
   - one two-way valve
   - one 20 mL syringe
   - two tubing clamps for transpiration experiments

7. The Vernier Barometer sensor can also be used to perform this experiment. If you already have a Barometer and wish to do this activity, you will need to order the following parts from Vernier Software:

   Pressure Sensor Accessories Kit (order code PS-ACC)

   Plastic 3-Way Pressure Sensor Valve (order code PSV)

8. The plastic tubing clamps used in the student procedure may be purchased from Vernier Software & Technology:

   Plastic Tubing Clamps (order code PTC: package of 100)

SAMPLE DATA

Table 1	
Test	Adjusted rate (kPa/min/cm^2)
Control	-2.39×10^{-3}
Light	-5.63×10^{-3}
Humidity	-0.52×10^{-3}
Wind	-0.16×10^{-3}

ANSWERS TO QUESTIONS

1. It is typically predicted that the light and wind will increase the rate of transpiration. This may not be apparent until after correction for surface area differences. Sometimes the wind, if too strong, may cause the leaves to droop or fold up, and in this case they may transpire less. Stomates may close to counter the dehydration. If this happens, discuss the nature of science experimentation, e.g., the expected may not always be the result. Usually, after correction for surface area, the high humidity plant will transpire less than a control. A student may question whether the light increased the temperature of the leaf. If the light was too close to the plant, temperature may indeed be a variable without a control.

2. Answers will vary—usually the light will produce the greatest rate of water loss. High light intensity increases water loss due to increased photosynthesis. Wind removes water vapor from the surface of the leaf more rapidly. It may increase the evaporation rate by increasing the gradient between water in the leaf air spaces and water vapor in the air.

3. Plants can increase or decrease water loss by

   - closing the stomata during water stress.
   - reducing the number of stomata.
   - waxy cuticles.
   - fleshy, thick leaves.
   - hairy surfaces.
   - reducing the overall leaf surface area.

Experiment
11A

Cell Respiration

Cell respiration refers to the process of converting the chemical energy of organic molecules into a form immediately usable by organisms. Glucose may be oxidized completely if sufficient oxygen is available by the following equation:

$$C_6H_{12}O_6 + 6O_2(g) \rightarrow 6\ H_2O + 6\ CO_2(g) + \text{energy}$$

All organisms, including plants and animals, oxidize glucose for energy. Often, this energy is used to convert ADP and phosphate into ATP. It is known that peas undergo cell respiration during germination. Do peas undergo cell respiration before germination? The results of this experiment will verify that germinating peas do respire. Using your collected data, you will be able to answer the question concerning respiration and non-germinating peas.

Using the O_2 Gas Sensor, you will monitor the oxygen consumed by peas during cell respiration. Both germinating and non-germinating peas will be tested. Additionally, cell respiration of germinating peas at two different temperatures will be tested.

OBJECTIVES

In this experiment, you will

- Use an O_2 Gas Sensor to measure concentrations of oxygen.
- Study the effect of temperature on cell respiration.
- Determine whether germinated and non-germinated peas respire.
- Compare the rates of cell respiration in germinated and non- germinated peas.

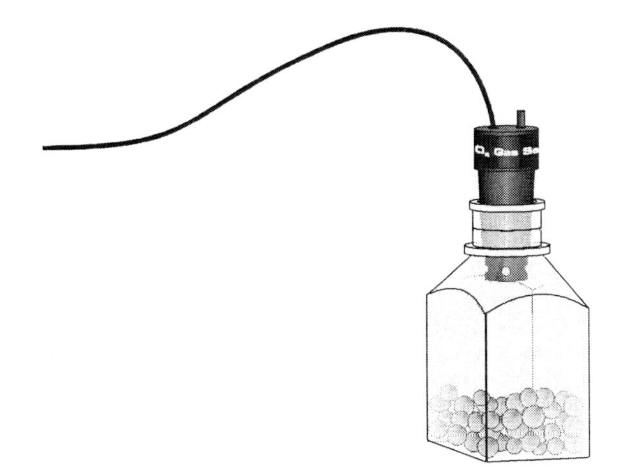

Figure 1

MATERIALS

computer
Vernier computer interface
Logger*Pro*
Vernier O_2 Gas Sensor
25 germinating peas
25 non-germinating peas
250 mL respiration chamber
ice cubes
1 L beaker
thermometer
two 100 mL beakers

PROCEDURE

1. Connect the O_2 Gas Sensor to the computer interface.

2. Prepare the computer for data collection by opening the file "11A Cell Resp O2" from the *Biology with Computers* folder of Logger*Pro*.

3. Obtain 25 germinating peas and blot them dry between two pieces of paper towel. Use the thermometer to measure the room temperature. Record the temperature in Table 1.

4. Place the germinating peas into the respiration chamber.

5. Place the O_2 Gas Sensor into the bottle as shown in Figure 1. Gently push the sensor down into the bottle until it stops. The sensor is designed to seal the bottle without the need for unnecessary force.

6. Wait two minutes, then begin collecting data by clicking Collect. Data will be collected for 10 minutes.

7. When data collection has finished, remove the O_2 Gas Sensor from the respiration chamber. Place the peas in a 100 mL beaker filled with cold water and an ice cube.

8. Fill the respiration chamber with water and then empty it. Thoroughly dry the inside of the respiration chamber with a paper towel.

9. Determine the rate of respiration:
   a. Click the Linear Fit button, , to perform a linear regression. A floating box will appear with the formula for a best fit line.
   b. Record the slope of the line, *m*, as the rate of respiration for germinating peas at room temperature in Table 2.
   c. Close the linear regression floating box.

10. Move your data to a stored run. To do this, choose Store Latest Run from the Experiment menu.

11. Obtain 25 non-germinating peas and place them in the respiration chamber

12. Repeat Steps 5 – 10 for the non-germinating peas.

Part II Germinating peas, cool temperatures

13. Place the respiration chamber in a 1 L beaker. Cover the outside of the chamber with ice.

14. Use the thermometer to measure the water temperature of the 100 mL beaker containing the germinating peas. Record the water temperature in Table 1.

15. Remove the peas from the cold water and blot them dry between two paper towels.

16. Repeat Steps 5 – 9 to collect data with the germinating peas at a cold temperature.

17. To print a graph showing all three data runs:
    a. Label all three curves by choosing Text Annotation from the Insert menu, and typing

"Room Temp Germinated" (or "Room Temp Non-germinated", or "Cold Germinated") in the edit box. Then drag each box to a position near its respective curve. Adjust the position of the arrow head.

b. Print a copy of the graph, with all three data sets and the regression lines displayed. Enter your name(s) and the number of copies of the graph you want.

DATA

Table 1	
Condition	Temperature (°C)
room	
cold water	

Table 2	
Peas	Rate of Respiration (%/min)
Germinating, room temperature	
Non-germinating, room temperature	
Germinating, cool temperature	

QUESTIONS

1. Do you have evidence that cell respiration occurred in peas? Explain.
2. What is the effect of germination on the rate of cell respiration in peas?
3. What is the effect of temperature on the rate of cell respiration in peas?
4. Why do germinating peas undergo cell respiration?

EXTENSIONS

1. Compare the respiration rate among various types of seeds.
2. Compare the respiration rate among seeds that have germinated for different time periods, such as 1, 3, and 5 days.
3. Compare the respiration rate among various types of small animals, such as insects or earthworms.

TEACHER INFORMATION

Cell Respiration

1. Allow the seeds to germinate for three days prior to the experiment. Prior to the first day, soak them in water overnight. On subsequent days, roll them in a moist paper towel and place the towel in a paper bag. Place the bag in a warm, dark place. Check each day to be sure the towels remain very moist. If time is short, the peas can be used after they have soaked overnight. For best results, allow them to germinate for the full three days.
2. To extend the life of the O_2 Gas Sensor, always store the sensor upright in the box it was shipped in.
3. The morning of the experiment fill a 1 L beaker with ice and water so that students will have cold water for Step 7. Students will also need access to ice for Steps 7 and 14.
4. The calibration stored in this experiment file works well for this experiment. The calibration is for the O_2 Gas Sensor (%).

SAMPLE RESULTS

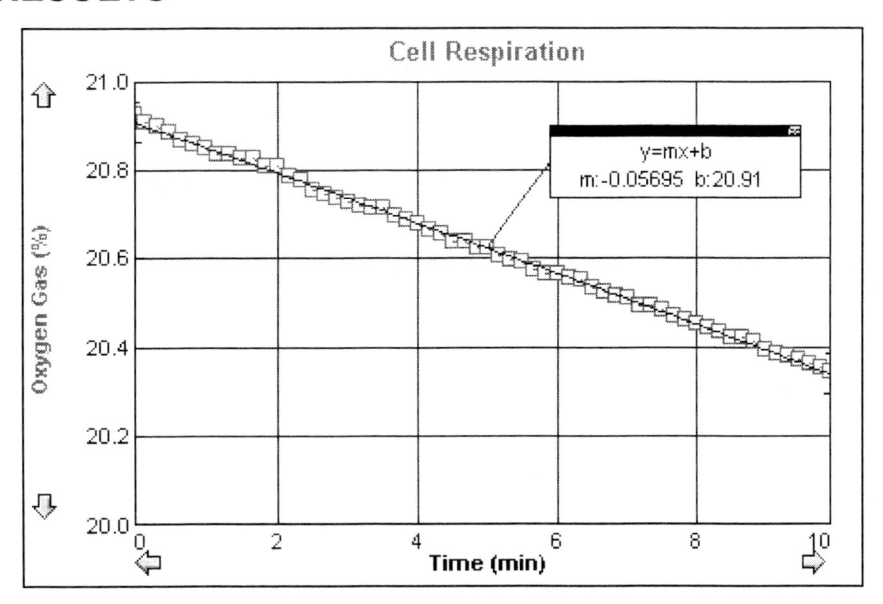

O_2 respired by germinated peas

Table 1	
Condition	Temperature (°C)
room	24
cold water	9

Table 2	
Peas	Rate of Respiration (%/min)
Germinating, room temperature	–0.057
Non-germinating, room temperature	–0.002
Germinating, cool temperature	–0.039

ANSWERS TO QUESTIONS

1. Yes, the oxygen concentration *vs.* time graph clearly indicates that oxygen is being consumed at a steady rate when germinating peas are in the respiration chamber.

2. Germination greatly accelerates the rate of cellular respiration. This reflects a higher rate of metabolic activity in germinating seeds. In most experiments, non-germinating seeds do not seem to be respiring. Occasionally, however, some respiration is detectable.

3. Warm temperatures increase the rate of respiration. This reflects a higher rate of metabolic activity in warm germinating seeds than in cool seeds.

4. It is necessary for germinating seeds to undergo cellular respiration in order to acquire the energy they need for growth and development. Unlike their mature relatives, seeds do not yet have the necessary photosynthetic abilities needed to produce their own energy sources.

Experiment

11B

Cell Respiration

Cell respiration refers to the process of converting the chemical energy of organic molecules into a form immediately usable by organisms. Glucose may be oxidized completely if sufficient oxygen is available by the following equation:

$$C_6H_{12}O_6 + 6O_2(g) \rightarrow 6\ H_2O + 6\ CO_2(g) + \text{energy}$$

All organisms, including plants and animals, oxidize glucose for energy. Often, this energy is used to convert ADP and phosphate into ATP. It is known that peas undergo cell respiration during germination. Do peas undergo cell respiration before germination? The results of this experiment will verify that germinating peas do respire. Using your collected data, you will be able to answer the question concerning respiration and non-germinating peas.

Using the CO_2 Gas Sensor, you will monitor the carbon dioxide produced by peas during cell respiration. Both germinating and non-germinating peas will be tested. Additionally, cell respiration of germinating peas at two different temperatures will be tested.

OBJECTIVES

In this experiment, you will

- Use a CO_2 Gas Sensor to measure concentrations of carbon dioxide.
- Study the effect of temperature on cell respiration.
- Determine whether germinated and non-germinated peas respire.
- Compare the rates of cell respiration in germinated and non- germinated peas.

Figure 1

MATERIALS

computer
Vernier computer interface
Logger*Pro*
Vernier CO_2 Gas Sensor
25 germinating peas
25 non-germinating peas
250 mL respiration chamber
ice cubes
1 L beaker
thermometer
two 100 mL beakers

PROCEDURE

1. Plug the CO_2 Gas Sensor into Channel 1 of the Vernier computer interface.

2. Prepare the computer for data collection by opening the "11B Cell Resp" file in the *Biology with Computers* folder.

3. Obtain 25 germinating peas and blot them dry between two pieces of paper towel. Use the thermometer to measure the room temperature. Record the temperature in Table 1.

4. Place the germinating peas into the respiration chamber.

5. Place the shaft of the CO_2 Gas Sensor in the opening of the respiration chamber. Gently twist the stopper on the shaft of the CO_2 Gas Sensor into the chamber opening. Do not twist the shaft of the CO_2 Gas Sensor or you may damage it.

6. Wait one minute, then begin measuring carbon dioxide concentration by clicking Collect. Data will be collected for 5 minutes.

7. Remove the CO_2 Gas Sensor from the respiration chamber. Place the peas in a 100 mL beaker filled with cold water and an ice cube. The cold water will prepare the peas for part II of the experiment.

8. Use a notebook or notepad to fan air across the openings in the probe shaft of the CO_2 Gas Sensor for 1 minute.

9. Fill the respiration chamber with water and then empty it. Thoroughly dry the inside of the respiration chamber with a paper towel.

10. Determine the rate of respiration:
    a. Move the mouse pointer to the point where the data values begin to increase. Hold down the left mouse button. Drag the mouse pointer to the end of the data and release the mouse button.
    b. Click the Linear Fit button, , to perform a linear regression. A floating box will appear with the formula for a best fit line.
    c. Record the slope of the line, *m*, as the rate of respiration for germinating peas at room temperature in Table 2.
    d. Close the linear regression floating box.

11. Move your data to a stored run. To do this, choose Store Latest Run from the Experiment menu.

12. Obtain 25 non-germinating peas and place them in the respiration chamber

13. Repeat Steps 5 – 11 for the non-germinating peas.

Part II Germinating peas, cool temperatures

14. Place the respiration chamber in a 1 L beaker. Cover the outside of the chamber with ice.

15. Use the thermometer to measure the water temperature of the 100 mL beaker containing the germinating peas. Record the water temperature in Table 1.

16. Remove the peas from the cold water and blot them dry between two paper towels.
17. Repeat Steps 5 – 11 to collect data with the germinating peas at a cold temperature.
18. To print a graph showing all three data runs:
    a. Label all three curves by choosing Text Annotation from the Insert menu, and typing "Room Temp Germinated" (or "Room Temp Non-germinated", or "Cold Germinated") in the edit box. Then drag each box to a position near its respective curve. Adjust the position of the arrowhead.
    b. Print a copy of the graph, with all three data sets and the regression lines displayed. Enter your name(s) and the number of copies of the graph you want.

DATA

Table 1	
Condition	Temperature (°C)
room	
cold water	

Table 2	
Peas	Rate of Respiration (ppm/min)
Germinating, room temperature	
Non-germinating, room temperature	
Germinating, cool temperature	

QUESTIONS

1. Do you have evidence that cell respiration occurred in peas? Explain.
2. What is the effect of germination on the rate of cell respiration in peas?
3. What is the effect of temperature on the rate of cell respiration in peas?
4. Why do germinating peas undergo cell respiration?

EXTENSIONS

1. Compare the respiration rate among various types of seeds.
2. Compare the respiration rate among seeds that have germinated for different time periods, such as 1, 3, and 5 days.
3. Compare the respiration rates of various small animal types, such as insects or earthworms.

Experiment

11B

TEACHER INFORMATION

Cell Respiration

1. Allow the seeds to germinate for three days prior to the experiment. Prior to the first day, soak them in water overnight. On subsequent days, roll them in a moist paper towel and place the towel in a paper bag. Place the bag in a warm, dark place. Check each day to be sure the towels remain very moist. If time is short, the peas can be used after they have soaked overnight. For best results, allow them to germinate for the full three days.
2. Heavy condensation buildup in the respiration chamber can interfere with readings from the CO_2 Gas Sensor. This can be a source of error if the peas are very wet when placed in the respiration chamber. Before placing the peas in the respiration chamber, blot them dry with a paper towel.
3. The stopper included with the CO_2 Gas Sensor is slit to allow easy application and removal from the probe. When students are placing the probe in the respiration chamber, they should gently twist the stopper into the chamber opening. Warn the students not to twist the probe shaft or they may damage the sensing unit.
4. The CO_2 Gas Sensor relies on the diffusion of gases into the probe shaft. Students should allow a couple of minutes between trials so that gases from the previous trial will have exited the probe shaft. Alternatively, the students can use a firm object such as a book or notepad to fan air through the probe shaft. This method is used in Step 8 of the student procedure.
5. The morning of the experiment fill a 1 L beaker with ice and water so that students will have cold water for Step 7. Students will also need access to ice for Step 14.
6. The calibration stored in this experiment file works well for this experiment. Initial readings that seem slightly high or low will still reflect an accurate change in gas levels.

SAMPLE RESULTS

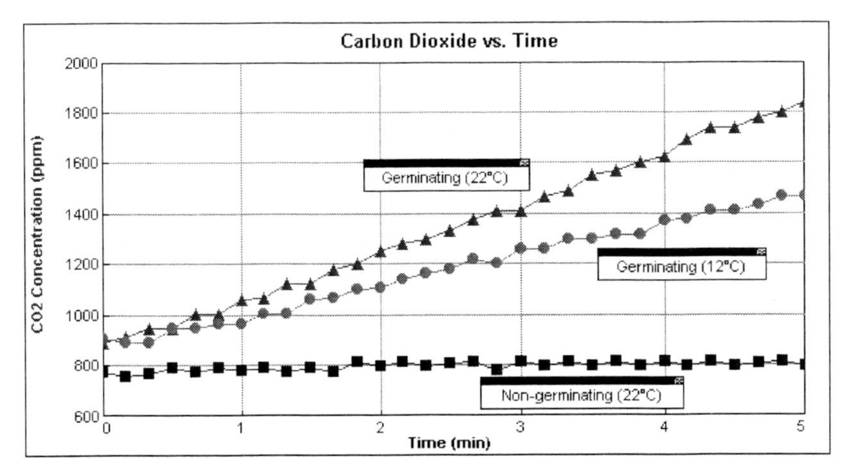

CO_2 respired by germinating and non-germinating peas

Table 1	
Condition	Temperature (°C)
room	22
cold water	10

Table 2	
Peas	Rate Respiration (ppm/min)
Germinating, room temperature	194.5
Non-germinating, room temperature	6.6
Germinating, cool temperature	122.1

ANSWERS TO QUESTIONS

1. Yes, the carbon dioxide concentration *vs.* time graph clearly indicates that carbon dioxide is being produced at a steady rate when germinating peas are in the respiration chamber.

2. Germination greatly accelerates the rate of cellular respiration. This reflects a higher rate of metabolic activity in germinating seeds. In most experiments, non-germinating seeds do not seem to be respiring. Occasionally, however, some respiration is detectable.

3. Warm temperatures increase the rate of respiration. This reflects a higher rate of metabolic activity in warm germinating seeds than in cool seeds.

4. It is necessary for germinating seeds to undergo cellular respiration in order to acquire the energy they need for growth and development. Unlike their mature relatives, seeds do not yet have the necessary photosynthetic abilities needed to product their own energy sources.

Experiment

11C

Cell Respiration

Cell respiration refers to the process of converting the chemical energy of organic molecules into a form immediately usable by organisms. Glucose may be oxidized completely if sufficient oxygen is available, by the following equation:

$$C_6H_{12}O_6 + 6O_2(g) \rightarrow 6\ H_2O + 6\ CO_2(g) + \text{energy}$$

All organisms, including plants and animals, oxidize glucose for energy. Often, this energy is used to convert ADP and phosphate into ATP.

To measure the rate of cell respiration, the pressure change due to the consumption of oxygen by peas will be measured. It is not possible to directly measure pressure changes due to oxygen, since the pressure sensor measures the total pressure change. Carbon dioxide is produced as oxygen is consumed. The pressure due to CO_2 might cancel out any change due to the consumption of oxygen. To eliminate this problem, a chemical will be added that will selectively remove CO_2. Potassium hydroxide, KOH, will chemically react with CO_2 by the following equation:

$$2\ KOH + CO_2 \rightarrow K_2CO_3 + H_2O$$

This will allow you to monitor pressure changes exclusively due to the consumption of oxygen.

A *respirometer* is the system used to measure cell respiration. Pressure changes in the respirometer are directly proportional to a change in the amount of gas in the respirometer, providing the volume and the temperature of the respirometer do not change. If you wish to compare the consumption of oxygen in two different respirometers, as we will in this experiment, you must keep the volume and temperature of the air equal in each respirometer.

Both germinating and non-germinating peas will be tested. Additionally, cell respiration of germinating peas at two different temperatures will be tested.

OBJECTIVES

In this experiment, you will

- Use a computer and a Gas Pressure Sensor to measure pressure changes.
- Study the effect of temperature on cell respiration.
- Determine whether germinated and non-germinated peas respire.
- Compare the rates of cell respiration in germinated and non-germinated peas.

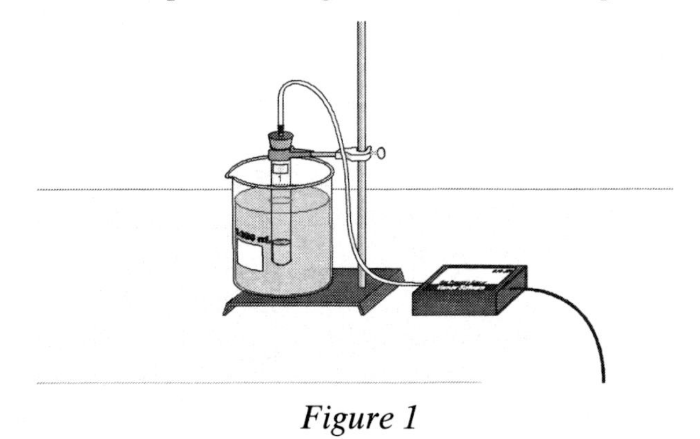

Figure 1

MATERIALS

computer
Vernier computer interface
Logger*Pro*
2 Vernier Gas Pressure Sensors
15% KOH in a dropper bottle
25 germinating peas
25 non-germinating peas
100 mL graduated cylinder
absorbent cotton
forceps
2 utility clamps
glass beads
ice
non-absorbent cotton
thermometer
test tube rack
timer with a second hand
three 18 ×150 mm test tubes
two 1-hole rubber stopper assemblies
two 1 L beakers
ring stand

PROCEDURE

1. Connect the plastic tubing to the valve on the Gas Pressure Sensor.
2. Connect the Gas Pressure Sensor to the computer interface. Prepare the computer for data collection by opening the file "11C Cell Resp (Pressure)" from the *Biology with Computers* folder of Logger*Pro*.

   To test whether germinating peas undergo cell respiration, you will need to

   - set up two water baths.
   - prepare a respirometer for the germinating peas.
   - prepare a second, control respirometer containing glass beads.
3. Set up two water baths, one at about 25°C and one at about 10°C. Obtain two 1 liter beakers and place about 800 mL of water in each. Add ice to attain the 10°C water bath.
4. To be sure the volumes of air in all respirometers are equal, you will need to measure the volume of the twenty-five peas that will be in the experimental respirometer. The control respirometer must have an equal volume of glass beads (or other non-oxygen consuming material) to make the air volume equal to the respirometer with germinating peas. Similarly, glass beads will be used to account for any volume difference between the germinating and non-germinating peas.
5. Obtain three test tubes and label them "T1", T2", and "T3". .
6. Place a 3 cm wad of absorbent cotton in the bottom of each test tube. Using a dropper pipette, carefully add a sufficient amount of KOH to the cotton to completely saturate it. Do not put so much that liquid can easily run out of the tube. Note: Do not allow any of the KOH to touch the sides of the test tube. The sides should be completely dry, or the KOH may damage the peas. **CAUTION:** Potassium hydroxide solution is caustic. Avoid spilling it on your clothes or skin.

7. Prepare the test tube containing germinating peas (T1):
   a. Add 50 mL of water to a 100 mL graduated cylinder.
   b. Place 25 germinating peas into the water.
   c. Measure the volume of the peas by water displacement. Record that volume in Table 1.
   d. Gently remove the peas from the graduated cylinder and blot them dry with a paper towel.
   e. Add a small wad of non-absorbent dry cotton to the bottom of the test tube to prevent the peas from touching the KOH saturated cotton.
   f. Add these germinating peas to the respirometer labeled "T1".

8. Prepare the test tube containing non-germinating peas (T2):
   a. Refill the graduated cylinder with 50 mL of water.
   b. Place 25 non-germinating peas into the water.
   c. Measure the volume of the peas by water displacement. Record the volume in Table 1.
   d. Add a sufficient number of glass beads to the non-germinating peas and water until they displace exactly the same volume of water as the germinating peas.
   e. Gently remove the peas and glass beads from the graduated cylinder and dry them with a paper towel.
   f. Add a small wad of dry non-absorbent cotton to the bottom of the test tube to prevent the peas from touching the KOH saturated cotton.
   g. Add the non-germinating peas and glass beads to the respirometer labeled "T2".

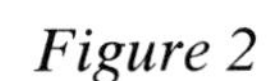

Figure 2

9. Prepare the test tube containing glass beads (T3):
   a. Refill the graduated cylinder with 50 mL of water.
   b. Add a sufficient number of glass beads to the water until they displace exactly the same volume of water as the germinating peas.
   c. Remove the glass beads from the graduated cylinder and dry them.
   d. Add a small wad of dry non-absorbent cotton to the bottom of the test tube to prevent the peas from touching the KOH saturated cotton.
   e. Add the glass beads to the respirometer labeled "T3".

Part I Germinating peas, room temperature

10. Insert a single-holed rubber-stopper into test tube T1 and T3. **Note**: *Firmly* twist the stopper for an *airtight* fit. Secure each test tube with a utility clamp and ring-stand as shown in Figure 1.

11. Arrange test tubes T1 and T3 in the warm water bath using the apparatus shown in Figure 1. Incubate the test tube for 10 minutes in the water bath. Be sure to keep the temperature of the water bath constant. If you need to add more hot or cold water, first remove about as much water as you will be adding, or the beaker may overflow. Use a basting bulb to remove excess water. Record the resulting temperature of the water bath once incubation has finished in Table 2.

    Note: Be sure the tubes are submerged to an equal depth, just up to the rubber stoppers. The temperature of the air in the tube must be constant for this experiment to work well.

12. When incubation has finished, connect the free-end of the plastic tubing to the connector in the rubber stopper as shown in Figure 3.

13. Click [▶ Collect] to begin data collection. Maintain the temperature of the water bath during the course of the experiment.

14. Data collection will end after 20 minutes. Monitor the pressure readings displayed in the live readouts on the toolbar. If the pressure exceeds 130 kPa, the pressure inside the tube will be too great and the rubber stopper is likely to pop off. Disconnect the plastic tubing from the Gas Pressure Sensor if the pressure exceeds 130 kPa.

15. The rate of respiration can be measured by examining the *slope* of the pressure change *vs.* time plot at the right of the screen. Calculate a linear regression for the pressure change vs. time graph:

    a. Click on the Pressure Change *vs.* Time graph to select it.
    b. Click the Linear Fit button, [Linear Fit], to perform a linear regression. A floating box will appear with the formula for a best fit line.
    c. Record the slope of the line, *m*, in Table 3 as the rate of oxygen consumption by germinating peas.
    d. Close the linear regression floating box.

Figure 3

16. Move your data to a stored data run. To do this, choose Store Latest Run from the Experiment menu.

Part II Non-germinating peas, room temperature

17. Disconnect the plastic tubing connectors from the rubber stoppers. Remove the rubber stopper from each test tube.

18. Repeat Steps 10 – 16, using test tubes T2 and T3.

Part III Germinating peas, cool temperatures

19. Disconnect the plastic tubing connectors from the rubber stoppers. Remove the rubber stopper from each test tube.

20. Repeat Steps 10 – 16, using test tubes T1 and T3 in a cold water bath.

21. If instructed by your teacher, make a printout of the graph with each of the three trials.

DATA

Table 1	
Peas	Volume (mL)
Germinating	
Non-germinating	

Table 2	
Water bath	Temperature (°C)
warm	
cool	

Table 3	
Peas	Rate of Respiration (kPa/min)
Germinated, room temperature	
Non-germinated, room temperature	
Germinated, cool temperature	

QUESTIONS

1. Do you have evidence that cell respiration occurred in peas? Explain.
2. What is the effect of germination on the rate of cell respiration in peas?
3. What is the effect of temperature on the rate of cell respiration in peas?
4. What was the role of the control respirometer in each series of experiments?
5. Why do germinating peas undergo cell respiration?

EXTENSIONS

1. Compare the respiration rate among various types of seeds.
2. Compare the respiration rate among seeds that have germinated for different time periods, such as 1, 3, and 5 days.
3. Compare the respiration rate among various types of small animals, such as insects or earthworms.

Experiment

11C

TEACHER INFORMATION

Cell Respiration

1. This experiment may take several 50 minute lab periods to complete. A good stopping place might be at the end of Step 16. If Part III is completed during a second or third lab period, you may want to cool the respirometers in the refrigerator prior to the class. This will save a substantial amount of equilibration time for students.

2. Allow the seeds to germinate for three days prior to the experiment. Prior to the first day, soak them in water overnight. On subsequent days, roll them in a moist paper towel and place the towel in a paper bag. Place the bag in a warm, dark place. Check each day to be sure the towels remain very moist. If time is short, the peas can be used after they have soaked overnight. For best results, allow them to germinate for the full three days.

3. To prepare the 15% KOH solution, add 75 grams of solid KOH to distilled water to make a total volume of 500 mL. If the solution will be stored for an extended time, it will be best to store it in a plastic container. Strong bases will damage glass containers. **HAZARD ALERT:** Corrosive solid; skin burns are possible; much heat evolves when added to water; very dangerous to eyes; wear face and eye protection when using this substance. Wear gloves. Hazard Code: B—Hazardous.

   The hazard information reference is: Flinn Scientific, Inc., *Chemical & Biological Catalog Reference Manual, 2000*, (800) 452-1261, www.flinnsci.com. See *Appendix D* of this book, *Biology with Computers*, for more information.

4. Emphasize to your students the importance of providing an airtight fit with all plastic-tubing connections and when closing valves or twisting the stopper into a test tube.

5. All of the pressure valves, tubing, and connectors used in this experiment are included with Vernier Gas Pressure Sensors shipped after February 15, 1998. These accessories are also helpful when performing respiration/fermentation experiments such as Experiments 12B, and 16B in this manual.

   If you purchased your Gas Pressure Sensor at an earlier date, Vernier has a Pressure Sensor Accessories Kit (PS-ACC) that includes all of the parts shown here for doing pressure-related experiments. Using this kit allows for easy assembly of a completely airtight system. The kit includes the following parts:

   - two ribbed, tapered valve connectors inserted into a No. 5 rubber stopper
   - one ribbed, tapered valve connectors inserted into a No. 1 rubber stopper
   - two Luer-lock connectors connected to either end of a piece of plastic tubing
   - one two-way valve
   - one 20 mL syringe
   - two tubing clamps for transpiration experiments

6. The accessory items used in this experiment are the #1 single hole stopper fitted with a tapered valve connector and the section of plastic tubing fitted with Luer-lock connectors.

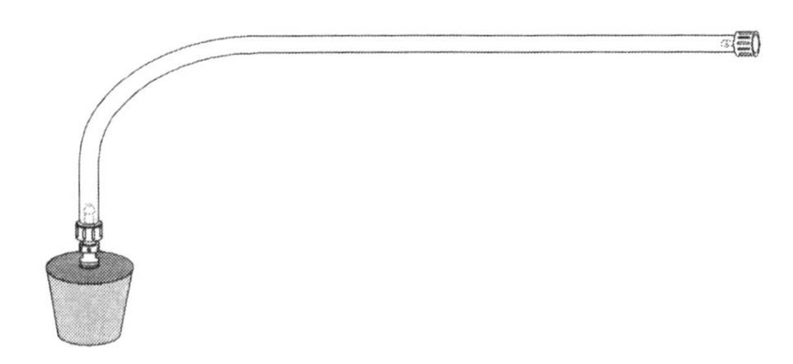

7. The length of plastic tubing connecting the rubber stopper assemblies to each gas pressure sensor must be the same for all groups. It is best to keep the length of tubing reasonably small to keep the volume of gas in the test tube low. **Note:** If pressure changes during data collection are too small, you may need to decrease the total gas volume in the system. Shortening the length of tubing used will help to decrease the volume.

SAMPLE RESULTS

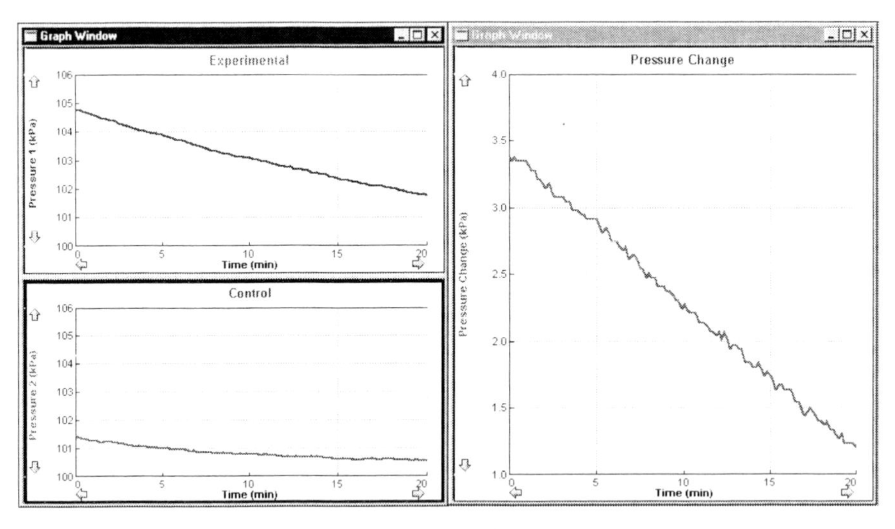

Figure 1

The graphs in Figure 1 illustrate cellular respiration of germinating and non-germinating peas at warm temperatures. The pressure change *vs.* time graph is calculated from the difference between the two left graphs.

Table 2	
Water bath	Temperature (°C)
warm	26
cool	10

Table 3	
Peas	Rate of O_2 consumption (kPa/min)
Germinating, warm	0.1124
Non-germinating, warm	0.0193
Germinating, cool	0.0699

ANSWERS TO QUESTIONS

1. Yes. The pressure change vs. time graph indicates that some gas is being removed at a constant rate from the respirometer when germinating seeds are present.

2. Germination greatly accelerates the rate of cellular respiration. This reflects a higher rate of metabolic activity in germinating seeds. In most experiments, non-germinating seeds do not seem to be respiring. Occasionally, however, some respiration is detectable.

3. Warm temperatures increase the rate of respiration. This reflects a higher rate of metabolic activity in warm germinating seeds than in cool seeds.

4. Gas in the control respirometer responded to temperature changes exactly as the experimental respirometer. By subtracting the control respirometer's pressure readings from the experimental respirometer's pressure readings, only the pressure change due to the removal of oxygen gas by seeds is detected. Any pressure change due to temperature fluctuations is eliminated.

   If one only looks at the graph of the experimental respirometer, it appears that both germinating and non-germinating seeds respire. This would be an erroneous conclusion, as the control respirometer's graph clearly indicates a change as well. The control would not be necessary only if the graph indicated no change in pressure throughout the experiment.

5. It is necessary for germinating seeds to undergo cellular respiration in order to acquire the energy they need for growth and development. Unlike their mature relatives, seeds do not yet have the necessary photosynthetic abilities needed to produce their own energy sources.

Experiment

11D

Cell Respiration

Cell respiration refers to the process of converting the chemical energy of organic molecules into a form immediately usable by organisms. Glucose may be oxidized completely if sufficient oxygen is available according to the following equation:

$$C_6H_{12}O_6 + 6O_2(g) \rightarrow 6\ H_2O + 6\ CO_2(g) + \text{energy}$$

All organisms, including plants and animals, oxidize glucose for energy. Often, this energy is used to convert ADP and phosphate into ATP. Peas undergo cell respiration during germination. Do peas undergo cell respiration before germination? Using your collected data, you will be able to answer the question regarding respiration and non-germinating peas.

Using the CO_2 Gas Sensor and O_2 Gas Sensor, you will monitor the carbon dioxide produced and the oxygen consumed by peas during cell respiration. Both germinating and non-germinating peas will be tested. Additionally, cell respiration of germinating peas at two different temperatures will be investigated.

OBJECTIVES

In this experiment, you will

- Use an O_2 Gas Sensor to measure concentrations of oxygen gas.
- Use a CO_2 Gas Sensor to measure concentrations of carbon dioxide gas.
- Study the effect of temperature on cell respiration.
- Determine whether germinating peas and non-germinating peas respire.
- Compare the rates of cell respiration in germinating and non-germinating peas.

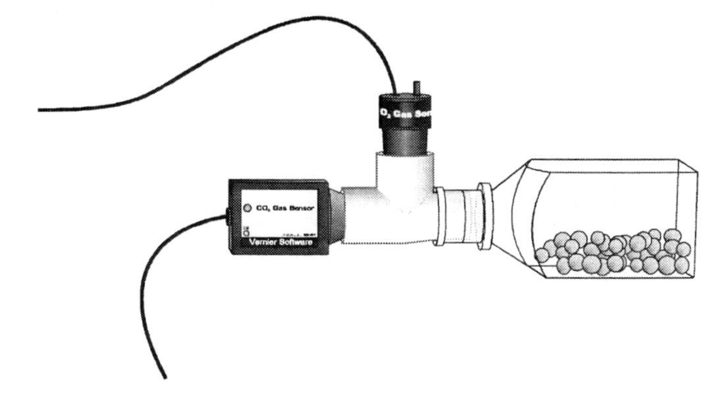

Figure 1

MATERIALS

computer
Vernier computer interface
Logger*Pro*
Vernier CO_2 Gas Sensor
Vernier O_2 Gas Sensor
CO_2–O_2 Tee
25 germinating peas
25 non-germinating peas
250 mL respiration chamber
ice cubes
1 L beaker
thermometer
two 100 mL beakers

PROCEDURE

1. Connect the CO_2 Gas Sensor to Channel 1 and the O_2 Gas Sensor to Channel 2 of the Vernier computer interface.

2. Prepare the computer for data collection by opening the file "11D Cell Respiration (CO2 and O2)" from the *Biology with Computers* folder of Logger*Pro*.

3. Obtain 25 germinating peas and blot them dry between two pieces of paper towel. Use the thermometer to measure the room temperature. Record the temperature in Table 1.

4. Place the germinating peas into the respiration chamber.

5. Insert the CO_2–O_2 Tee into the neck of the respiration chamber. Place the O_2 Gas Sensor into the CO_2–O_2 Tee as shown in Figure 1. Insert the sensor snugly into the Tee. The O_2 Gas Sensor should remain vertical throughout the experiment. Place the CO_2 Gas Sensor into the Tee directly across from the respiration chamber as shown in Figure 1. Gently twist the stopper on the shaft of the CO_2 Gas Sensor into the chamber opening. Do not twist the shaft of the CO_2 Gas Sensor or you may damage it.

6. Wait four minutes for readings to stabilize, then begin collecting data by clicking Collect. Collect data for ten minutes and click Stop.

7. When data collection has finished, remove the CO_2–O_2 Tee from the respiration chamber. Place the peas in a 100 mL beaker filled with cold water and ice.

8. Fill the respiration chamber with water and then empty it. Thoroughly dry the inside of the respiration chamber with a paper towel.

9. Determine the rate of respiration:
   a. Click anywhere on the CO_2 graph to select it. Click the Linear Fit button, , to perform a linear regression. A floating box will appear with the formula for a best fit line.
   b. Record the slope of the line, *m*, as the rate of respiration for germinating peas at room temperature in Table 2.
   c. Close the linear regression floating box.
   d. Repeat Steps 9a – c for the O_2 graph.

10. Move your data to a stored run. To do this, choose Store Latest Run from the Experiment menu.

11. Obtain 25 non-germinating peas and place them in the respiration chamber

12. Repeat Steps 5 – 10 for the non-germinating peas.

Part II Germinating peas, cool temperatures

13. Place the respiration chamber in a 1 L beaker. Cover the outside of the chamber with ice.

14. Use the thermometer to measure the water temperature of the 100 mL beaker containing the germinating peas. Record the water temperature in Table 1.

15. Remove the peas from the cold water and blot them dry between two paper towels.

16. Repeat Steps 5 – 9 to collect data with the germinating peas at a cold temperature.

17. To print a graph of concentration *vs.* volume showing all three data runs:
    a. Click anywhere on the CO_2 graph. Label all three curves by choosing Text Annotation from the Insert menu, and typing "Room Temp Germinated" (or "Room Temp Non-germinated", or "Cold Germinated") in the edit box. Then drag each box to a position near its respective curve. Adjust the position of the arrow head.
    b. Print a copy of the graph, with all three data sets and the regression lines displayed. Enter your name(s) and the number of copies of the graph you want.
    c. Click on the O_2 graph and repeat the process to print a copy of the O_2 graph.

DATA

Table 1	
Condition	Temperature (°C)
Room	
Cold water	

Table 2		
Peas	CO_2 Rate of respiration (ppt/min)	O_2 Rate of consumption (ppt/min)
Germinating, room temperature		
Non-germinating, room temperature		
Germinating, cool temperature		

QUESTIONS

1. Do you have evidence that cell respiration occurred in peas? Explain.
2. What is the effect of germination on the rate of cell respiration in peas?
3. What is the effect of temperature on the rate of cell respiration in peas?
4. Why do germinating peas undergo cell respiration?

EXTENSIONS

1. Compare the respiration rate among various types of seeds.
2. Compare the respiration rate among seeds that have germinated for different time periods, such as 1, 3, and 5 days.
3. Compare the respiration rate among various types of small animals, such as insects or earthworms.

TEACHER INFORMATION

Experiment **11D**

Cell Respiration

1. Allow the seeds to germinate for three days prior to the experiment. Prior to the first day, soak them in water overnight. On subsequent days, roll them in a moist paper towel and place the towel in a paper bag. Place the bag in a warm, dark place. Check each day to be sure the towels remain very moist. If time is short, the peas can be used after they have soaked overnight. For best results, allow them to germinate for the full three days.
2. The O_2 Gas Sensor should always be stored upright in the box in which it was shipped.
3. The morning of the experiment, fill a 1 L beaker with ice and water so that students will have cold water for Step 7. Students will also need access to ice for Steps 7 and 13.
4. The calibrations stored in this experiment file for both sensors work well for this experiment. Initial readings that seem slightly high or low will still reflect an accurate change in gas levels.
5. The stopper included with the CO_2 Gas Sensor is slit to allow easy application and removal from the probe. When students are placing the probe in the respiration chamber, they should gently twist the stopper into the chamber opening. Warn the students not to twist the probe shaft or they may damage the sensing unit.

SAMPLE RESULTS

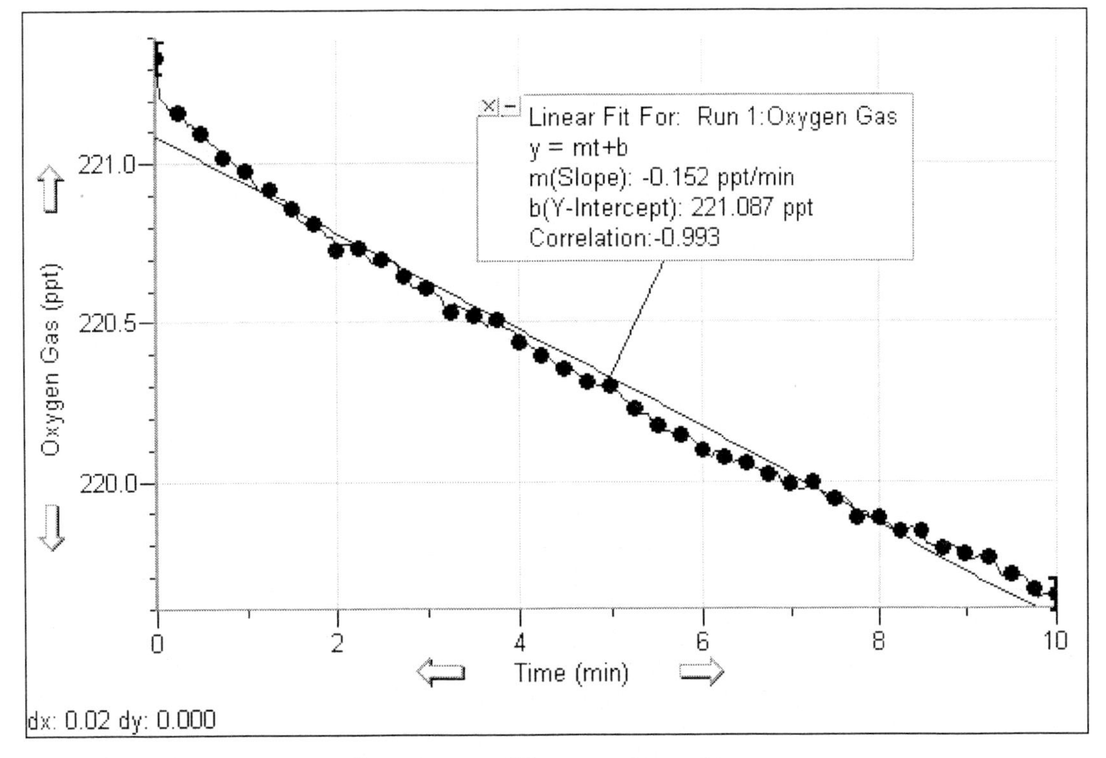

O_2 consumed by germinated peas

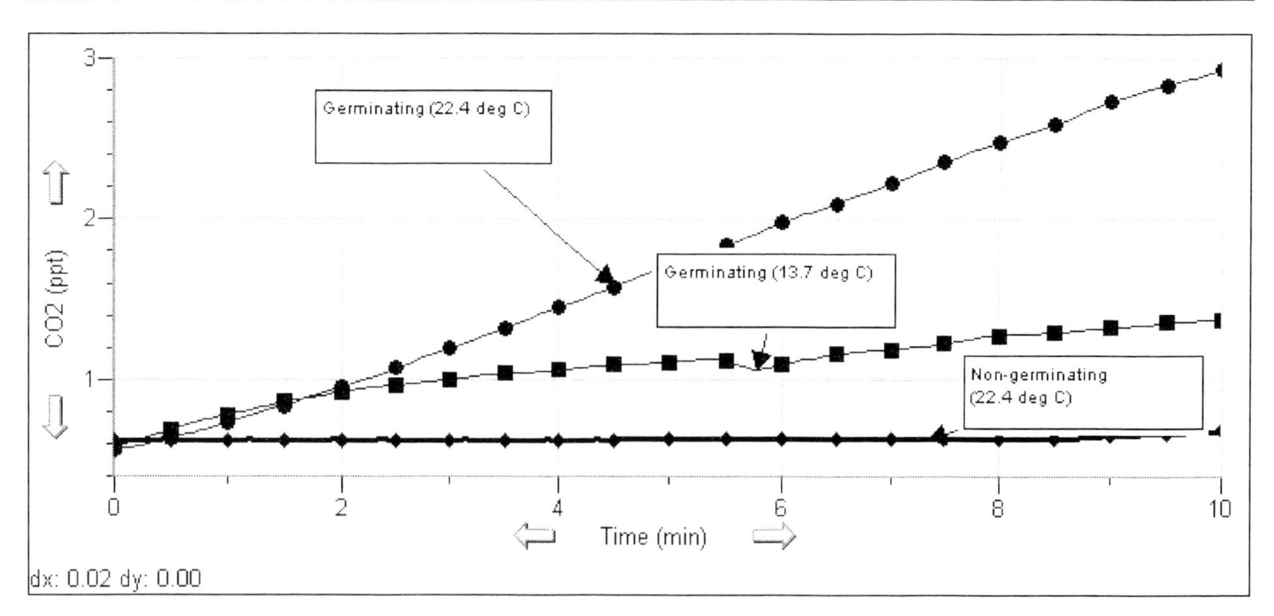

CO_2 levels of germinating and non-germinating peas

Table 1	
Condition	Temperature (°C)
room	22.4
cold water	13.7

Table 2		
Peas	CO_2 rate of respiration (ppt/min)	O_2 rate of consumption (ppt/min)
Germinating, room temperature	0.249	–0.152
Non-germinating, room temperature	0.003	–0.002
Germinating, cool temperature	0.066	–0.087

ANSWERS TO QUESTIONS

1. Yes, the oxygen concentration *vs.* time graph clearly indicates that oxygen is being consumed at a steady rate when germinating peas are in the respiration chamber. The carbon dioxide concentration *vs.* time graph indicates that carbon dioxide is being produced at a steady rate.

2. Germination greatly accelerates the rate of cellular respiration. This reflects a higher rate of metabolic activity in germinating seeds. In most experiments, non-germinating seeds do not seem to be respiring.

3. Warm temperatures increase the rate of respiration. This reflects a higher rate of metabolic activity in warm germinating seeds than in cooler seeds.

4. It is necessary for germinating seeds to undergo cellular respiration in order to acquire the energy they need for growth and development. Unlike their mature relatives, seeds do not yet have the necessary photosynthetic abilities needed to produce their own energy sources.

Experiment
12A

Respiration of Sugars by Yeast

Yeast are able to metabolize some foods, but not others. In order for an organism to make use of a potential source of food, it must be capable of transporting the food into its cells. It must also have the proper enzymes capable of breaking the food's chemical bonds in a useful way. Sugars are vital to all living organisms. Yeast are capable of using some, but not all sugars as a food source. Yeast can metabolize sugar in two ways, *aerobically,* with the aid of oxygen, or *anaerobically*, without oxygen.

In this lab, you will try to determine whether yeast are capable of metabolizing a variety of sugars. When yeast respire aerobically, oxygen gas is consumed and carbon dioxide, CO_2, is produced. You will use a CO_2 Gas Sensor to monitor the production of carbon dioxide as yeast respire using different sugars. The four sugars that will be tested are glucose (blood sugar), sucrose (table sugar), fructose (fruit sugar), and lactose (milk sugar).

OBJECTIVES

In this experiment, you will

- Use a CO_2 Gas Sensor to measure concentrations of carbon dioxide.
- Determine the rate of respiration by yeast while using different sugars.
- Determine which sugars can be used as a food source by yeast.

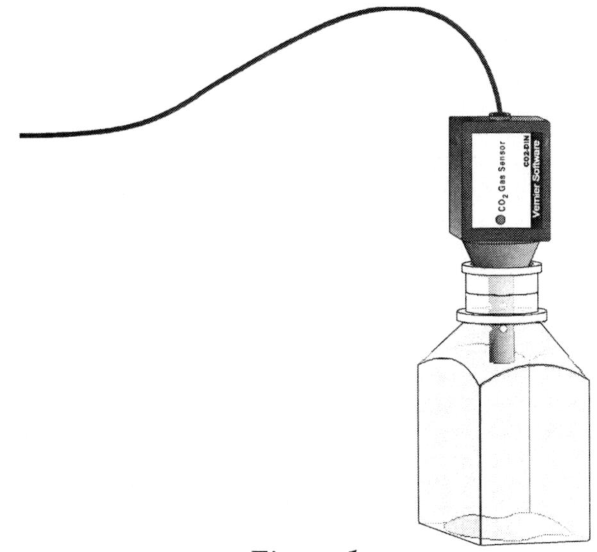

Figure 1

MATERIALS

computer
Vernier computer interface
Logger *Pro*
Vernier CO_2 Gas Sensor
250 mL respiration chamber
5% glucose, sucrose, lactose, and fructose sugar solutions
600 mL beaker (for water bath)
Beral pipettes
hot and cold water
thermometer
four 10 × 100 mm test tube
yeast suspension

PROCEDURE

1. Prepare a water bath for the yeast. A water bath is simply a large beaker of water at a certain temperature. This ensures that the yeast will remain at a constant and controlled temperature. To prepare the water bath, obtain some warm and cool water from your teacher. Combine the warm and cool water in the 600 mL beaker until it reaches 38 – 40°C. The beaker should be filled with about 300 – 400 mL water. Leave the thermometer in the water bath during the course of the experiment to monitor the temperature of the water bath.

2. Obtain five test tubes and label them G, S, F, L, and W.

3. Obtain the four sugar solutions: glucose, sucrose, fructose, and lactose.
   a. Place 2 mL of the glucose solution in test tube G.
   b. Place 2 mL of the sucrose solution in test tube S.
   c. Place 2 mL of the fructose solution in test tube F.
   d. Place 2 mL of the lactose solution in test tube L.
   e. Place 2 mL of distilled water in test tube W.

4. Obtain the yeast suspension. Gently swirl the yeast suspension to mix the yeast that settles to the bottom. Put 2 mL of yeast into each of the five test tubes. Gently swirl each test tube to mix the yeast into the solution.

5. Set the five test tubes into the water bath.

6. Incubate the test tubes for 10 minutes in the water bath. Keep the temperature of the water bath constant. If you need to add more hot or cold water, first remove as much water as you will add, or the beaker may overflow. Use a beral pipet to remove excess water. While the test tubes are incubating, proceed to Step 7.

7. Connect the CO_2 Gas Sensor to the computer interface. Prepare the computer for data collection by opening the file "12A Yeast Respiration" from the *Biology with Computers* folder of Logger*Pro*.

8. When incubation is finished, use a beral pipet to place 1 mL of the solution in test tube G into the 250 mL respiration chamber. Note the temperature of the water bath and record as the actual temperature in Table 1.

9. Quickly place the shaft of the CO_2 Gas Sensor in the opening of the respiration chamber. Gently twist the stopper on the shaft of the CO_2 Gas Sensor into the chamber opening. Do not twist the shaft of the CO_2 Gas Sensor or you may damage it.

10. Begin measuring carbon dioxide concentration by clicking [▶ Collect]. Data will be collected for 4 minutes.

11. When data collection has finished, remove the CO_2 Gas Sensor from the respiration chamber. Fill the respiration chamber with water and then empty it. Make sure that all yeast have been removed. Thoroughly dry the inside of the chamber with a paper towel.

12. Determine the rate of respiration:
    a. Move the mouse pointer to the point where the data values begin to increase. Hold down the left mouse button. Drag the pointer to the end of the data and release the mouse button.
    b. Click on the Linear Fit button, ,to perform a linear regression. A floating box will appear with the formula for a best fit line.
    c. Record the slope of the line, *m*, as the rate of respiration in Table 1.
    d. Close the linear regression floating box.
    e. Share your data with the class by recording the sugar type and respiration rate on the board.

13. Move your data to a stored run. To do this, choose Store Latest Run from the Experiment menu.

14. Use a notebook or notepad to fan air across the openings in the probe shaft of the CO_2 Gas Sensor for 1 minute.

15. Repeat Steps 8 – 14 for the other four test tubes.

DATA

Table 1		
Sugar Tested	Actual Temperature (°C)	Respiration Rate (ppm/min)
Glucose		
Sucrose		
Fructose		
Lactose		
Water (control)		

Table 2: Class Averages	
Sugar Tested	Respiration Rate (ppm/min)
Glucose	
Sucrose	
Fructose	
Lactose	
Water	

PROCESSING THE DATA

1. When all other groups have posted their results on the board, calculate the average rate of respiration for each solution tested. Record the average rate values in Table 2.

2. On Page 2 of the experiment file, make a bar graph of rate of respiration *vs*. sugar type. The rate values should be plotted on the y-axis, and the sugar type on the x-axis. Use the rate values from Table 2.

QUESTIONS

1. Considering the results of this experiment, do yeast equally utilize all sugars? Explain.
2. Hypothesize why some sugars were not metabolized while other sugars were.
3. Why do you need to incubate the yeast before you start collecting data?
4. Yeast live in many different environments. Make a list of some locations where yeast might naturally grow. Estimate the possible food sources at each of these locations.

TEACHER INFORMATION

Respiration of Sugars by Yeast

1. To prepare the yeast solution, dissolve 7 g (1 package) of dried yeast for every 100 mL of water. Incubate the suspension in 37 – 40°C water for at least 10 minutes.

2. After the 10 minute incubation period, transfer the yeast to dispensing tubes. Each group will need about 12 mL of yeast.

3. To prepare the 5% sugar solutions, add 5 g of sugar per 100 mL of solution.

4. The stopper included with the CO_2 Gas Sensor is slit to allow easy application and removal from the probe. When students are placing the probe in the respiration chamber, they should gently twist the stopper into the chamber opening. Warn the students not to twist the probe shaft or they may damage the sensing unit.

5. The CO_2 Gas Sensor relies on the diffusion of gases into the probe shaft. Students should allow a couple of minutes between trials so that gases from the previous trial will have exited the probe shaft. Alternatively, the students can use a firm object such as a book or notepad to fan air through the probe shaft. This method is used in Step 14 of the student procedure.

6. The stored calibration for the CO_2 Gas Sensor works very well for this experiment.

SAMPLE RESULTS

Table 1

Sugar Tested	Actual Temperature (°C)	Respiration Rate (ppm/min)
Glucose	37 °C	823.44
Sucrose	37 °C	812.55
Fructose	37 °C	568.15
Lactose	37 °C	75.52
Water (control)	37 °C	65.54

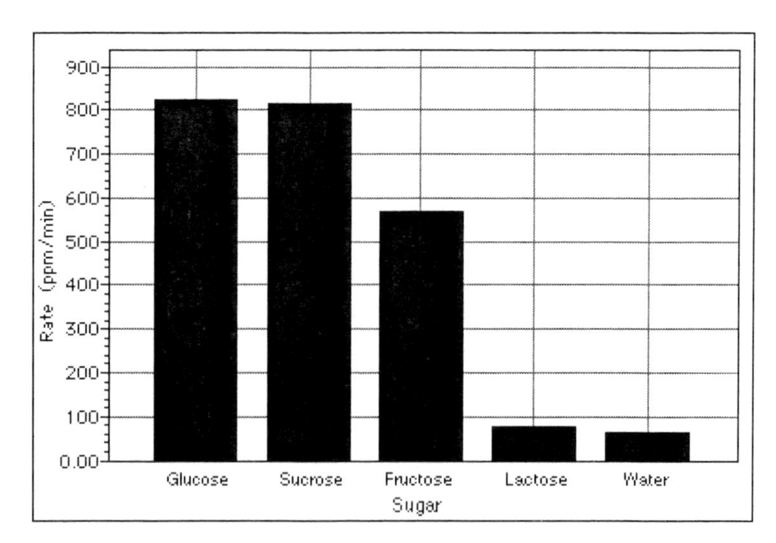

Rate of respiration of four sugars

ANSWERS TO QUESTIONS

1. Yeast cannot utilize all of the sugars equally well. While glucose, sucrose, and fructose can all be metabolized by yeast, lactose is not utilized at all.

2. Yeast may not have the proper enzymes to either transport lactose across its cell membrane, or it may not have the enzyme needed to convert it from a disaccharide to a monosaccharide.

3. It takes the yeast a few minutes to transport the sugar into the cell, to respire at a constant rate, and to reach the proper temperature.

4. Some yeast live on other organisms. If they are warm blooded, they may be near the optimal temperature for yeast respiration, 37°C. Many yeast live in soils. The temperature of soils may easily be measured at different times of the year.

Experiment

12B

Sugar Fermentation in Yeast

Yeast are able to metabolize some foods, but not others. In order for an organism to make use of a potential source of food, it must be capable of transporting the food into its cells. It must also have the proper enzymes capable of breaking the food's chemical bonds in a useful way. Sugars are vital to all living organisms. Yeast are capable of using some, but not all sugars as a food source. Yeast can metabolize sugar in two ways, *aerobically,* with the aid of oxygen, or *anaerobically*, without oxygen.

In this lab, you will try to determine whether yeast are capable of metabolizing a variety of sugars. Although the aerobic fermentation of sugars is much more efficient, in this experiment we will have yeast ferment the sugars anaerobically. When the yeast respire aerobically, oxygen gas is consumed at the same rate that CO_2 is produced—there would be no change in the gas pressure in the test tube. When yeast ferment the sugars anaerobically, however, CO_2 production will cause a change in the pressure of a closed test tube, since no oxygen is being consumed. We can use this pressure change to monitor the fermentation rate and metabolic activity of the organism.

The fermentation of glucose can be described by the following equation:

$$\underset{\text{glucose}}{C_6H_{12}O_6} \rightarrow 2\ \underset{\text{ethanol}}{CH_3CH_2OH} + 2\ \underset{\text{carbon dioxide}}{CO_2} + \text{energy}$$

Note that alcohol is a byproduct of this fermentation.

OBJECTIVES

In this experiment, you will

- Use a Gas Pressure Sensor to measure the pressure change caused by carbon dioxide released during fermentation.
- Determine the rate of fermentation.
- Determine which sugars yeast can metabolize.

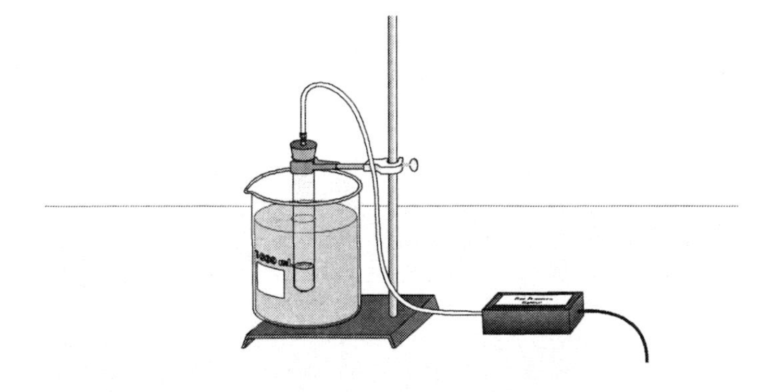

Figure 1

MATERIALS

computer
Vernier computer interface
Logger*Pro*
Vernier Gas Pressure Sensor
l-hole rubber stopper assembly
plastic tubing with Luer-lock fitting
5% glucose, sucrose, lactose, and one other sugar solution
ring stand
18 × 150 mm test tube
1 L beaker (for water bath)
basting bulb or Beral pipette
hot and cold water
test tube rack
thermometer
yeast suspension
vegetable oil in dropper bottle
utility clamp

PROCEDURE

1. Connect the Gas Pressure Sensor to the computer interface. Prepare the computer for data collection by opening the file "12B Fermentation (Pressure)" from the *Biology with Computers* folder of Logger*Pro*.
2. Connect the plastic tubing to the valve on the Gas Pressure Sensor.
3. Prepare a water bath for the yeast. A water bath is simply a large beaker of water at a certain temperature. This ensures that the yeast will remain at a constant and controlled temperature. To prepare the water bath, obtain some warm and cool water from your teacher. Combine the warm and cool water into the 1 liter beaker until it reaches 38 – 40°C. The beaker should be filled with about 600 – 700 mL water. Place the thermometer in the water bath to monitor the temperature during the experiment.
4. Obtain two test tubes and label them 1 and 2.
5. Your team will test two of the four sugar solutions. Obtain two of the four sugar solutions: glucose, sucrose, lactose, and one other sugar solution, as directed by your instructor. Place 2.5 mL of the first sugar solution into test tube 1 and 2.5 mL of the second sugar solution into test tube 2. Record which solutions you tested in Table 1.
6. Set the test tubes into the water bath.
7. Obtain the yeast suspension. Gently swirl the yeast suspension to mix the yeast that settles to the bottom. Using a 10 mL pipette or graduated cylinder, transfer 2.5 mL of yeast into test tube 1. Gently mix the yeast into the sugar solution. Be gentle with the yeast—they are living organisms!
8. In the test tube, place enough vegetable oil to completely cover the surface of the yeast/glucose mixture as shown in Figure 3. Be careful to not get oil on the inside wall of the test tube. Set the test tube in the water bath.
9. Insert the single-holed rubber-stopper into the test tube. **Note**: *Firmly* twist the stopper for an *airtight* fit. Secure the test tube with a utility clamp and ring-stand as shown in Figure 1.

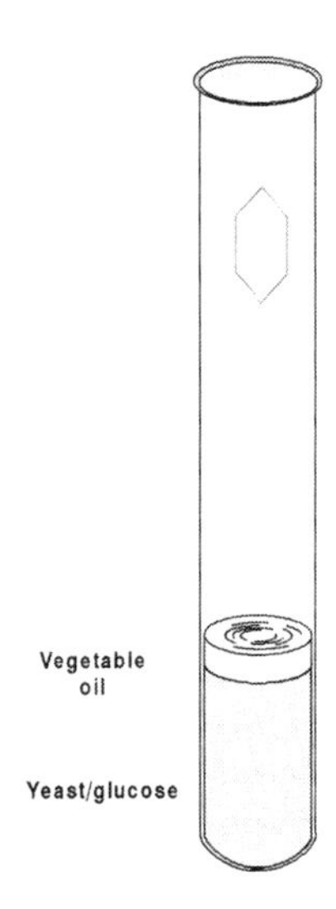

Figure 2

10. Incubate the test tube for 10 minutes in the water bath. Be sure to keep the temperature of the water bath constant. If you need to add more hot or cold water, first remove about as much water as you will be adding, or the beaker may overflow. Use a basting bulb to remove excess water.

Note: Be sure that most of the test tube is completely covered by the water in the water bath. The temperature of the air in the tube must be constant for this experiment to work well.

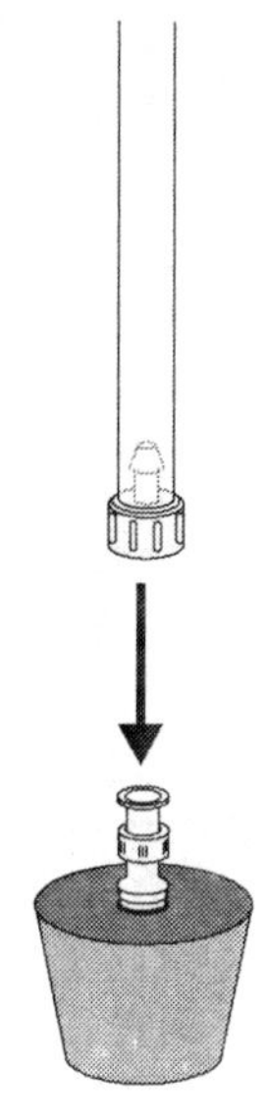

Figure 3

11. When incubation has finished, connect the free end of the plastic tubing to the connector in the rubber stopper as shown in Figure 3.
12. Click [Collect] to begin data collection. Maintain the temperature of the water bath during the course of the experiment. Data will be collected for 15 minutes.
13. Monitor the pressure readings displayed in the meter. If the pressure exceeds 130 kPa, the pressure inside the tube will be too great and the rubber stopper is likely to pop off. Disconnect the plastic tubing from the Gas Pressure Sensor if the pressure exceeds 130 kPa.
14. When data collection has finished, disconnect the plastic tubing connector from the rubber stopper. Remove the rubber stopper from the test tube and discard the contents in a waste beaker.
15. From the Experiment menu, choose Store Latest Run. This stores the data so it can be used later, but it will still be displayed while you collect data for the second test tube.
16. Repeat Steps 7 – 14 using test tube 2.
17. Click on the Linear Fit button, [Linear Fit]. Click [OK]. A best-fit linear regression line will be shown for each of your runs. In Table 1, record the value of the slope, *m*, for each of the test tubes. (The linear regression statistics are displayed in a floating box for each of the data sets.)
18. To print a graph of pressure *vs.* time showing both data runs:
    a. Label both curves by choosing Text Annotation from the Insert menu, and typing the sugar you tested in the edit box. Repeat for the second sugar tested. Then drag each box to a position near its respective curve.
    b. Print a copy of the graph, with both data sets and the regression lines displayed. Enter your name(s) and the number of copies of the graph you want.

DATA

Table 1

	Type of Sugar	Rate of Fermentation (kPa/min)
Test tube 1		
Test tube 2		

Table 2: Class Averages	
Sugar Tested	Fermentation Rate (kPa/min)
Control	

PROCESSING THE DATA

1. Share your data with the rest of the class by recording the sugar type you tested and the rate of fermentation on the board.
2. Using the class data, calculate the average rates of fermentation for each of the sugar types tested. Record the average rates in Table 2, along with the names of the 4 sugars tested.
3. On Page 2 of the experiment file, use the class data in Table 2, make a bar graph of rate of fermentation vs. sugar type. The rate values should be plotted on the y-axis, and the sugar type on the x-axis.

QUESTIONS

1. Considering the results of this experiment, can yeast utilize all of the sugars equally well? Explain.
2. Hypothesize why some sugars were not metabolized while other sugars were.
3. Why do you need to incubate the yeast before you start monitoring air pressure?
4. Yeast live in many different environments. Make a list of some locations where yeast might naturally grow. Estimate the food sources of each of these locations.

TEACHER INFORMATION

Sugar Fermentation in Yeast

In this experiment, students will test the ability of yeast to respire anaerobically using various sugar sources.

1. To prepare the yeast solution, dissolve 7 g (1 package) of dried yeast for every 100 mL of water. Incubate the suspension in 37 – 40°C water for at least 10 minutes.

2. After the 10-minute incubation period, transfer the yeast to dispensing tubes. Each group will need about 11 mL of yeast.

3. To prepare the 5% sugar solutions, add 5 g of sugar per 100 mL of solution.

4. Fructose is a good sugar to have available, as is galactose. Students may want to test fruit juices. If so, do not dilute these with water.

5. Emphasize to your students the importance of providing an airtight fit with all plastic-tubing connections and when closing valves or twisting the stopper into a test tube.

6. All of the pressure valves, tubing, and connectors used in this experiment are included with Vernier Gas Pressure Sensors shipped after February 15, 1998. These accessories are also helpful when performing respiration/fermentation experiments such as Experiments 11C, and 16B in this manual.

   If you purchased your Gas Pressure Sensor at an earlier date, Vernier has a Pressure Sensor Accessories Kit (PS-ACC) that includes all of the parts shown here for doing pressure-related experiments. Using this kit allows for easy assembly of a completely airtight system. The kit includes the following parts:

   - two ribbed, tapered valve connectors inserted into a No. 5 rubber stopper
   - one ribbed, tapered valve connectors inserted into a No. 1 rubber stopper
   - two Luer-lock connectors connected to either end of a piece of plastic tubing
   - one two-way valve
   - one 20 mL syringe
   - two tubing clamps for transpiration experiments

7. The accessory items used in this experiment are the #1 single hole stopper fitted with a tapered valve connector and the section of plastic tubing fitted with Luer-lock connectors.

8. The length of plastic tubing connecting the rubber stopper assemblies to each gas pressure sensor must be the same for all groups. It is best to keep the length of tubing reasonably

small to keep the volume of gas in the test tube low. **Note:** If pressure changes during data collection are too small, you may need to decrease the total gas volume in the system. Shortening the length of tubing used will help to decrease the volume.

9. If the Vernier Gas Pressure Sensor or Biology Gas Pressure Sensor is unavailable, the Vernier Pressure Sensor may be used as an alternative.

SAMPLE RESULTS

The following data may be different from students' results. The actual values depend upon the viability and concentration of the yeast, among other factors.

Table 1		
Test	Type of Sugar	Rate of Fermentation (kPa/min)
1	glucose	1.998
2	sucrose	1.662
3	lactose	0.006
4	fructose	0.468

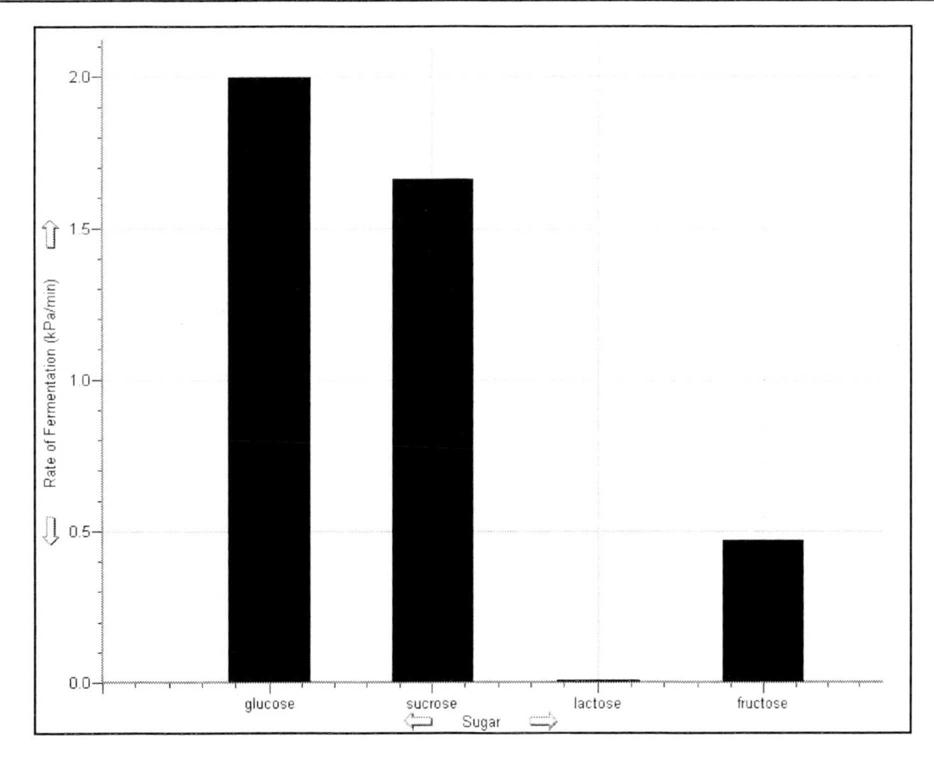

Fermentation rates for each of the four sugars

ANSWERS TO QUESTIONS

1. Yeast cannot utilize all of the sugars equally well. While glucose, sucrose, and fructose all can be metabolized by yeast, lactose is not utilized at all.

2. Yeast may not have the proper enzymes to either transport lactose across its cell membrane, or it may not have the enzyme needed to convert it from a disaccharide to a monosaccharide.

3. The yeast need to be incubated so that the oxygen in the test tube will be completely consumed. If the yeast respire aerobically, no pressure change occurs, because much oxygen is consumed as CO_2 is produced. It also takes a few minutes for the yeast to transport the sugar into the cell, to respire at a constant rate, and to reach the proper temperature.

4. Some yeast live on other organisms. If they are warm blooded, they may be near the optimal temperature for yeast respiration, 37°C. Many yeast live in soils. The temperature of soils may easily be measured at different times of the year.

Experiment 13

Population Dynamics

As organisms reproduce, die, and move in and out of an area, their populations fluctuate. In a closed population, organisms do not move in and out of an area, so their populations change only through *natality*, or births, and *mortality*, or deaths of individuals. Yeast are very small, rapidly reproducing organisms. They can experience dramatic population changes over a relatively short time. In this experiment, a population of yeast will be given a small amount of food and placed in a closed environment. Organic materials will not enter or leave the environment—only inorganic gases will be allowed to be exchanged. The population of yeast can be monitored by measuring the *turbidity*, or cloudiness, of the medium that contains the yeast.

To measure the yeast population, you will be using the Colorimeter shown in Figure 1. In this device, light from an LED light source will pass through the medium containing yeast and strike a photocell. Photons of light that strike a yeast cell will be reflected away from the photocell and will make the medium appear more turbid. The computer-interfaced Colorimeter monitors the light received by the photocell as either an *absorbance* or a *percent transmittance* value. The absorbance value is proportional the population of yeast present in the medium.

Figure 1

OBJECTIVES

In this experiment, you will

- Use a computer and Colorimeter to monitor a closed population of yeast.
- Use a microscope to monitor a closed population of yeast.
- Compare the population estimates obtained using the two different techniques.
- Practice making dilutions for population counts.

MATERIALS

computer
Vernier computer interface
Logger *Pro*
Vernier Colorimeter
18 × 150 mm test tubes
cuvette and lid
two 5 mL pipettes or 10 mL graduated cylinders
dropper pipet or Beral pipet
glass-marking pencil
graph paper
microscope
microscope slide and cover slip
test-tube rack
cotton swabs

PRE-LAB ACTIVITY

Your instructor will be placing a small amount of yeast in a closed environment containing food at the start of the experiment. Use the Draw Prediction feature found under the Analyze menu to predict how the yeast population might change over a long time period

PROCEDURE

Prepare the computer and Colorimeter

If instructed to do so by your teacher, one team should prepare the computer and Colorimeter for use. The equipment will be located on a resource table for use by all of the class teams.

1. Connect the Colorimeter to the computer interface. Prepare the computer for data collection by opening the file "13 Population Dynamics" from the *Biology with Computers* folder of Logger*Pro*.
2. You are now ready to calibrate the Colorimeter. Prepare a *blank* by filling a cuvette 3/4 full with distilled water. To correctly use a Colorimeter cuvette, remember:
   - All cuvettes should be wiped clean and dry on the outside with a tissue.
   - Handle cuvettes only by the top edge of the ribbed sides.
   - All solutions should be free of bubbles.
   - Always position the cuvette with its reference mark facing toward the white reference mark at the top of the cuvette slot on the Colorimeter.

3.Calibrate the Colorimeter.

   a. Open the Colorimeter lid.
   b. Holding the cuvette by the upper edges, place it in the cuvette slot of the Colorimeter. Close the lid.
   c. If your Colorimeter has a CAL button, Press the < or > button on the Colorimeter to select a wavelength of 565 nm (Green) for this experiment. Press the CAL button until the red LED begins to flash. Then release the CAL button. When the LED stops flashing, the calibration is complete. Proceed directly to Step 12. If your Colorimeter does not have a CAL button, continue with this step to calibrate your Colorimeter.

   First Calibration Point

   d. Choose Calibrate ▸ CH1: Colorimeter (%T) from the Experiment menu and then click Calibrate Now.
   e. Turn the wavelength knob on the Colorimeter to the "0% T" position.
   f. Type "0" in the edit box.
   g. When the displayed voltage reading for Reading 1 stabilizes, click Keep.

   Second Calibration Point

   h. Turn the knob of the Colorimeter to the Green LED position (565 nm).
   i. Type "100" in the edit box.
   j. When the displayed voltage reading for Reading 2 stabilizes, click Keep, then click Done.

Measure the yeast population with the Colorimeter (Day 0)

Each team should perform the following steps.

4. Obtain a 2.5 mL yeast sample and control sample from your instructor. Add 2.5 mL of distilled water to the sample to dilute it 50%.

5. Mix and transfer 2.5 mL of the diluted yeast into a clean, dry cuvette. Place a cuvette lid on the cuvette. Put an identifying mark on the bottom of the cuvette. Use this same cuvette each time you take readings with the Colorimeter.

6. You are now ready to collect absorbance data for the yeast. Quickly perform these steps:

   a. Mix the cuvette contents until all air bubbles are removed from the clear sides of the cuvette.
   b. Wipe the outside of the cuvette with a tissue and place it into the Colorimeter.
   c. Close the lid and wait for the absorbance value in the meter to stabilize.
   d. Record the absorbance value in Table 2.
   e. Remove the cuvette from the Colorimeter.

7. Repeat Steps 5 – 6 for the control test tube.

8. Record your absorbance values for the experimental test tube and the control test tube on the board, as instructed by your teacher. Discard the highest and lowest values and average the remaining absorbance values from all the class teams. Record the class average in Table 2.

Measure the yeast population with a Microscope (Day 0)

9. Mix the yeast in the cuvette.

10. Using a dropper pipet, withdraw a small amount of culture and transfer one drop (or the amount specified by your instructor) onto a clean microscope slide.

11. Place a clean cover slip over the culture. Do not allow air bubbles to get trapped. The liquid should barely fill the area under the coverslip, but should not ooze out.

12. Place the slide on a microscope and focus it under low power. Refocus near the center of the slide under high (40× – 45×) power. If there are too many cells to count, you will need to make a dilution. If a dilution must be made, follow the steps below. If not, proceed to Step 13.

    a. Mix the yeast in the test tube used to make the latest dilution and transfer 0.5 mL of yeast into a clean, dry test tube.
    b. Add 4.5 mL of water to the 0.5 mL of yeast. This will make a 1/10 dilution.
    c. Mix the contents thoroughly.
    d. Record the dilution in your notebook.
    e. Repeat Step 10.

13. Record the final dilution of yeast. If undiluted, then record 1/1. If one dilution was made, record 1/10. If two dilutions were made, then record 1/100 (1/10 × 1/10).

14. Have one team partner count the number of yeast cells in one field of view. Be sure to count each cell—a yeast bud counts as a cell. If you see a clump of cells, estimate the number of cells in the clump. Record the appearance of the yeast cells and the odor of the tube in your notebook.

15. Calculate the count of yeast in the original sample. To do this, divide the final count of yeast by the dilution factor. Divide this value by 1/2—the dilution factor from Step 5—to obtain the final result. Record the result in Table 2.

16. Move the slide to a different field of view. Have another team member count the number of yeast cells in this field of view. Repeat this for every member of the team. Count the yeast in at least 6 fields.

17. Discard the highest and lowest count. Average the remaining values from all of the team members and record the average value in Table 2 and on the board.

18. Repeat Steps 10 – 17 for the control test tube.

Measure the yeast population (Day 1–8)

19. Repeat Steps 4 – 18 for both test tubes each day, except the weekends, until the 9th day.

20. Your instructor may give you data from other classes to increase the number of data points in this experiment. If so, extend Table 2 in your notebook.

DATA

Table 1	
Prediction	Plot

Table 2								
Day	Absorbance				Microscopic Count			
	Team data		Class average		Team data		Class average	
	Expt	Control	Expt	Control	Expt	Control	Expt	Control
0								
1								
2								
3								
4								
5								
6								
7								
8								

PROCESSING THE DATA

1. On Page 2 of the experiment file, manually enter the absorbance values for the experimental data.
   a. Enter the absorbance values for your team in Column 2 of Data Set 1of the data table.
   b. Enter the absorbance values for the control data your team measured in Column 3 of the data table.
   c. Enter the class average experimental absorbance values in Column 4 of the data table.
   d. Enter the class average control absorbance values in Column 5 of the data table.
2. Use Text Annotation from the Insert menu to appropriately identify each of the four different lines on the graph.
3. On Page 3 of the experiment file, manually enter the count values for the experimental data.
   a. Enter the count values for your team in Column 2 Data Set 2 of the data table.
   b. Enter the count values for the control data your team measured in Column 3 of the data table.
   c. Enter the class average experimental microscopic count values in Column 4 of the data table.
   d. Enter the class average control microscopic values in Column 5 of the data table.
4. Use Text Annotation from the Insert menu to appropriately identify each of the four lines on the graph.

QUESTIONS

1. Compare the plot of yeast population obtained from the absorbance readings with that from a microscopic count. Describe the similarities and differences among the plots.
2. How do your results compare to that obtained from the class average? If there are any differences, explain how the differences might have occurred.
3. What was the purpose of the control test tube?
4. How do the results from the control test tube compare with those from the experimental test tube? What information does this provide?
5. Could you tell whether a yeast cell was dead or alive during this experiment? How might this affect your results?
6. What factors encouraged the growth of yeast in this experiment? Explain.
7. What factors limited the growth of yeast in this experiment? Explain.
8. How did your prediction compare with your results? How did your prediction compare with the results from the class average? Explain.
9. How did the odor in the tube change throughout the experiment? What do you think was responsible for any changes?

EXTENSIONS

1. Design and perform an experiment to determine the effect temperature has on the growth rate of yeast.

2. Design and perform an experiment to determine the effect different growth media have on the rate of growth for yeast. Consider growing them in plain water or in media with specific sugars as a food source.

TEACHER INFORMATION

Population Dynamics

1. Students should work in teams of at least three members each.
2. It is best to begin this experiment early in the week. Otherwise, the period of rapid growth may occur during the weekend.
3. The instructor will need to prepare the cultures and aliquot samples for students prior to each class, or the amount of student class time this experiment will require may be extreme.
4. Prepare a culture of rapidly dividing yeast prior to the experiment. Rehydrate about 1/2 teaspoon of yeast in 50 mL of apple cider or other yeast media on the day before class.

   Although apple cider is readily available and easy to obtain, you may want to consider making alternate media. If so, prepare one that is fairly clear.
5. For each 500 mL of media required, do the following:
   a. Place 500 mL apple cider or other broth into each of two 1 L flasks. Fit the flasks with a cap or cotton plug.
   b. Sterilize the broth for 15 minutes at 15 psi in an autoclave or pressure cooker.
   c. Allow the tubes to slowly cool.
   d. When cool, inoculate one of the two flasks with 1-mL of the actively growing yeast culture. The other flask will be for the control media.
6. Sterilize a sufficient number of screw cap test tubes. Sterilize the tubes for 15 minutes at 15 psi in an autoclave or pressure cooker.

   Prepare two tubes for each day. One tube is for the yeast culture and one tube is for the control. Each tube should have sufficient media for every team in all of the classes that will perform the experiment that day. Each team will require 2.5-mL of media. Prepare a few extra tubes in case some of the tubes become contaminated.
7. Assuming that 25 mL of broth is sufficient for each day, prepare the media as follows:
   a. Label nine (or more) screw cap test tubes E0 through E9. These will be Experimental tubes.
   b. Mix and sterilely transfer 25 mL of the inoculated media into each of the Experimental screw-cap test tubes and loosely fit the cap onto the tube.
   c. Label nine (or more) screw cap test tubes C0 through C9. These will be Control tubes.
   d. Sterilely transfer 25 mL of the sterile media into each of the Control screw-cap test tubes and loosely fit the cap onto the tube.
   e. Place each tube in a secure location at room temperature. Select a warm location where the temperature will not vary. If placed in an incubator at 35 – 40°C, the growth rate may be too rapid. Use a test-tube shaker, if available.
8. Prior to class each day, remove one inoculated Experimental tube and one Control tube for use by all of your classes. Check them for contamination—select another tube if it is found to be contaminated. Aliquot 2.5 mL of each solution into clean, dry, labeled test tubes for each team. Place the tubes in a refrigerator until the beginning of class.

9. The first day (Day 0) of this experiment may take an entire laboratory period. Students may need to become familiar with counting techniques and taking measurements with a colorimeter. If students do not finish counting cells with a microscope, the yeast sample may be stored in the refrigerator until the next day.
10. On Days 1 – 8, it will take approximately 15 – 20 minutes to take absorbance readings and to make cell counts. One team member should measure the absorbance while other members count the yeast during this time period.
11. Each team will need two 5 mL pipets each day and possibly four test tubes.
12. Each team member should perform both the microscopic and computer-assisted measurements at some time during the 9 day period.
13. Have students use the same interface, colorimeter, and cuvette throughout the 9 days of the experiment. One unit should be set up on a resource table for class use. Once the colorimeter is calibrated, it can be used throughout the day by all of the classes.
14. If you have multiple classes performing this experiment, share the data among all of the classes. This will increase the granularity of the plot.
15. Be sure that students count individual cells and not clumps of cells.
16. You may want to have students plot the population growth on linear and semi-log paper.
17. The largest source of error is when students do not mix the contents of their tube thoroughly. They should invert the tube several times.
18. The student's procedure specifies that 1 drop of culture be placed on a microscope slide. Check to be sure this is the appropriate amount of culture to use, as coverslips come in a variety of sizes. Liquid should barely fill the area under the coverslip, but not ooze out. Be sure to have students use the same size of coverslip every day as well as the same style of dropper or pipette.
19. If a micropipette is available, you may want to use it to deliver a precise amount of culture onto the microscope slide. Experiment to determine an appropriate amount of liquid to match your coverslip. 30 µL will work well for a 25 × 22 mm coverslip.

SAMPLE RESULTS

Table 1: Class Averages					
Date	Absorbance	Count	Date	Absorbance	Count
2/14/00 8:48 AM	0.000	0	2/17/00 3:00 PM	0.348	39
2/15/00 9:00 AM	0.012	20	2/18/00 8:00 AM	0.334	44
2/15/00 3:00 PM	0.074	34	2/18/00 2:30 PM	0.362	40
2/16/00 9:00 AM	0.234	72	2/21/00 8:00 AM	0.351	28
2/16/00 3:00 PM	0.282	65	2/21/00 3:00 PM	0.313	20
2/17/00 9:00 AM	0.344	50			

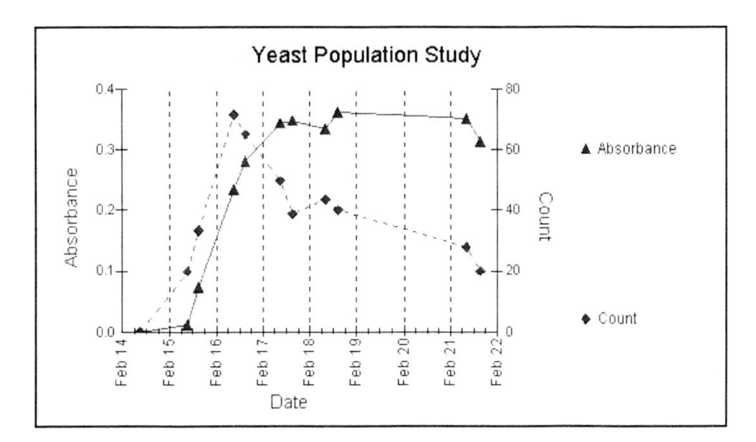

ANSWERS TO QUESTIONS

1. The plots should be similar in shape, especially at the beginning of the experiment. As cells die, some may lyse. This will cause the direct count to indicate a decreasing population compared to the results obtained by turbidity readings. Lysed cells do not remain intact but continue to affect turbidity. As seen in the above graph, the general shape of the two plots reflect this. They both increase, peak, and begin to decrease. The direct count decreases more rapidly than does the count by absorbance.

2. The class average results are usually more uniform than any one team's data. By averaging the data, counting errors are smoothed out. A team's data may fluctuate due to lack of experience with counting techniques or different counting techniques by team members.

3. The control tube allow students to measure the effect of the medium on measurements each day. It also indicates whether the initial culture media was contaminated.

4. No growth should be observed in the control tube. The count and absorbance measurements did not change in the control tube throughout the experiment. Thus, any change in the daily measurements were due to the yeast population.

5. It is not easy to tell if a cell is dead or alive. Some students may observe cellular debris from lysed cells, but few will identify the debris as a cell. If dead cells are counted in each daily measurement, the cell count will appear higher than it really is. The peak in the graph might occur earlier and reduce to a lower value if only living cells were counted.

6. Factors that might encourage the growth of yeast at the start of this experiment include:
   - Food in the medium.
   - A suitable temperature for growth.
   - Few waste products at the start of the experiment.
   - No competition from other types of organisms.
   - Plenty of space for yeast to grow.

7. Factors that might limit the growth of yeast at the end of this experiment include:
   - Lack of food in the medium or waste products accumulating in the medium.
   - Competition from other yeast cells for food.
   - Competition from other types of organisms, if contaminated.

8. Results may vary.

9. By the end of the experiment, alcohol was produced as a waste product. This gives yeast a "fermented" smell.

Experiment

14

Interdependence of Plants and Animals

Plants and animals share many of the same chemicals throughout their lives. In most ecosystems, O_2, CO_2, water, food and nutrients are exchanged between plants and animals. In this lab, you will be designing your own experiments to determine the relationships between two organisms—a plant (Elodea) and an animal (a snail).

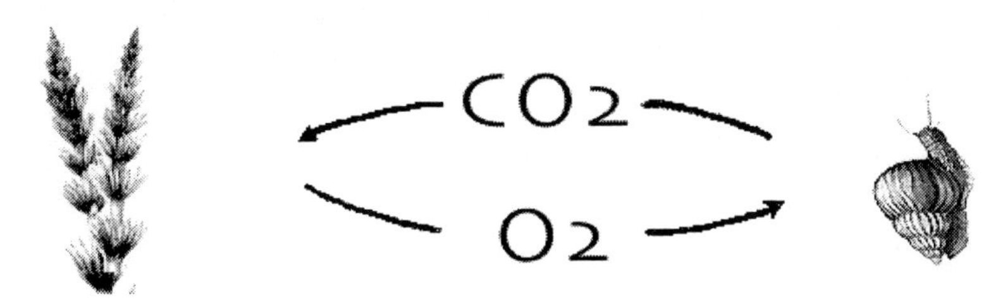

Figure 1

Several hypotheses have been discussed in the past about possible relationships between the Elodea and the snail. You will test to determine how oxygen and CO_2 are exchanged among Elodea plants, snails, and the water both exist in.

To perform the necessary tests, you will need to determine the presence of CO_2. An easy way to do this is to monitor the pH of the pond water. If CO_2 dissolves in water, it forms carbonic acid, H_2CO_3, and the pH decreases. If CO_2 is removed from pond water, the amount of carbonic acid goes down and the pH increases. One can use a computer to monitor the pH and determine whether CO_2 is released into the pond water or is taken from the water. Dissolved oxygen (DO) can be monitored with the aid of a Dissolved Oxygen Probe. Increases or decreases in the amount of dissolved oxygen can be rapidly assessed with this probe.

OBJECTIVES

In this experiment, you will

- Use a computer and a Dissolved Oxygen Probe to measure the dissolved oxygen in water.
- Use a computer and a pH Sensor to measure the pH of water.
- Use pH measurements to make inferences about the amount of CO_2 dissolved in water.
- Determine whether snails consume or produce oxygen and CO_2 in water.
- Determine whether plants consume or produce oxygen and CO_2 in the light.
- Determine whether plants consume or produce oxygen and CO_2 in the dark.

MATERIALS

computer
Vernier computer interface
Logger *Pro*
Vernier Dissolved Oxygen Probe
Vernier pH Sensor
eight 25 × 150 mm screw top test tubes
distilled wash water
250 mL beaker
Aluminum foil
Parafilm
pond snails
pond water
sprigs of elodea
two test tube racks

PRE-LAB PROCEDURE

Important: Prior to each use, the Dissolved Oxygen Probe must warm up for a period of 10 minutes as described below. If the probe is not warmed up properly, inaccurate readings will result. Perform the following steps to prepare the Dissolved Oxygen Probe.

1. Prepare the Dissolved Oxygen Probe for use.
   a. Remove the blue protective cap.
   b. Unscrew the membrane cap from the tip of the probe.
   c. Using a pipet, fill the membrane cap with 1 mL of DO Electrode Filling Solution.
   d. Carefully thread the membrane cap back onto the electrode.
   e. Place the probe into a container of water.

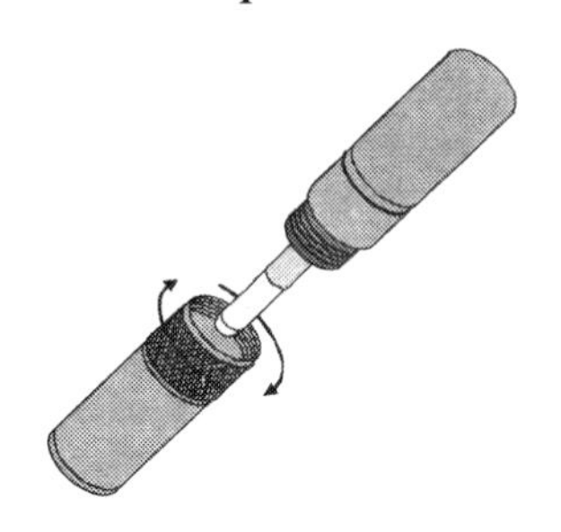

Remove membrane cap

Add electrode filling solution

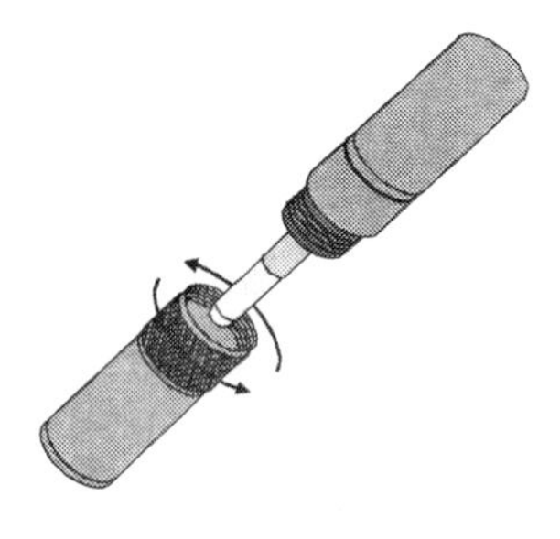

Replace membrane cap

2. Plug the Dissolved Oxygen Probe into Channel 2 of the Vernier interface. Connect the pH Sensor to Channel 1.

3. Prepare the computer for data collection by opening the file "14 Plants and Animals" from the *Biology with Computers* folder of Logger*Pro*. Two meters will be displayed. The left one is set to read pH while the right one is set for dissolved oxygen in mg/L.

4. It is necessary to warm up the Dissolved Oxygen Probe for 10 minutes before taking readings. To warm up the probe, leave it connected to the interface, with Logger *Pro* running, for 10 minutes. The probe must stay connected at all times to keep it warmed up. If disconnected for a few minutes, it will be necessary to warm up the probe again.

5. You are now ready to calibrate the Dissolved Oxygen Probe.
   - If your instructor directs you to use the calibration stored in the experiment file, then proceed to Step 6.
   - If your instructor directs you to perform a new calibration for the Dissolved Oxygen Probe, follow this procedure.

   Zero-Oxygen Calibration Point

   a. Choose Calibrate ▸ CH2: Dissolved Oxygen (mg/L) from the Experiment menu and then click Calibrate Now.
   b. Remove the probe from the water and place the tip of the probe into the Sodium Sulfite Calibration Solution. **Important:** No air bubbles can be trapped below the tip of the probe or the probe will sense an inaccurate dissolved oxygen level. If the voltage does not rapidly decrease, tap the side of the bottle with the probe to dislodge any bubbles. The readings should be in the 0.2 to 0.5 V range.

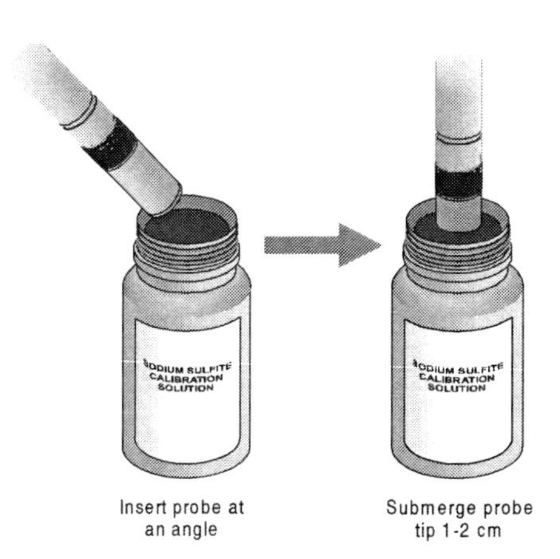

Insert probe at an angle

Submerge probe tip 1-2 cm

c. Type "0" (the known value in mg/L) in the edit box.
d. When the displayed voltage reading for Reading 1 stabilizes, click Keep.

Saturated DO Calibration Point

e. Rinse the probe with distilled water and gently blot dry.
f. Unscrew the lid of the calibration bottle provided with the probe. Slide the lid and the grommet about 2 cm onto the probe body.

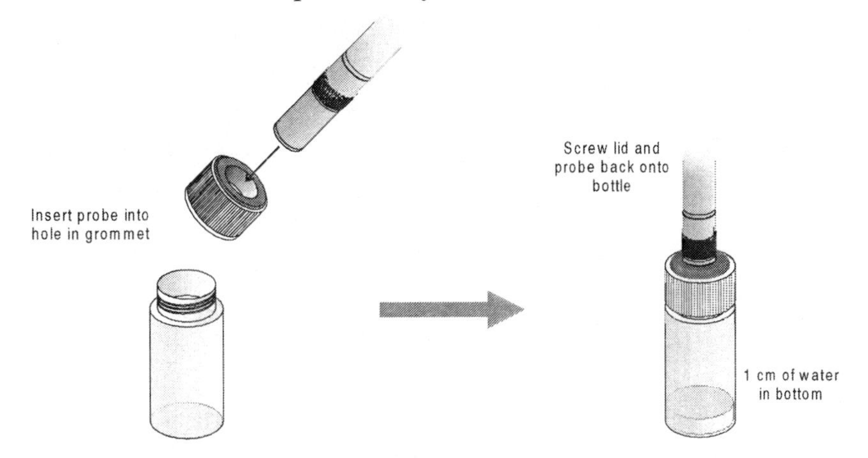

g. Add water to the bottle to a depth of about 1 cm and screw the bottle into the cap, as shown. **Important:** Do not touch the membrane or get it wet during this step. Keep the probe in this position for about a minute.
h. Type the correct saturated dissolved-oxygen value (in mg/L) from Table 3 (for example, "8.66") using the current barometric pressure and air temperature values. If you do not have the current air pressure, use Table 4 to estimate the air pressure at your altitude.
i. When the displayed voltage reading for Reading 2 stabilizes (readings should be above 2.0 V), click Keep.
j. Choose Calibration Storage from the calibration pull-down menu. Choose Experiment File (calibration stored with the current document). Click Done.
k. From the File menu, select Save As and save the current experiment file with a new name.

PROCEDURE

Day 1

6. Obtain and wear an apron.
7. Obtain and label eight test tubes 1 – 8.
8. Set test tubes 1 – 4 in one test tube rack and test tubes 5 – 8 in a second test tube rack.
9. Fill each tube with pond water.
10. Place one snail each in test tubes 2, 4, 6, and 8.
11. Place one sprig of elodea in test tubes 3, 4, 7, and 8. Each set of four tubes should appear similar to those in Figure 5.

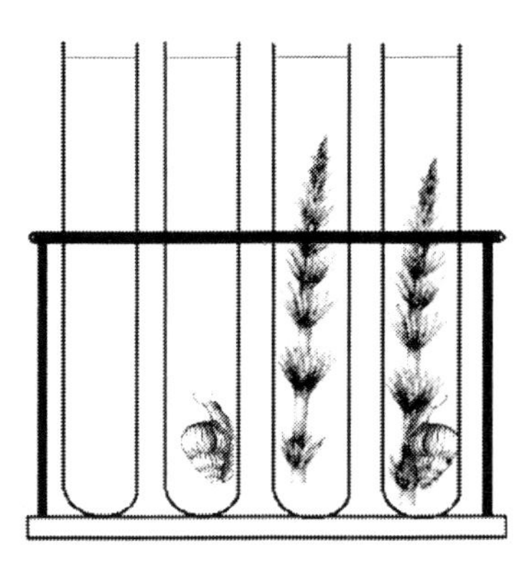

Figure 5

12. Wrap test tubes 5 – 8 in aluminum foil to make each light tight.

13. Remove the pH Sensor from the storage bottle. Rinse the probe thoroughly with distilled water. Place the probe into test tube 1 and gently swirl to allow water to move past the probe's tip. When the reading displayed in the meter stabilizes, record the pH value in Table 1.

14. Repeat Step 13 for each of the other seven test tubes.

15. When all of the pH readings have been taken, rinse the pH Sensor and return it to the pH storage bottle.

16. Place the Dissolved Oxygen Probe into test tube 1 so that it is submerged half the depth of the water. Gently and continuously move the probe up and down a distance of about 1 cm in the tube. This allows water to move past the probe's tip. Note: Do not agitate the water, or oxygen from the atmosphere will mix into the water and cause erroneous readings.

17. When the dissolved oxygen reading stabilizes (~30 seconds), record its value in Table 1.

18. Repeat Steps 16 – 17 for each of the other seven test tubes.

19. When all of the dissolved oxygen readings have been taken, rinse the Dissolved Oxygen Probe and return it to the distilled water beaker.

20. Completely fill each test tube with pond water and tighten the cap onto the tube. Do not allow any air bubbles to remain in any of the test tubes. Unscrew each cap slightly, so that they just barely open. Wrap each tube with Parafilm so that they do not leak water. The Parafilm will expand, if necessary, to accommodate any pressure build-up in a tube. No oxygen or carbon dioxide should enter or leave a tube.

21. Place test tubes 1 – 8 near the light source, as directed by your instructor.

22. Predict how the pH and dissolved oxygen will change in each tube. Write a short statement that explains your reasoning. Be specific about the roles of both the snail and elodea. Be prepared to discuss your reasoning in class on Day 2.

Day 2

23. Open the file saved on Day 1. Repeat Steps 1 – 6 to set up the pH Sensor and Dissolved Oxygen Probe.

24. Repeat Steps 13 – 21 to take pH and DO readings for each of the test tubes.

25. Now, the elodea will use the environment established by the snail and the snail will use the environment established by the elodea. Remove the snail from test tube 2 and the elodea from test tube 3. Place the snail in test tube 3 and the elodea in test tube 2. **Note**: Try not to aerate the water during the transfer.

26. Remove the snail from test tube 6 and the elodea from test tube 7. Place the snail in test tube 7 and the elodea in test tube 6.

27. Measure the pH and DO of test tubes 1 – 3 and test tubes 5 – 7. Record the results in Table 2. These values should be similar to those measured before the transfer. If not, the water may have been mixed too vigorously with the atmospheric air.

28. Completely fill test tubes 1 – 3 and test tubes 5 – 7 with water and tighten the cap onto each tube, as in Step 20. Wrap each slightly opened test tube with Parafilm. Place test tubes 1 – 3 and test tubes 5 – 7 near the light source, as in Step 21.

29. Return the snails and elodea from test tubes 4 and 8, as directed by your instructor. Clean and return the test tubes.

Day 3

30. Repeat Steps 13 – 19 for test tubes 1 – 3 and 5 – 7. Record the results in Table 2.

31. Return the snails and elodea, as directed by your instructor. Clean and return the test tubes.

DATA

Table 1

Test Tube	pH Day 1	pH Day 2	ΔpH	DO Day 1	DO Day 2	ΔDO
1						
2						
3						
4						
5						
6						
7						
8						

Table 2						
Test Tube	pH Day 2	pH Day 3	ΔpH	DO Day 2	DO Day 3	ΔDO
1						
2						
3						
5						
6						
7						

PROCESSING THE DATA

1. Calculate the change in pH for Tables 1 – 2. Record your results in Tables 1 – 2.
2. Calculate the change in dissolved oxygen for Tables 1 – 2. Record your results in Tables 1 – 2.

QUESTIONS

1. Consider the snails. Comparing the readings from day 1 to day 2, answer the following questions:
   a. Do snails produce or consume CO_2 when in the light?
   b. Do snails produce or consume oxygen when in the light?
   c. Which test tubes allow you to answer this? Which test tube is the experimental test tube? Which test tube is the control test tube?
   d. Do snails produce or consume CO_2 when in the dark?
   e. Do snails produce or consume oxygen when in the dark?
   f. Which test tubes allow you to answer this? Which test tube is the experimental test tube? Which test tube is the control test tube?
2. Consider the elodea. Comparing the readings from day 1 to day 2, answer the following questions:
   a. Do elodea produce or consume CO_2 when in the light?
   b. Do elodea produce or consume oxygen when in the light?
   c. Which test tubes allow you to answer this? Which test tube is the experimental test tube? Which test tube is the control test tube?
   d. Do elodea produce or consume CO_2 when in the dark?
   e. Do elodea produce or consume oxygen when in the dark?
   f. Which test tubes allow you to answer this? Which test tube is the experimental test tube? Which test tube is the control test tube?
3. Consider the elodea placed in the snail's water on days 2–3. Comparing the readings from day 2 to day 3, answer the following questions:
   a. Do elodea produce or consume CO_2 when in the light?
   b. Do elodea produce or consume oxygen when in the light?

   c. Which test tubes allow you to answer this? Which test tube is the experimental test tube? Which test tube is the control test tube?
   d. Do elodea produce or consume CO_2 when in the dark?
   e. Do elodea produce or consume oxygen when in the dark?
   f. Which test tubes allow you to answer this? Which test tube is the experimental test tube? Which test tube is the control test tube?

4. Consider the snail placed in the elodea's water on days 2–3. Comparing the readings from day 2 to day 3, answer the following questions:

   a. Do snails produce or consume CO_2 when in the light?
   b. Do snails produce or consume oxygen when in the light?
   c. Which test tubes allow you to answer this? Which test tube is the experimental test tube? Which test tube is the control test tube?
   d. Do snails produce or consume CO_2 when in the dark?
   e. Do snails produce or consume oxygen when in the dark?
   f. Which test tubes allow you to answer this? Which test tube is the experimental test tube? Which test tube is the control test tube?

5. Summarize the relationship between snails and plants in a pond. Explain your reasoning.

6. Interpret the results of Test Tube 4 and Test Tube 8. Compare your findings to the results obtained from Table 2.

7. How do your conclusions compare to your predictions in Step 25?

CALIBRATION TABLES

Table 3: 100% Dissolved Oxygen Capacity (mg/L)

	770 mm	760 mm	750 mm	740 mm	730 mm	720 mm	710 mm	700 mm	690 mm	680 mm	670 mm	660 mm
0°C	14.76	14.57	14.38	14.19	13.99	13.80	13.61	13.42	13.23	13.04	12.84	12.65
1°C	14.38	14.19	14.00	13.82	13.63	13.44	13.26	13.07	12.88	12.70	12.51	12.32
2°C	14.01	13.82	13.64	13.46	13.28	13.10	12.92	12.73	12.55	12.37	12.19	12.01
3°C	13.65	13.47	13.29	13.12	12.94	12.76	12.59	12.41	12.23	12.05	11.88	11.70
4°C	13.31	13.13	12.96	12.79	12.61	12.44	12.27	12.10	11.92	11.75	11.58	11.40
5°C	12.97	12.81	12.64	12.47	12.30	12.13	11.96	11.80	11.63	11.46	11.29	11.12
6°C	12.66	12.49	12.33	12.16	12.00	11.83	11.67	11.51	11.34	11.18	11.01	10.85
7°C	12.35	12.19	12.03	11.87	11.71	11.55	11.39	11.23	11.07	10.91	10.75	10.59
8°C	12.05	11.90	11.74	11.58	11.43	11.27	11.11	10.96	10.80	10.65	10.49	10.33
9°C	11.77	11.62	11.46	11.31	11.16	11.01	10.85	10.70	10.55	10.39	10.24	10.09
10°C	11.50	11.35	11.20	11.05	10.90	10.75	10.60	10.45	10.30	10.15	10.00	9.86
11°C	11.24	11.09	10.94	10.80	10.65	10.51	10.36	10.21	10.07	9.92	9.78	9.63
12°C	10.98	10.84	10.70	10.56	10.41	10.27	10.13	9.99	9.84	9.70	9.56	9.41
13°C	10.74	10.60	10.46	10.32	10.18	10.04	9.90	9.77	9.63	9.49	9.35	9.21
14°C	10.51	10.37	10.24	10.10	9.96	9.83	9.69	9.55	9.42	9.28	9.14	9.01
15°C	10.29	10.15	10.02	9.88	9.75	9.62	9.48	9.35	9.22	9.08	8.95	8.82
16°C	10.07	9.94	9.81	9.68	9.55	9.42	9.29	9.15	9.02	8.89	8.76	8.63
17°C	9.86	9.74	9.61	9.48	9.35	9.22	9.10	8.97	8.84	8.71	8.58	8.45
18°C	9.67	9.54	9.41	9.29	9.16	9.04	8.91	8.79	8.66	8.54	8.41	8.28
19°C	9.47	9.35	9.23	9.11	8.98	8.86	8.74	8.61	8.49	8.37	8.24	8.12
20°C	9.29	9.17	9.05	8.93	8.81	8.69	8.57	8.45	8.33	8.20	8.08	7.96
21°C	9.11	9.00	8.88	8.76	8.64	8.52	8.40	8.28	8.17	8.05	7.93	7.81
22°C	8.94	8.83	8.71	8.59	8.48	8.36	8.25	8.13	8.01	7.90	7.78	7.67
23°C	8.78	8.66	8.55	8.44	8.32	8.21	8.09	7.98	7.87	7.75	7.64	7.52
24°C	8.62	8.51	8.40	8.28	8.17	8.06	7.95	7.84	7.72	7.61	7.50	7.39
25°C	8.47	8.36	8.25	8.14	8.03	7.92	7.81	7.70	7.59	7.48	7.37	7.26
26°C	8.32	8.21	8.10	7.99	7.89	7.78	7.67	7.56	7.45	7.35	7.24	7.13
27°C	8.17	8.07	7.96	7.86	7.75	7.64	7.54	7.43	7.33	7.22	7.11	7.01
28°C	8.04	7.93	7.83	7.72	7.62	7.51	7.41	7.30	7.20	7.10	6.99	6.89
29°C	7.90	7.80	7.69	7.59	7.49	7.39	7.28	7.18	7.08	6.98	6.87	6.77
30°C	7.77	7.67	7.57	7.47	7.36	7.26	7.16	7.06	6.96	6.86	6.76	6.66

Table 4: Approximate Barometric Pressure at Different Elevations

Elevation (m)	Pressure (mm Hg)	Elevation (m)	Pressure (mm Hg)	Elevation (m)	Pressure (mm Hg)
0	760	800	693	1600	628
100	748	900	685	1700	620
200	741	1000	676	1800	612
300	733	1100	669	1900	604
400	725	1200	661	2000	596
500	717	1300	652	2100	588
600	709	1400	643	2200	580
700	701	1500	636	2300	571

Experiment

14

TEACHER INFORMATION

Interdependence of Plants and Animals

1. The Dissolved Oxygen Probe must be calibrated the first day of use. Follow the pre-lab procedure to prepare and calibrate the Dissolved Oxygen Probe. To save time, you may wish to calibrate the probe and record the calibration values on paper. The students can then skip the pre-lab procedure and they will have the calibration values available for manual entry in case the values stored in the program are lost.
2. At the end of class instruct the students to leave Logger *Pro* running. This will keep power going to the probes. When the next group of students come in, they can begin at Step 6 of the procedure. They can skip the pre-lab procedure because the initial group of students has completed all of the setup. Have the last group of students for the day shut everything off and put things away.
2. The pond water should be adjusted to pH 7 before class begins. Use 0.1 M NaOH or 0.1 M HCl to adjust the pH. Be sure the elodea are fresh and healthy.
3. Florescent lamps should be used as a source of light. They should be on for the entire 24 hour period, set a few inches from the tubes. If the tubes are water tight, as they should be, test tube racks are not necessary. Students can place them horizontally on a table and the light can be lowered until it is just above the tubes.
4. Wrap the test tubes thoroughly in aluminum foil if they require darkness, or place them in a darkened part of the room. If there are not a sufficient number of test tube racks for these, place the set of four wrapped tubes in a small beaker.
5. Between classes, store the Dissolved Oxygen Probes in a beaker of distilled water. At the end of the day, be sure to empty out the electrode filling solution in the Dissolved Oxygen Probe and rinse the inside of the membrane cap with distilled water.
6. If you have a pH System, but do not have a Dissolved Oxygen Probe, the experiment may be modified to indirectly investigate carbon dioxide levels using only the pH System.
7. When taking dissolved oxygen readings, the students should allow ample time for the readings to stabilize. In some instances this can take 60 seconds.
8. Each student team should use the same set of equipment to make measurements each day.
9. When setting up the Dissolved Oxygen Probe, be sure to remove the blue plastic cap from the end of the probe. The cap is made of a soft plastic material and easily slides off the probe end.
10. Elodea canadensisis a good alternative for those who live in any area to which it is illegal to ship Elodea. Other aquatic plants may work equally well.

ANSWERS TO QUESTIONS

1. Consider the snails.

   a. Snails produce CO_2 when in the light. The pH decreased, meaning the acidity increased. Higher acidity means more CO_2 is dissolved. A snail was the only organism present to produce the CO_2.
   b. Snails consume O_2 when in the light. The DO decreased, so less O_2 is dissolved. A snail was the only organism present to consume the O_2.
   c. Experimental Test Tube: 2. Control Test Tube: 1.
   d. Snails produce CO_2 when in the dark. The pH decreased, so the acidity increased. Higher acidity means more CO_2 is dissolved. A snail was the only organism present to produce the CO_2.
   e. Snails consume O_2 in the dark. The DO decreased, so less O_2 is dissolved. A snail was the only organism present to consume the O_2.
   f. Experimental Test Tube: 6. Control Test Tube: 5.

2. Consider the Elodea.

   a. Elodea consume CO_2 when in the light. The pH increased, so the acidity decreased. Lower acidity means less CO_2 is dissolved. Elodea was the only organism present to consume the CO_2.
   b. Elodea produce O_2 when in the light. The DO increased, so more O_2 is dissolved. Elodea were the only organism present to make the O_2.
   c. Experimental Test Tube: 3. Control Test Tube: 1.
   d. Elodea produce CO_2 in the dark. The pH decreased, so the acidity increased. Higher acidity means more CO_2 is dissolved. Elodea were the only organism present to produce the CO_2.
   e. Elodea consume O_2 in the dark. The DO decreased, so less O_2 is dissolved. Elodea was the only organism present to consume the O_2.
   f. Experimental Test Tube: 7. Control Test Tube: 5.

3. Consider the elodea placed in the snail's water on days 2–3.

   a. Elodea consumes the CO_2 that snails release when in the light. CO_2 increased when a snail was alone in the pond water, yet it decreased when elodea replaced the snail. Some of the CO_2 used by the plant must have come from the snail. The increase in pH was greater for elodea in Tube 2 of Table 2 than in Tube 3, Table 1, so more CO_2 was consumed from water the snail was in.
   b. Elodea produces O_2 when in the light, as above. The DO increased, so more O_2 is dissolved.
   c. Experimental Test Tube: 2. Control Test Tube: 1.
   d. The pH change in Test Tube 6 from Day 2 to Day 3 was negative. This indicates that the plant did not remove the CO_2; rather, it was added as the plant respired. Elodea did not consume CO_2 in the dark, and did not use the CO_2 from the snail.
   e. Elodea consumed O_2 when in the dark. The DO decreased, so the plant respired when in the dark.
   f. Experimental Test Tube: 6. Control Test Tube: 5.

Experiment 15

Biodiversity and Ecosystems

Biodiversity is critical in any self-sustaining environment. Complex and diverse ecological systems are made up of many organisms and a huge variety of interactions. Simple ecosystems have few organisms, few interactions, and are often fragile. All ecosystems, whether diverse or sparse, involve an intimate interaction of living things with their abiotic environment. Biodiversity implies variety, and variety in an ecosystem often ensures a greater chance of survival in a changing world.

The Earth is losing its biodiversity at a worrisome rate. Humans simplify ecosystems for many reasons: to increase the agricultural base, to make way for cities and industrial zones, or for aesthetic reasons, such as making lawns and gardens. This practice has direct effects upon many abiotic factors within an environment. The air temperatures found in cities, for instance, are usually significantly higher than that in surrounding, non-urbanized areas. Such cities are said to produce *heat islands*. An area's biodiversity has profound effects upon the physical and biological makeup of an ecosystem.

OBJECTIVES

In this experiment, you will

- Examine how biodiversity affects an environment's temperature.
- Determine how animal diversity changes in different environments.
- Work with your classmates to compare biodiversity in areas with different plant patch sizes.

MATERIALS

computer
Vernier computer interface
Temperature Probe
Logger *Pro*
notebook
meter stick
string or twine

PROCEDURE

1. Choose two sites, one that is diverse with a fair variety of different types of plants. Call this Site A. Find a simple site, such as a grassy lawn. Call this Site B. Two such sites might be like those shown here:

Site A

Site B

2. Connect the Temperature Probe to the computer interface. Prepare the computer for data collection by opening the file "15 Biodiversity" from the *Biology with Computers* folder of Logger*Pro*.

Site A:

3. Using a meter stick, measure out a one-square meter area at Site A. Mark the area with string or twine.

Identifying organisms

4. Examine this area closely. Your group will need to make several decisions:
   - How will we identify an organism? What is a grass organism? One blade? A group from one set of roots? A patch of grass?
   - How will we count similar organisms? If there are thousands of one type of organism in your area, do you count each one, or find ways of estimating? What ways of estimating are reasonably accurate? Since many animals and birds move in and out of an area, over what time period will you count organisms?
   - How will we identify a plant patch? A patch of one type of plant is an area of similar plants that are physically separate from another area of similar plants. Patches need not be the same size. You will need to distinguish one patch from another.

   Discuss how your group will determine each of these.

   Record your decisions and the rationale for your choices.

5. Now, record information about the different living organisms in your area. Include estimates of the following:
   - The type of organism. The actual name is not important. You might write something like:
     - a maple tree
     - tall (20-30 cm) grass with wide blades
     - cut grass, 3 cm long
     - beetle #1 (0.5 cm long, black)
     - bee #1 (yellow and black stripes about 1 cm long)
     - a fly flew in and out of the area
   - the approximate number of each type of organism.
   - the number of plant patches in the 1 m^2 area.

6. Using a soil borer or a trowel, examine a small sample of soil. Record the depth of the humus in your soil in Table 1. Humus is made of decaying organic matter and is usually darker in color than the non-organic soil. Note any animals in your sample.

7. Examine your data from Steps 5 and 6 and record each of the following in Table 1:

   a. The number of different animals in your area

   b. The number of different plants in your area

   c. The number of plant patches in your area

Identifying physical factors

8. Place a meter stick vertically in your area. One end of the stick (reading 0 cm) should be on the soil. If leaf litter exists, move it aside and place the stick on the dirt. The other end (100 cm) should be in the air, so that the stick is as vertical as possible.

9. Place the temperature probe at the top of the meter stick (at a height of 100 cm).

10. You are ready to begin the measurements. Start measuring by clicking Collect. Note that a new button, Keep, is available.

11. Allow the temperature reading to stabilize then click Keep. This button tells the computer that you want to record a measurement.

12. A text box will appear. Enter the height of the sensor. At a height of 100 cm, type "100" in the text box and press ENTER.

13. Move the sensor down to 90 cm. Repeat Steps 11 and 12.

14. Continue measuring the temperature at heights of 80, 70, 60, 50, 40, 30, 20, 10, and 0 cm. To do so, repeat the Step-13 procedure. When all measurements have been made click Stop.

15. Obtain the temperature values from the table and record them in Table 2.

16. Move your data to a stored data run. To do this, choose Store Latest Run from the Experiment menu. This will allow you to keep the first plot on the screen while you are measuring Site B.

Site B

17. Repeat Steps 5 – 15 at Site B. Save your data when finished.

Class Data

18. Obtain the data found by each team in your class and record their values in Table 3. Calculate the averages for each column and record your results in the bottom row of Table 3.

DATA

Table 1

Count in your area	Site A (diverse)	Site B (sparse)
Number of different animals		
Number of different plants		
Number of plant patches		
Humus depth (cm)		

Table 2											
Height (cm)	100	90	80	70	60	50	40	30	20	10	0
Site A Temperature (°C)											
Site B Temperature (°C)											

Table 3								
Team	# of Animals		# of Plants		# of Patches		Humus depth (cm)	
	Site A	Site B	Site A	Site B	Site A	Site B	Site A	Site B
1								
2								
3								
4								
5								
6								
7								
8								
9								
10								
Average								

QUESTIONS

1. How would you compare Sites A and B? Include a comparison of the biodiversity at each site.
2. Which site supported a *larger* animal population? Which site supported a *more diverse* animal population? Hypothesize why this might be so.
3. Examine your data in Table 3. How does the average number of plants compare to the average number of plant patches?
4. What seems to be a more meaningful indicator of biodiversity—a count of the number of plants, or a count of the number of plant patches? Explain your answer.
5. How would each group's definition of what a plant was affect the results in Table 3 and the answer to Question 4?

6. Compare your plots of temperature *vs.* height for the two sites. If you performed Extension 1, compare plots of the measurements for the two sites.
7. Using your answer to Question 6, summarize how living organisms affect both biotic and abiotic factors in an ecosystem.
8. Which ecosystem, the complex one found in Site A or the simple one found in Site B, requires a greater expenditure of human resources to maintain? Explain your answer.
9. If each ecosystem experienced a fundamental environmental change, which would be more likely to survive? Explain your reasoning.
10. Summarize your conclusions of this field experiment.

EXTENSIONS

1. What other environmental measurements (in addition to temperature) could be taken that would help one understand how abiotic variables are affected by living organisms? Design an experiment to measure these variables.
2. Write an essay that discusses how an animal might perceive a patch site.
3. Make two possible food pyramids from your data, one from Site A and the other from Site B. Discuss how biodiversity affects the two food pyramids.

Experiment

15

TEACHER INFORMATION

Biodiversity and Ecosystems

1. If possible, this lab should be performed on a warm, dry day.
2. Students will need class time to discuss how they will define their plants and plant patches. This should be done while (or after) the students visualize their study areas. Since groups will be sharing data, it is best if the class comes to a common agreement prior to data collection.

   A patch is an area where plants are separate from other plants. This can be a single organism or a large number of similar organisms. Students will need to determine how to distinguish one patch from another.
3. Have student groups select a wide variety of locations for Site A. This will provide ample information when the groups share data. The area should have some plants at least 0.5 m tall.
4. A lawn or playing field is a good choice for Site B. They usually are part of a very simple ecosystems.
5. Allow sufficient time at each site so that students observe animals that move into and out of their site. Their site includes a vertical column as high as they can observe.
6. It is strongly recommended that students perform Extension 1. In the data below, light and relative humidity were measured in addition to temperature. Students may need class time to decide what additional measurements to take. You will need to advise students what sensors and probes you have available.

SAMPLE RESULTS

The following data may be different from students' results. The data below came from the two sites in Figure 1. In this discussion, Extension 1 was performed using a light and relative humidity sensor.

Table 1

Count in your area	Site A	Site B
Number of different animals	8	0
Number of different plants	9	2
Number of plant patches	14	4
Humus depth (cm)	5	1

Table 2						
	Site A (diverse)			Site B (sparse)		
Height (cm)	Temp (°C)	Relative humidity	Light	Temp (°C)	Relative humidity	Light
100	27.8	35.0	0.95	31.3	13.9	0.95
90	27.8	35.0	0.91	31.2	14.0	0.95
80	27.4	34.9	0.85	31.2	14.1	0.95
70	27.2	34.8	0.71	31.2	14.0	0.95
60	26.5	39.9	0.59	31.2	14.1	0.93
50	25.5	38.3	0.44	31.2	14.1	0.88
40	25.2	43.9	0.39	31.3	15.0	0.83
30	24.7	43.7	0.25	31.4	16.4	0.73
20	24.2	42.2	0.19	31.5	20.6	0.60
10	23.9	43.6	0.21	31.6	22.1	0.32
0	20.4	61.0	0.030	33.4	61.3	0.05

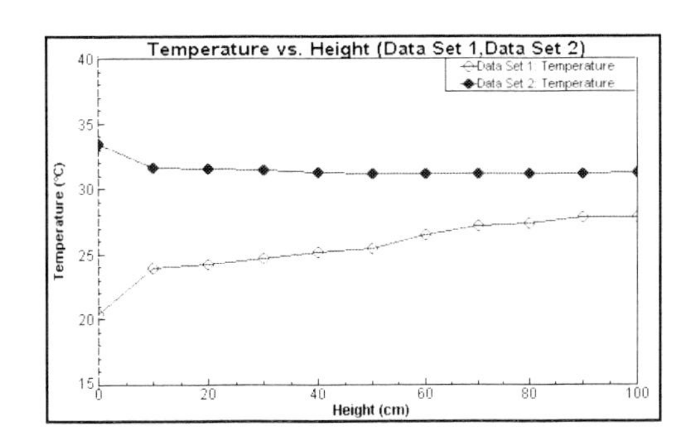

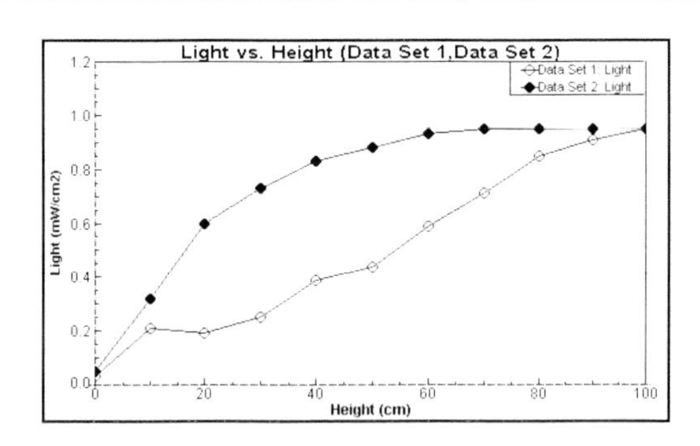

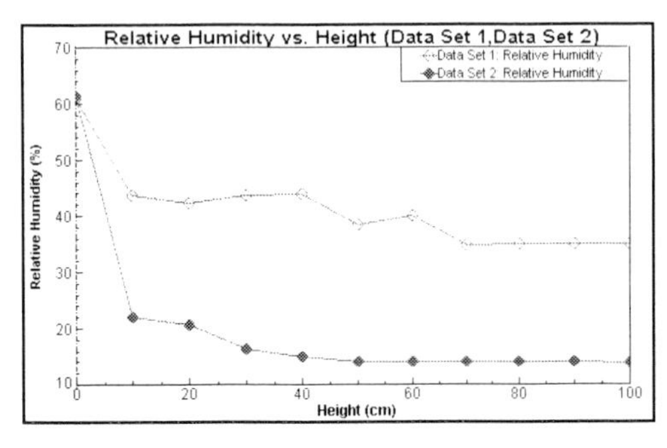

These plots were obtained from the sites in Figure 1. At Site A, the relative humidity remains high, the temperatures are quite cool, and the light reflected is lower due to the plant life.
At Site B, a grassy lawn, the relative humidity is high only at ground level, the temperatures are consistently high, and the light reflected is consistently higher. This provides a much more narrow habitat range than in Site A.

ANSWERS TO QUESTIONS

1. Answers will vary. In this sample, 17 types of organisms were found at Site A, while only two were at Site B. The biodiversity at Site A was greater than at Site B.

2. Site A supported more kinds and a higher population of animals. The range of habitats available to animals was greater, as was the potential food supply.

3. Answers may vary, although generally, the number of different patches of plants is greater in the more diverse area. The number of individual plants in either area may be quite high, however.

4. In the study area measured above, there are thousands of blades of grass, but grass was one of only two types of plants present at site B. A count of the number of plant patches seems to be a more meaningful indicator of biodiversity than a count of the number of plants.

5. Using different definitions will introduce extra variables into the experiment. For instance, if two groups examined the same area and obtained vastly different numbers, the two group's combined data would appear to be totally unrelated. If one group's method of defining plants and/or patches was used to measure all of the sites, the numbers could be compared in a meaningful way.

6. In this discussion, Extension 1 was performed using a light and relative humidity sensor.
   a. At Site A, the temperatures are quite cool due to the plant life. At Site B, a grassy lawn, the temperatures are consistently high. They may be highest at the ground level, since there is no shade.
   b. The light increases dramatically when there is little vegetation, and is not as intense when plants absorb light energy.
   c. At Site A, the relative humidity remains high due to the plant life. At Site B, a grassy lawn, the relative humidity is high only at ground level. It plummets drastically just above the height of the grass.

7. A diverse habitat provides a wide range of environments for plants and animals to live. The hot, dry conditions found in Site B are inhospitable to life compared to the moist, cool environment found in Site A. The plants at Site A tend to
   - provide shade, keeping the animals, the plant roots, the soil, and some plants cool.
   - trap the moisture, so that it does not evaporate into the atmosphere as rapidly.
   - provide a rich and more varied food supply.
   - keep the temperatures moderate and consistent under the shade.

8. Many resources were used to maintain a grassy lawn. Fossil fuels were combusted, metals processed for the mower, and human time and energy were constantly applied to maintain a simple ecosystem. Few human energy or resources were expended to maintain Site A—it was left alone for over 30 years.

9. Site A is more resilient to change. For instance, if a disease came though that destroyed one specie of grass, Site B might be devastated, while Site A would only be slightly affected. Because of the many different types of plants and animals present, the disappearance of any one organism would not be devastating at Site A.

Experiment

16A

Effect of Temperature on Respiration

Temperature changes have profound effects upon living things. Enzyme-catalyzed reactions are especially sensitive to small changes in temperature. Because of this, the metabolism of *poikilotherms,* organisms whose internal body temperature is determined by their surroundings, are often determined by the surrounding temperature. Bakers who use yeast in their bread making are very aware of this. Yeast is used to leaven bread (make it rise). Yeast leavens bread by fermenting sugar, producing carbon dioxide, CO_2, as a waste product. Some of the carbon dioxide is trapped by the dough and forms small "air" pockets that make the bread light. If the yeast is not warmed properly, it will not be of much use as a leavening agent; the yeast cells will burn sugar much too slowly. In this experiment, you will watch yeast cells respire (burn sugar) at different temperatures and measure their rates of respiration. Each team will be assigned three temperatures by your teacher and will share their results with other class members.

You will observe the yeast under aerobic conditions and monitor the change in concentration of carbon dioxide released by the yeast. When yeast burn sugar under anaerobic conditions, ethanol (ethyl alcohol) and carbon dioxide are released in a process called fermentation. When yeast burn sugar under aerobic conditions, water and carbon dioxide are released in a process known as oxidative respiration. Thus, the metabolic activity of yeast may be measured by monitoring the concentration of carbon dioxide produced during a specific period of time.

OBJECTIVES

In this experiment, you will

- Use a CO_2 Gas Sensor to measure the concentration of carbon dioxide produced during respiration.
- Determine the rate of respiration of yeast at different temperatures.

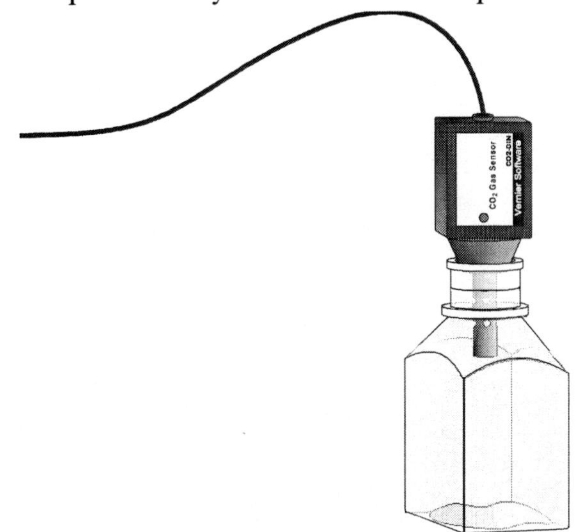

Figure 1

MATERIALS

computer
Vernier computer interface
Logger *Pro*
Vernier CO_2 Gas Sensor
250 mL respiration chamber
5% sugar solution
yeast suspension
three 600 mL beakers
Beral pipettes
hot and cold water
thermometer
three 10 × 100 mm test tube

PROCEDURE

1. Prepare a water bath for the yeast. A water bath is simply a large beaker of water at a certain temperature. This ensures that the yeast will remain at a constant and controlled temperature. To prepare the water bath, obtain some warm and cool water from your teacher. If you have been assigned very cold temperatures, you will need to get some ice from your teacher. Combine the warm and cool water into the 600 mL beaker until it reaches the assigned temperature. The beaker should be filled with about 300 – 400 mL water. Leave the thermometer in the water bath during the course of the experiment to monitor the temperature of the water bath.

2. Obtain three test tubes and label them according to the temperatures you were assigned.

3. Place 2 mL of sugar solution in each test tube.

4. Obtain the yeast suspension. Gently swirl the yeast suspension to mix the yeast that settles to the bottom. Put 2 mL of yeast into each of the test tubes. Gently swirl each test tube to mix the yeast into the solution.

5. Set one test tube in each of the water baths. Incubate the test tubes for 10 minutes in the water baths. Keep the temperature of each water bath constant. If you need to add more hot or cold water, first remove as much water as you will add, or the beaker may overflow. Use a Beral pipet to remove excess water. While the test tubes are incubating proceed to Step 6.

6. Connect the CO_2 Gas Sensor to the computer interface. Prepare the computer for data collection by opening the file "16A Temp-Respiration" from the *Biology with Computers* folder of Logger*Pro*.

7. When incubation is finished, use a Beral pipet to place 1 mL of the sugar/yeast solution from the first assigned temperature into the 250 mL respiration chamber. Note the temperature of the water bath when you transfer the yeast and record the temperature in Table 1.

8. Quickly place the shaft of the CO_2 Gas Sensor in the opening of the respiration chamber. Gently twist the stopper on the shaft of the CO_2 Gas Sensor into the chamber opening. Do not twist the shaft of the CO_2 Gas Sensor or you may damage it.

9. Hold the respiration chamber down in the water bath during data collection. Begin measuring carbon dioxide concentration by clicking Collect.

10. Data collection will end after 4 minutes. Remove the CO_2 sensor from the respiration chamber.

11. Determine the rate of respiration:
    a. Move the mouse pointer to the point where the data values begin to increase. Hold down the left mouse button. Drag the pointer to the point where the data ceases to rise and release the mouse button.
    b. Click on the Linear Fit button, , to perform a linear regression. A floating box will appear with the formula for a best fit line.
    c. Record the slope of the line, *m*, as the rate of respiration in Table 1.
    d. Close the linear regression floating box.
12. Move your data to a stored run. To do this, choose Store Latest Run from the Experiment menu.
13. Fill the respiration chamber with water and then empty it. Make sure that all yeast have been removed from the chamber. Thoroughly dry the inside with a paper towel.
14. Use a notebook or notepad to fan air across the openings in the probe shaft of the CO_2 gas sensor for 1 minute.
15. Repeat Steps 7 - 14 for the other two assigned temperatures.

DATA

Table 1

Temperature tested (°C)	Respiration rate (ppm/min)

Table 2: Class Averages	
Temperature (°C)	Respiration rate (ppm/min)
5 – 10°C	
10 – 15°C	
15 – 20°C	
20 – 25°C	
25 – 30°C	
30 – 35°C	
35 – 40°C	
40 – 45°C	
45 – 50°C	

PROCESSING THE DATA

1. Share your data with the class by recording on the board the temperatures you tested and the rate of respiration you calculated for each.
2. Using the class data, calculate the average for each temperature range and record in Table 2.
3. On Page 2 of the experiment file, create a graph of the rate of respiration *vs.* temperature using the class data in Table 2. The respiration rate values should be plotted on the y-axis, and the temperature values plotted on the x-axis.

QUESTIONS

1. On the basis of the results of this experiment, does temperature affect the rate of respiration of yeast? Explain.
2. Is there an *optimal* temperature for yeast respiration? How does this compare to the temperature yeast are at bread "rises"? You may have to consult a cookbook to answer this.
3. Why does the respiration rate decrease at very high temperatures?
4. It is sometimes said that the metabolism of poikilotherms doubles with every 10°C increase in temperature. Does your data support this statement? Explain.
5. Do yeast always live in conditions where their consumption of energy is optimal? Explain.
6. Yeast live in many different environments. Make a list of some locations where yeast might naturally grow. Estimate the temperatures of each of these locations and compare them to your results.

Experiment

16A

TEACHER INFORMATION

Effect of Temperature on Respiration

In this experiment, students will test the ability of yeast to respire aerobically at different temperatures.

1. To prepare the yeast solution, dissolve 7 g (1 package) of dried yeast for every 100 mL of water. Incubate the suspension in 37 – 40°C water for at least 10 minutes.

2. After the 10 minute incubation period, transfer the yeast to dispensing tubes. Each group will need about 12 mL of yeast.

3. To prepare the 5% sugar solution add 5 g of sucrose per 100 mL of solution.

4. Assign one of the following sets of temperatures to each lab group:

Group 1	Group 2	Group 3
5 – 10°C	20 – 25°C	35 – 40°C
10 – 15°C	25 – 30°C	40 – 45°C
15 – 20°C	30 – 35°C	45 – 50°C

   These are the temperatures that they will be responsible for testing. Try to get each temperature tested an equal number of times. It is important to be in the temperature range, but not critical to be at a specific temperature.

5. The stopper included with the CO_2 Gas Sensor is slit to allow it to be easily added or removed from the probe. When students are placing the probe in the respiration chamber, they should gently twist the stopper into the chamber opening. Warn the students not to twist the probe shaft or they may damage the sensing unit.

6. The CO_2 Gas Sensor is dependent on the diffusion of gases into the probe shaft. Students should allow a couple of minutes between trials so that gases from the previous trial will have diffused out of the probe shaft. Alternatively, the students can use a firm object such as a book or notepad to fan air through the probe shaft. This method is used in Step 14 of the student procedure.

7. The morning of the experiment fill a 1 L beaker with ice and water so students will have cold water for Step 1 of the procedure. Students will also need ice for the water baths.

8. The CO_2 Gas Sensor requires a 90 second period to warm up and read accurately. The students should plug in the CO_2 Gas Sensor at least 90 seconds before data is collected. Alternatively, you can plug in the sensors prior to the start of class.

9. The CO_2 Gas Sensor was calibrated at the time of manufacturing, so you should have no need to calibrate it. The only instance where you may need to recalibrate the sensor is if it appears to be giving unusual readings. Refer to the CO_2 Gas Sensor probe booklet to calibrate the sensor.

SAMPLE RESULT

Table 2: Class Averages	
Temperature (°C)	Respiration Rate (ppm/min)
5 – 10°C	101.4
10 – 15°C	105.2
15 – 20°C	246.6
20 – 25°C	546.1
25 – 30°C	628.8
30 – 35°C	681.3
35 – 40°C	738.6
40 – 45°C	787.8
45 – 50°C	595.8
50 – 55°C	402.6

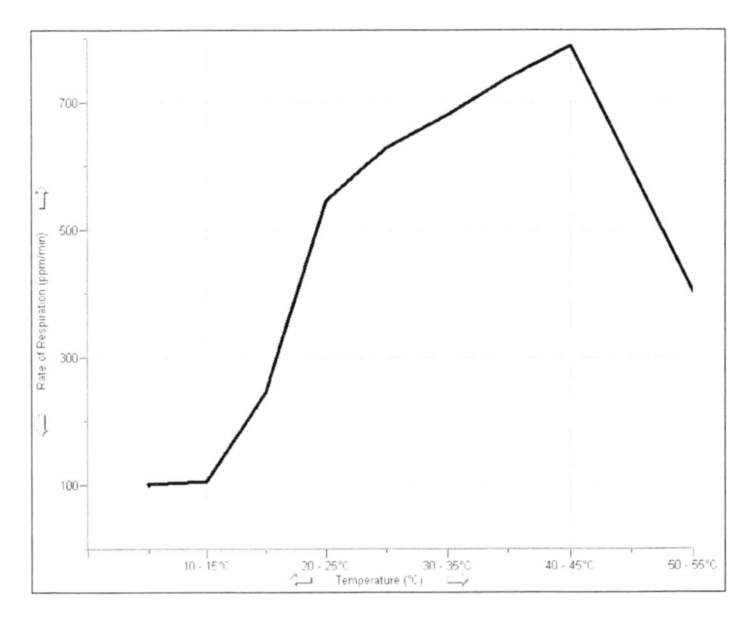

Class Averages

ANSWERS TO QUESTIONS

1. Yes, temperature does affect the rate of respiration in yeast. The rate should increase from 10°C to 40°C. At 50°C and above, the rates will decrease.

2. The temperature yielding the highest rate of respiration should be in the 30 – 40°C range.

3. The respiration rate at high temperatures will be low because high temperatures are lethal to yeast. Enzymes move too fast and denature at high temperatures.

4. Results may vary.

5. No, yeast do not always live in conditions where their consumption of energy is optimal. Higher temperature conditions may not really be optimal under all conditions. Overpopulation, competition for resources such as food, and waste production may accompany optimal energy consumption. Also, the majority of habitats available to yeast are not at this temperature, so their distribution is much greater.

6. Lists may vary. Some yeast live on other organisms. If the other organisms are warm blooded, they may be near the optimal temperature for yeast respiration, 37°C. Many yeast live in soils. The temperature of soils may easily be measured at different times of the year.

Experiment

16B

Effect of Temperature on Fermentation

Temperature changes have profound effects upon living things. Enzyme-catalyzed reactions are especially sensitive to small changes in temperature. Because of this, the metabolism of *poikilotherms,* organisms whose internal body temperature is determined by their surroundings, are often determined by the surrounding temperature. Bakers who use yeast in their bread making are very aware of this. Yeast is used to leaven bread (make it rise). Yeast leavens bread by fermenting sugar, producing carbon dioxide, CO_2, as a waste product. Some of the carbon dioxide is trapped by the dough and forms small "air" pockets that make the bread light. If the yeast is not warmed properly, it will not be of much use as a leavening agent; the yeast cells will burn sugar much too slowly. In this experiment, you will watch yeast cells ferment (burn sugar in the absence of oxygen) at different temperatures and measure their rates of fermentation. Each team will be assigned one temperature and will share their results with other class members.

You will observe the yeast under anaerobic conditions and monitor the change in air pressure due to CO_2 released by the yeast. When yeast burn sugar under anaerobic conditions, ethanol (ethyl alcohol) and CO_2 are released as shown by the following equation:

$$\underset{\text{glucose}}{C_6H_{12}O_6} \rightarrow 2\ \underset{\text{ethanol}}{CH_3CH_2OH} + 2\ \underset{\text{carbon dioxide}}{CO_2} + \text{energy}$$

Thus, the metabolic activity of yeast may be measured by monitoring the pressure of gas in the test tube. If the yeast were to respire aerobically, there would be no change in the pressure of gas in the test tube, because oxygen gas would be consumed at the same rate as CO_2 is produced.

OBJECTIVES

In this experiment, you will

- Use a Gas Pressure Sensor to measure the change in pressure due to carbon dioxide released during respiration.
- Determine the rate of fermentation of yeast at different temperatures.

MATERIALS

computer
Vernier computer interface
Logger *Pro*
Vernier Gas Pressure Sensor
5% glucose solution
1 liter beaker (for water bath)
Beral pipet, graduated
10 mL graduated cylinder
18 × 150 mm test tube
yeast suspension
vegetable oil in a dropper bottle
1-hole rubber stopper assembly
plastic tubing w/Luer-lock fitting
pipet pump or bulb
ring stand
thermometer
basting bulb or Beral pipet
utility clamp
warm and cool water

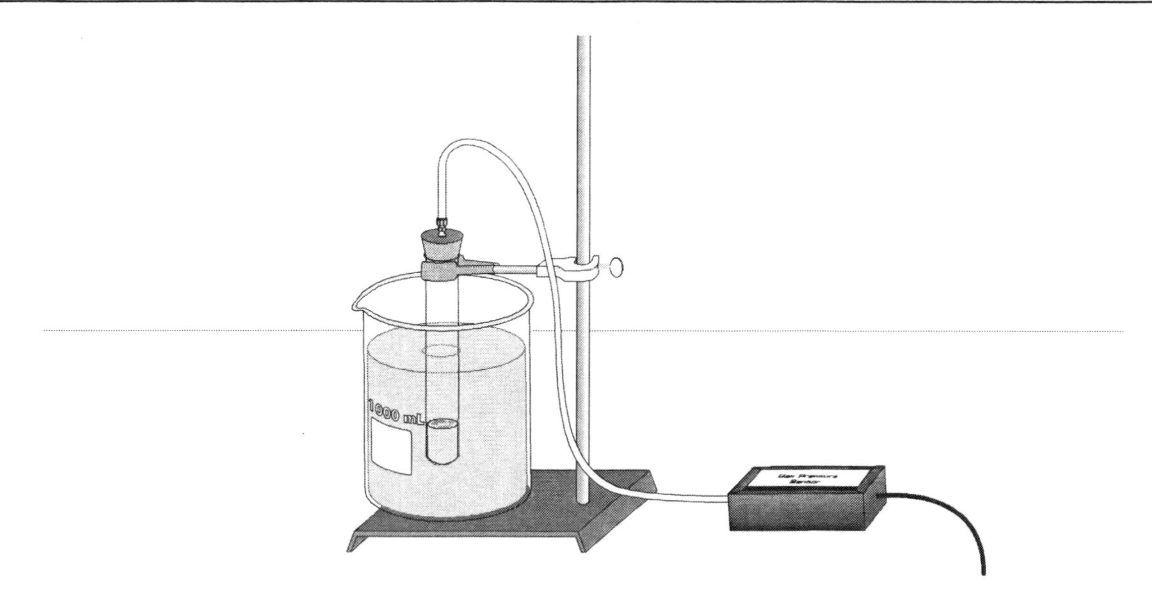

Figure 1

PROCEDURE

1. Connect the Gas Pressure Sensor to the computer interface. Prepare the computer for data collection by opening the file "16B Temp-Ferment (Pressure)" from the *Biology with Computers* folder of Logger*Pro*.

2. Connect the plastic tubing to the valve on the Gas Pressure Sensor.

3. Prepare a water bath for the yeast. A water bath is simply a large beaker of water at a certain temperature. This ensures that the yeast will remain at a constant and controlled temperature. To prepare the water bath, obtain some warm and cool water from your teacher. Combine the warm and cool water into the 1 liter beaker until it reaches the temperature you were assigned. The beaker should be filled with about 600 – 700 mL water. Place the thermometer in the water bath to monitor the temperature during the experiment.

4. Place 2.5 mL of the glucose solution into a clean 18 × 150 mm test tube.

5. Obtain the yeast suspension. Gently swirl the yeast suspension to mix the yeast that settles to the bottom. Using a 10 mL pipette or graduated cylinder, transfer 2.5 mL of yeast into the test tube. Gently mix the yeast into the sugar solution. Be gentle with the yeast—they are living organisms!

6. In the test tube, place enough vegetable oil to completely cover the surface of the yeast/glucose mixture as shown in Figure 2. Be careful to not get oil on the inside wall of the test tube. Set the test tube in the water bath.

7. Insert the single-holed rubber-stopper into the test tube. **Note**: *Firmly* twist the stopper for an *airtight* fit. Secure the test tube with a utility clamp and ring-stand as shown in Figure 1.

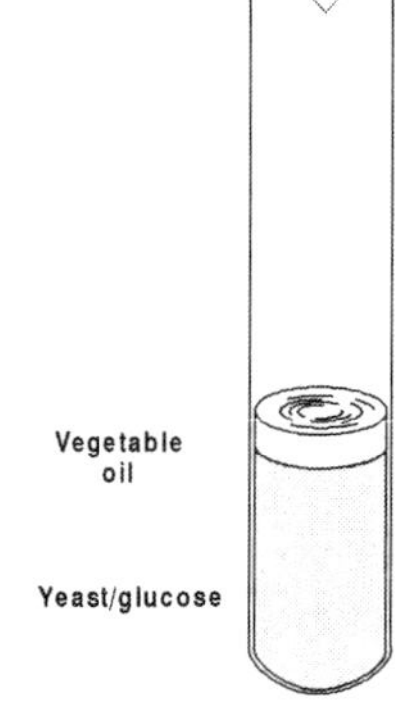

Figure 2

8. Incubate the test tube for 10 minutes in the water bath. Be sure to keep the temperature of the water bath constant. If you need to add more hot or cold water, first remove about as much water as you will be adding, or the beaker may overflow. Use a basting bulb to remove excess water.

   Note: Be sure that most of the test tube is completely covered by the water in the water bath. The temperature of the air in the tube must be constant for this experiment to work well.

9. When incubation has finished, connect the free-end of the plastic tubing to the connector in the rubber stopper as shown in Figure 3.

10. Click [Collect] to begin data collection. Maintain the temperature of the water bath during the course of the experiment. Data will be collected for 15 minutes.

11. Monitor the pressure readings displayed in the meter. If the pressure exceeds 130 kPa, the pressure inside the tube will be too great and the rubber stopper is likely to pop off. Disconnect the plastic tubing from the Gas Pressure Sensor if the pressure exceeds 130 kPa.

12. When data collection has finished, disconnect the plastic tubing connector from the rubber stopper. Remove the rubber stopper from the test tube and discard the contents in a waste beaker.

Figure 3

13. Find the rate of fermentation:

    a. Move the mouse pointer to the point where the pressure values begin to increase. Hold down the mouse button. Drag the mouse pointer to the position where the data values cease to rise and release the mouse button.
    b. Click the Linear Fit button, [Linear Fit], to perform a linear regression. A floating box will appear with the formula for a best fit line.
    c. Record the slope of the line, m, as the rate of fermentation in Table 1.
    d. Close the linear regression floating box.
    e. Share your data with other classmates by recording the temperature and the rate of fermentation on the board.

PROCESSING THE DATA

1. Record the data from the other teams in Table 2.

2. On Page 2 of the experiment file, create a graph of fermentation rate *vs*. temperature using the data recorded in Table 2. The fermentation rate values should be plotted on the y-axis, and the temperature on the x-axis.

DATA

Table 1: Team Data	
Temperature (°C)	Rate of Fermentation (kPa/min)

Table 2: Class Data	
Temperature (°C)	Rate of Fermentation (kPa/min)

QUESTIONS

1. On the basis of the results of this experiment, does temperature affect the rate of fermentation of yeast? Explain.
2. Is there an *optimal* temperature for yeast to ferment sugar? How does this compare to the temperature yeast are at when "rising" while making bread? You may have to consult a cookbook to answer this.
3. Why does the fermentation rate decrease at very high temperatures?
4. It is sometimes said that the metabolism of poikilotherms doubles with every 10°C increase in temperature. Does your data support this statement? Explain.
5. What is the purpose of the oil on top of the water?
6. Do yeast always live in conditions where their consumption of energy is optimal? Explain.
7. Why do you need to incubate the yeast before you start monitoring air pressure?
8. Yeast live in many different environments. Make a list of some locations where yeast might naturally grow. Estimate the temperatures of each of these locations and compare them to your results.

Effect of Temperature on Fermentation

1. To prepare the yeast solution, dissolve 7 g (1 package) of dried yeast per 100 mL of water. Incubate the suspension in 37 – 40°C water for at least 10 minutes.

2. Emphasize to your students the importance of providing an airtight fit with all plastic-tubing connections and when closing valves or twisting the stopper into a test tube.

3. All of the pressure valves, tubing, and connectors used in this experiment are included with Vernier Gas Pressure Sensors shipped after February 15, 1998. These accessories are also helpful when performing respiration/fermentation experiments such as Experiments 11C, and 12B in this manual.

   If you purchased your Gas Pressure Sensor at an earlier date, Vernier has a Pressure Sensor Accessories Kit (PS-ACC) that includes all of the parts shown here for doing pressure-related experiments. Using this kit allows for easy assembly of a completely airtight system. The kit includes the following parts:

   - two ribbed, tapered valve connectors inserted into a No. 5 rubber stopper
   - one ribbed, tapered valve connectors inserted into a No. 1 rubber stopper
   - two Luer-lock connectors connected to either end of a piece of plastic tubing
   - one two-way valve
   - one 20 mL syringe
   - two tubing clamps for transpiration experiments

5. The accessory items used in this experiment are the #1 single hole stopper fitted with a tapered valve connector and the section of plastic tubing fitted with Luer-lock connectors.

6. The length of plastic tubing connecting th: rubber stopper assemblies to each gas pressure sensor must be the same for all groups. It is best to keep the length of tubing reasonably small to keep the volume of gas in the test tube low. **Note:** If pressure changes during data collection are too small, you may need to decrease the total gas volume in the system. Shortening the length of tubing used will help to decrease the volume.

7. To prepare the 5% glucose sugar solution, add 5 grams glucose per 100 mL of solution. Sucrose may be substituted for glucose, if desired.

8. Temperatures of 10, 20, 30, 40, 50, and 60°C are suggested. Assign one temperature to each student team.

9. If you have turkey basters, they are excellent tools for removing water from the water bath as the temperature of the water bath is being maintained. Beral pipettes also work well.
10. Students can practice maintaining the water bath temperature during the incubation period.
11. If time permits, extend the experiment length to 30 minutes. This can be done by increasing the time between samples from 10 to 20 seconds.

SAMPLE RESULTS

The following data may be different from students' results. The actual values depend upon the viability and age of the yeast, among other factors.

Class Data	
Temperature (°C)	Rate of Fermentation (kPa/min)
10	0.0517
20	0.0709
30	0.3991
40	0.7912
50	0.5298
60	0.0051

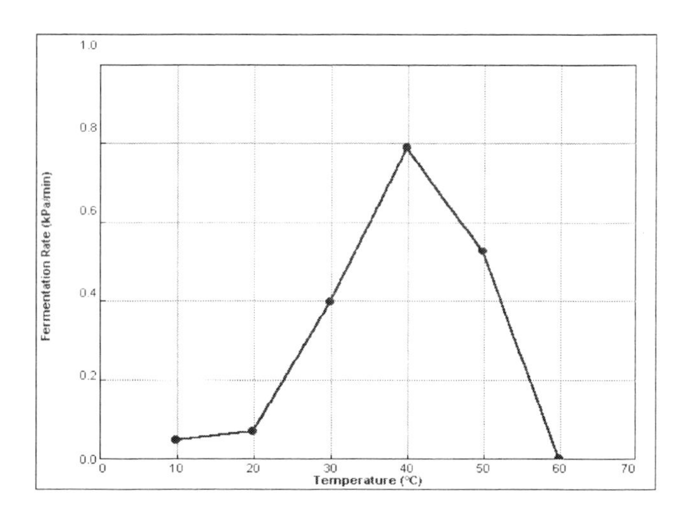

ANSWERS TO QUESTIONS

1. Yes, temperature does affect the rate of fermentation in yeast. The rate should increase from 10°C to 40°C. At 50°C and above, the rates will decrease.
2. The temperature yielding the highest rate of respiration should be in the 40°C range.
3. The fermentation rate at high temperatures will be low because high temperatures are lethal to yeast. Enzymes move too fast and denature at high temperatures.
4. Results may vary.
5. The oil prevents oxygen from the air in the test tube from entering the solution. This helps keep the yeast anaerobic. It also helps keep the water from evaporating.
6. No, yeast do not always live in conditions where their consumption of energy is maximal. Higher temperature conditions may not really be optimal under all conditions. Overpopulation, competition for resources such as food, and waste production may accompany maximal energy consumption. Also, the majority of habitats available to yeast are not at this temperature, so their distribution is much greater.
7. The yeast need to be incubated so that the oxygen in the test tube will be completely consumed. If the yeast respire aerobically, no pressure change occurs, because much oxygen is consumed as CO_2 is produced. It also takes a few minutes for the yeast to transport the

sugar into the cell, to respire at a constant rate, and to reach the proper temperature.

8. Lists may vary. Some yeast live on other organisms. If the other organisms are warm blooded, they may be near the optimal temperature for yeast respiration, 37°C. Many yeast live in soils. The temperature of soils may easily be measured at different times of the year.

Experiment
17

Aerobic Respiration

Aerobic cellular respiration is the process of converting the chemical energy of organic molecules into a form immediately usable by organisms. Glucose may be oxidized completely if sufficient oxygen is available, by the following reaction:

$$C_6H_{12}O_6 + 6O_2(g) \rightarrow 6\ H_2O + 6\ CO_2(g) + \text{energy}$$

All organisms, including plants and animals, oxidize glucose for energy. Often, this energy is used to convert ADP and phosphate into ATP. In this experiment, the rate of cellular respiration will be measured by monitoring the consumption of oxygen gas.

Many environmental variables might affect the rate of aerobic cellular respiration. Temperature changes have profound effects upon living things. Enzyme-catalyzed reactions are especially sensitive to small changes in temperature. Because of this, the metabolism of *ectotherms,* organisms whose internal body temperature is determined by their surroundings, are often determined by the surrounding temperature. In this experiment, you will determine the effect temperature changes have on the aerobic respiration of yeast.

OBJECTIVES

In this experiment, you will

- Use a computer to measure changes in dissolved oxygen concentration.
- Study the effect of temperature on cellular respiration.
- Make a plot of the rate of cellular respiration as a function of temperature.

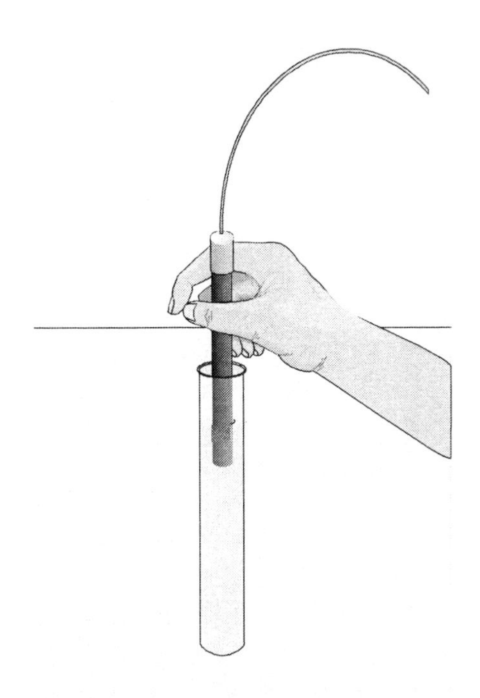

Figure 1

MATERIALS

computer
Vernier computer interface
Vernier Dissolved Oxygen Probe
Vernier Temperature Probe
Logger*Pro*
two 18 × 150 mm test tubes or vials
1% glucose solution
two 10 × 100 mm test tubes
25 mL graduated cylinder
two 600 mL beakers (for water bath)
2 mL pipette or 10 mL graduated cylinder
pipette bulb or pump
ring stand (optional)
test tube rack
utility clamp
warm and cool water
wash bottle with distilled water
yeast solution

PRE-LAB PROCEDURE

Important: Prior to each use, the Dissolved Oxygen Probe must warm up for a period of 10 minutes as described below. If the probe is not warmed up properly, inaccurate readings will result. Perform the following steps to prepare the Dissolved Oxygen Probe.

1. Prepare the Dissolved Oxygen Probe for use.
   a. Remove the blue protective cap.
   b. Unscrew the membrane cap from the tip of the probe.
   c. Using a pipet, fill the membrane cap with 1 mL of DO Electrode Filling Solution.
   d. Carefully thread the membrane cap back onto the electrode.
   e. Place the probe into a container of water.

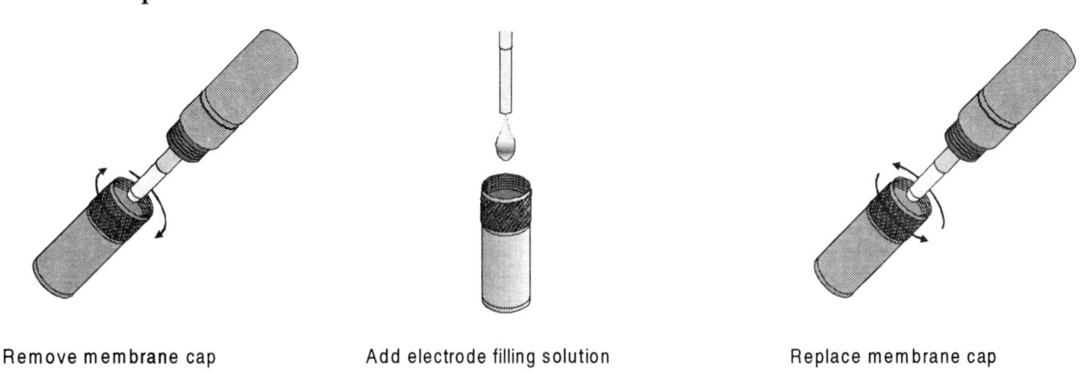

Remove membrane cap   Add electrode filling solution   Replace membrane cap

2. Plug the Dissolved Oxygen Probe into Channel 1 of the Vernier interface. Connect the Temperature Probe to Channel 2.
3. Prepare the computer for data collection by opening the file "17 Aerobic Respiration" from the *Biology with Computers* folder of Logger*Pro*.
4. It is necessary to warm up the Dissolved Oxygen Probe for 10 minutes before taking readings. To warm up the probe, leave it connected to the interface, with Logger *Pro* running, for 10 minutes. The probe must stay connected at all times to keep it warmed up. If disconnected for a few minutes, it will be necessary to warm up the probe again.
5. You are now ready to calibrate the Dissolved Oxygen Probe.
   - If your instructor directs you to use the calibration stored in the experiment file, then proceed to Step 6.

- If your instructor directs you to perform a new calibration for the Dissolved Oxygen Probe, follow this procedure.

Zero-Oxygen Calibration Point

a. Choose Calibrate ▸ CH1: Dissolved Oxygen (mg/L) from the Experiment menu and then click Calibrate Now.

b. Remove the probe from the water and place the tip of the probe into the Sodium Sulfite Calibration Solution. **Important:** No air bubbles can be trapped below the tip of the probe or the probe will sense an inaccurate dissolved oxygen level. If the voltage does not rapidly decrease, tap the side of the bottle with the probe to dislodge any bubbles. The readings should be in the 0.2 to 0.5 V range.

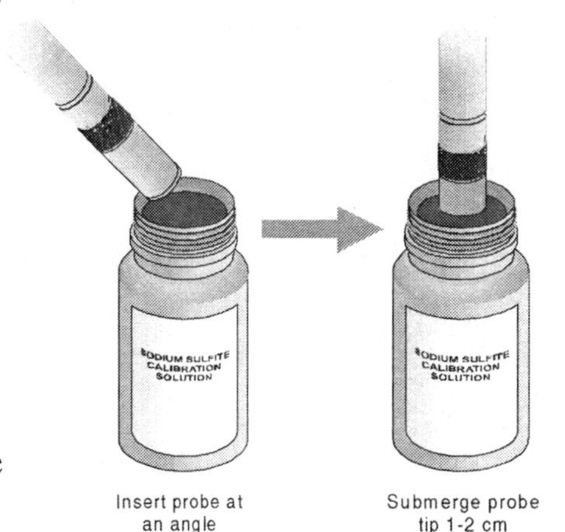

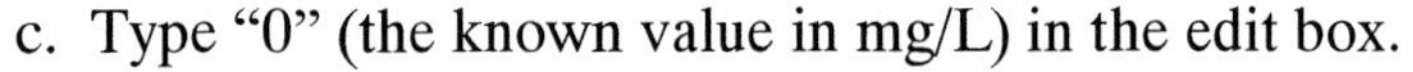

c. Type "0" (the known value in mg/L) in the edit box.

d. When the displayed voltage reading for Reading 1 stabilizes, click Keep.

Saturated DO Calibration Point

e. Rinse the probe with distilled water and gently blot dry.

f. Unscrew the lid of the calibration bottle provided with the probe. Slide the lid and the grommet about 2 cm onto the probe body.

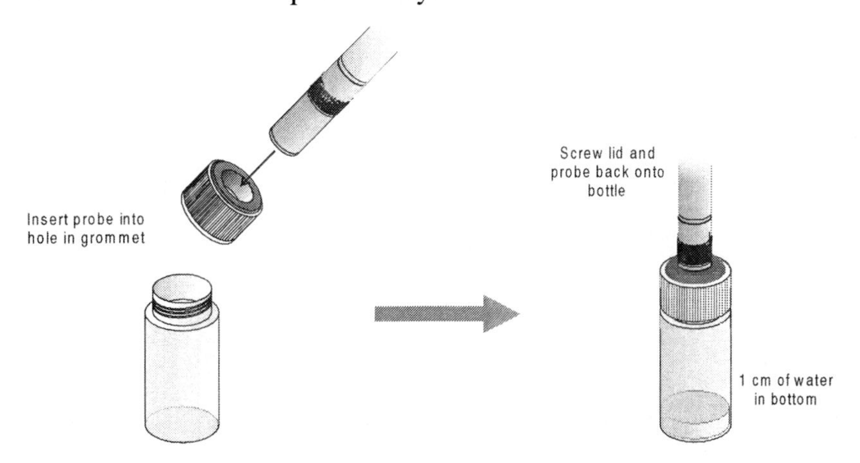

g. Add water to the bottle to a depth of about 1 cm and screw the bottle into the cap, as shown. **Important:** Do not touch the membrane or get it wet during this step. Keep the probe in this position for about a minute.

h. Type the correct saturated dissolved-oxygen value (in mg/L) from Table 2 (for example, "8.66") using the current barometric pressure and air temperature values. If you do not have the current air pressure, use Table 3 to estimate the air pressure at your altitude.

i. When the displayed voltage reading for Reading 2 stabilizes (readings should be above 2.0 V), click Keep.

j. Choose Calibration Storage from the calibration pull-down menu. Choose Experiment File (calibration stored with the current document). Click Done.

k. From the File menu, select Save As and save the current experiment file with a new name.

PROCEDURE

6. Obtain and wear goggles.

7. Set up two water baths, one at room temperature, approximately 20 – 25°C, and one at approximately 10°C. A water bath is simply a large beaker of water at the correct temperature. Obtain two 600 mL beakers and place about 200 mL of water in each. Add ice to attain 10°C in the cool water bath.

8. Obtain two clean, small test tubes. Label them "1" and "2". Place 2 mL of the yeast solution into each test tube.

9. Obtain two clean, large test tubes and label them "A", and "B". Place 20 mL of the glucose solution into each test tube.

10. Place test tube 1 and test tube A in the warm water bath and place test tube 2 and test tube B in the cool water bath.

11. After test tube 1 and test tube A reach the temperature of the water bath,
    a. Seal test tube A and vigorously mix the contents for about 60 seconds to aerate the solution.
    b. Return test tube A to the water bath and insert the Dissolved Oxygen Probe as shown in Figure 1. Place the Temperature Probe into the water bath.
    c. Gently and continuously stir the solution with the probe for 2 – 3 minutes, or until the dissolved oxygen reading is stable.
    d. Gently mix test tube 1 to suspend the yeast.
    e. Remove the probe from test tube A and add 2 mL of yeast solution from test tube 1.
    f. Place the Dissolved Oxygen Probe into test tube A.

12. Gently and continuously stir the solution with the probe.

13. After 10 seconds, start data collection by clicking Collect.

14. If the dissolved oxygen concentration drops below 1.0 mg/L, stop the experiment by clicking Stop. Otherwise data collection will stop after 2 minutes.

15. When data collection stops remove the Dissolved Oxygen Probe. Click on the plot of temperature *vs.* time with the mouse pointer. Click the Statistics button, , and record the mean temperature in Table 1. Close the statistics box by clicking in the top left corner of the box.

16. Determine the rate of aerobic respiration.
    a. Click on the plot of dissolved oxygen *vs.* time with the mouse pointer.
    b. Click the Linear Fit button, , to perform a linear regression. A floating box will appear with the formula for a best fit line.
    c. Record the slope of the line, m, as the rate of O_2 consumption in Table 1.

17. Remove the Dissolved Oxygen Probe from test tube A and rinse with water until clean.

18. Repeat Steps 11 – 17, using test tubes "2" and "B", which are in the cool water bath. Be sure the temperatures of the water bath and both solutions have equilibrated. **Note**: One of the team members may want to start Step 19 at this time.

19. Prepare two more water baths, one at 40°C and one at 60°C.

20. Repeat Steps 11 – 17, using test tubes "1", "A" and the water bath at 40°C.

21. Repeat Steps 11 – 17, using test tubes "2", "B" and the water bath at 60°C.

22. Clean all of the glassware and probes.

23. If time permits,

    a. Prepare two more water baths, one at 30°C and one at 50°C.
    b. Repeat Steps 11 – 17, using test tubes "1", "A" and the water bath at 30°C.
    c. Repeat Steps 11 – 17, using test tubes "2", "B" and the water bath at 50°C.

24. Clean all of the glassware and probes. Place the Dissolved Oxygen Probe in a beaker of distilled water.

25. On Page 2 of the experiment file, create a graph of rate of oxygen consumption *vs.* temperature. The rate values from Table 1 should be plotted on the y-axis, and the temperature on the x-axis.

DATA

Table 1

Assigned Temperature (°C)	Mean Temperature (°C)	Rate of O_2 consumption (mg O_2/L/min)
10		
20		
30		
40		
50		
60		

QUESTIONS

1. Do you have evidence that aerobic cellular respiration occurs in yeast? Explain.

2. How did temperature affect the rate of respiration in yeast? Explain.

3. Is there an *optimal* temperature for yeast respiration?

4. It is sometimes said that the metabolism of ectotherms doubles with every 10°C increase in temperature. Do your data support this statement? Explain.

5. What will happen to the yeast after the oxygen concentration in the solution drops to zero? Why will they continue to respire sugar? Explain.

EXTENSIONS

1. Design an experiment to measure and compare the effect of other environmental conditions on the respiration rate of yeast.
2. Design an experiment to measure and compare the respiration rate of a different aquatic organism at different temperatures. What changes to this experimental procedure might be necessary?
3. Carry out the experiment you designed in Step 1 or Step 2.
4. Compare the initial concentration of dissolved oxygen you measured at every temperature. What trends do you see? How does this compare to what you expected? Explain.

CALIBRATION TABLES

Table 2: 100% Dissolved Oxygen Capacity (mg/L)												
	70 mm	**60 mm**	**50 mm**	**40 mm**	**30 mm**	**20 mm**	**10 mm**	**00 mm**	**90 mm**	**80 mm**	**70 mm**	**60 mm**
0°C	14.76	14.57	14.38	14.19	13.99	13.80	13.61	13.42	13.23	13.04	12.84	12.65
1°C	14.38	14.19	14.00	13.82	13.63	13.44	13.26	13.07	12.88	12.70	12.51	12.32
2°C	14.01	13.82	13.64	13.46	13.28	13.10	12.92	12.73	12.55	12.37	12.19	12.01
3°C	13.65	13.47	13.29	13.12	12.94	12.76	12.59	12.41	12.23	12.05	11.88	11.70
4°C	13.31	13.13	12.96	12.79	12.61	12.44	12.27	12.10	11.92	11.75	11.58	11.40
5°C	12.97	12.81	12.64	12.47	12.30	12.13	11.96	11.80	11.63	11.46	11.29	11.12
6°C	12.66	12.49	12.33	12.16	12.00	11.83	11.67	11.51	11.34	11.18	11.01	10.85
7°C	12.35	12.19	12.03	11.87	11.71	11.55	11.39	11.23	11.07	10.91	10.75	10.59
8°C	12.05	11.90	11.74	11.58	11.43	11.27	11.11	10.96	10.80	10.65	10.49	10.33
9°C	11.77	11.62	11.46	11.31	11.16	11.01	10.85	10.70	10.55	10.39	10.24	10.09
10°C	11.50	11.35	11.20	11.05	10.90	10.75	10.60	10.45	10.30	10.15	10.00	9.86
11°C	11.24	11.09	10.94	10.80	10.65	10.51	10.36	10.21	10.07	9.92	9.78	9.63
12°C	10.98	10.84	10.70	10.56	10.41	10.27	10.13	9.99	9.84	9.70	9.56	9.41
13°C	10.74	10.60	10.46	10.32	10.18	10.04	9.90	9.77	9.63	9.49	9.35	9.21
14°C	10.51	10.37	10.24	10.10	9.96	9.83	9.69	9.55	9.42	9.28	9.14	9.01
15°C	10.29	10.15	10.02	9.88	9.75	9.62	9.48	9.35	9.22	9.08	8.95	8.82
16°C	10.07	9.94	9.81	9.68	9.55	9.42	9.29	9.15	9.02	8.89	8.76	8.63
17°C	9.86	9.74	9.61	9.48	9.35	9.22	9.10	8.97	8.84	8.71	8.58	8.45
18°C	9.67	9.54	9.41	9.29	9.16	9.04	8.91	8.79	8.66	8.54	8.41	8.28
19°C	9.47	9.35	9.23	9.11	8.98	8.86	8.74	8.61	8.49	8.37	8.24	8.12
20°C	9.29	9.17	9.05	8.93	8.81	8.69	8.57	8.45	8.33	8.20	8.08	7.96
21°C	9.11	9.00	8.88	8.76	8.64	8.52	8.40	8.28	8.17	8.05	7.93	7.81
22°C	8.94	8.83	8.71	8.59	8.48	8.36	8.25	8.13	8.01	7.90	7.78	7.67
23°C	8.78	8.66	8.55	8.44	8.32	8.21	8.09	7.98	7.87	7.75	7.64	7.52
24°C	8.62	8.51	8.40	8.28	8.17	8.06	7.95	7.84	7.72	7.61	7.50	7.39
25°C	8.47	8.36	8.25	8.14	8.03	7.92	7.81	7.70	7.59	7.48	7.37	7.26
26°C	8.32	8.21	8.10	7.99	7.89	7.78	7.67	7.56	7.45	7.35	7.24	7.13
27°C	8.17	8.07	7.96	7.86	7.75	7.64	7.54	7.43	7.33	7.22	7.11	7.01
28°C	8.04	7.93	7.83	7.72	7.62	7.51	7.41	7.30	7.20	7.10	6.99	6.89
29°C	7.90	7.80	7.69	7.59	7.49	7.39	7.28	7.18	7.08	6.98	6.87	6.77
30°C	7.77	7.67	7.57	7.47	7.36	7.26	7.16	7.06	6.96	6.86	6.76	6.66

Table 3: Approximate Barometric Pressure at Different Elevations					
Elevation (m)	**Pressure (mm Hg)**	**Elevation (m)**	**Pressure (mm Hg)**	**Elevation (m)**	**Pressure (mm Hg)**
0	760	800	693	1600	628
100	748	900	685	1700	620
200	741	1000	676	1800	612
300	733	1100	669	1900	604
400	725	1200	661	2000	596
500	717	1300	652	2100	588
600	709	1400	643	2200	580
700	701	1500	636	2300	571

Experiment
17

TEACHER INFORMATION

Aerobic Respiration

1. To prepare the yeast solution, dissolve 7 g (1 package) of dried yeast per 100 mL of water. Incubate the suspension in 37 – 40°C water for at least 10 minutes.

2. The Dissolved Oxygen Probe must be calibrated the first day of use. Follow the pre-lab procedure to prepare and calibrate the Dissolved Oxygen Probe. Be sure to remove the blue shipping cap from the Dissolved Oxygen Probe. To save time, you may wish to calibrate the probe and record the calibration values on paper. The students can then skip the pre-lab procedure and they will have the calibration values available for manual entry in case the values stored in the program are lost.

3. At the end of class instruct the students to leave Logger *Pro* running. This will keep power going to the probes. When the next group of students come in, they can begin at Step 6 of the procedure. They can skip the pre-lab procedure because the initial group of students has completed all of the setup. Have the last group of students for the day shut everything off and put things away.

4. To prepare the 1% glucose sugar solution, add 1 g of glucose per 100 mL of solution. Sucrose may be substituted for glucose, if desired.

5. Temperatures of 10, 20, 30, 40, 50, and 60°C are suggested. If time does not permit measuring respiration at each temperature, you may want to assign temperatures to each student team. Data can be shared among different teams.

6. Students must stir the yeast-sugar solution constantly with the Dissolved Oxygen Probe, or erroneous readings will result.

7. Thoroughly rinse any yeast from the membrane cap of the Dissolved Oxygen Probe after data has been collected.

8. Between classes store the Dissolved Oxygen Probe in a beaker of distilled water. The probe should remain plugged into an active interface box. At the end of the day, be sure to empty out the electrode filling solution in the probe membrane cap. Rinse the inside of the membrane cap and the glass electrode with distilled water.

9. When setting up the Dissolved Oxygen Probe, be sure to remove the blue plastic cap from the end of the probe. The cap is made of a soft plastic material and easily slides off the probe end.

SAMPLE RESULTS

Table 1	
Actual Temperature (°C)	Rate of O_2 consumption (mg O_2/L/min)
8.0	0.00
20.4	1.69
30.0	3.96
35.5	5.97
45.0	8.01
55.0	3.57
58.0	0.76

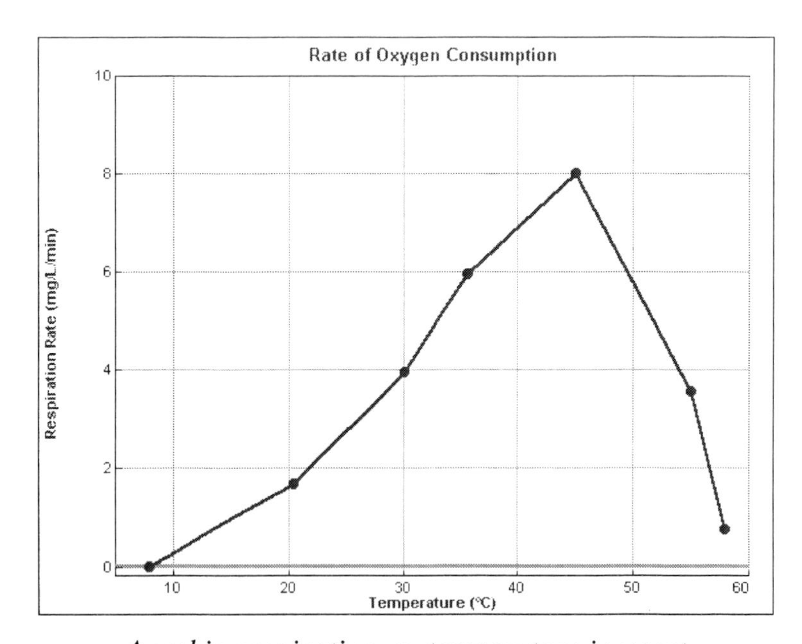

Aerobic respiration vs. temperature in yeast

ANSWERS TO QUESTIONS

1. Yes, there is evidence that aerobic cellular respiration occurred in yeast. The concentration of oxygen steadily decreased to zero with time in most experiments.
2. Temperature did affect the rate of sugar respiration in yeast. The rate increased from 10°C to 45°C. At 50°C and above, the rates decreased.
3. The temperature yielding the highest rate of respiration should be in the 40°C – 50°C range.
4. Results may vary. The generalization that the metabolism of poikilotherms doubles with every 10°C increase in temperature is generally true at temperatures below 40°C in this experiment.
5. When the oxygen concentration in the solution drops to zero, the yeast will respire anaerobically, since they are facultative anaerobes. The rate of respiration should decrease considerably, however.

Experiment
18

Acid Rain

Acid rain is a topic of much concern in today's world. As carbon dioxide gas, CO_2, dissolves in water droplets of unpolluted air, the following reaction occurs:

$$CO_2 + H_2O \leftrightarrow H_2CO_3$$

H_2CO_3 is a weak acid that causes the rain from unpolluted air to be slightly acidic. This source of "acid rain" is not usually considered to be a pollutant, since it is natural and usually does not alter the pH of rain water very much. Oxides of sulfur dissolve in water droplets to cause more serious problems. Sulfur dioxide dissolves to produce sulfurous acid, H_2SO_3, by the equation:

$$SO_2 + H_2O \leftrightarrow H_2SO_3$$

This source of sulfur dioxide can occur naturally, as from volcanic gases. More often, however, sulfur dioxide is considered a pollutant, since it is a by-product of fossil-fuel combustion.

The acidity of a solution can be expressed using the pH scale, which ranges from 0 to 14. Solutions with a pH above 7 are basic, solutions with pH below 7 are acidic, and a neutral solution has a pH of 7. In Part I of this experiment, you will study how the pH of water changes when CO_2 is dissolved in water. In Part II, you will study the effect sulfuric acid has on the pH of different water types.

OBJECTIVES

In this experiment, you will

- Use a computer to measure pH.
- Study the effect of dissolved CO_2 on the pH of distilled water.
- Study the effect on pH of dissolving H_2SO_4 in various waters.
- Learn why some bodies of water are more vulnerable to acid rain than others.

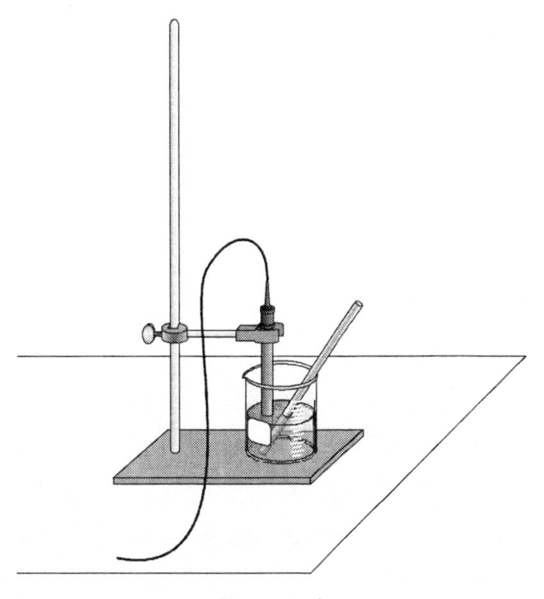

Figure 1

MATERIALS

computer
Vernier computer interface
Vernier pH Sensor
Logger*Pro*
250 mL and 100 mL beaker
buffer solution
dilute H_2SO_4
ring stand and utility clamp
straw
wash bottle with distilled water
water from a lake
water from the ocean (optional)

PROCEDURE

Part I CO_2 and Water

As carbon dioxide gas, CO_2, dissolves into water droplets suspended in the atmosphere, it changes the rainwater's pH. Here you will test to see if CO_2 will affect the pH of distilled water. The source of CO_2 will be your breath.

1. Obtain and wear goggles.

2. Connect the pH Sensor to the computer interface. Prepare the computer for data collection by opening the file "18a Acid Rain" from the *Biology with Computers* folder of Logger*Pro*.

3. Use a rinse bottle to thoroughly rinse the pH electrode as shown by your teacher. Catch the rinse water in a beaker.

4. Wash a 100 mL beaker with tap water and dry it with a paper towel. Note: All glassware must be clean in this experiment! Put 50 mL of distilled water into this clean beaker. Lower the pH electrode into the distilled water and swirl the water around the electrode briefly.

5. Begin data collection by clicking Collect. After the computer registers the initial pH, use a straw to blow your breath into the distilled water for 1 minute. Your breath contains CO_2. The computer will save readings as you blow.

6. Click the Statistics button, STAT, to determine the maximum and minimum pH values. Record the maximum and minimum pH in Table 1.

Part II Simulating Acid Rain Using H_2SO_4

7. Rinse the pH electrode thoroughly with distilled water.

8. Wash and dry the 100 mL beaker. Get a new 50 mL portion of distilled water. Place the pH electrode into the distilled water and secure in place with a ring stand and utility clamp.

9. Prepare the computer for data collection by opening the file "18b Acid Rain" from the *Biology with Computers* folder of Logger*Pro*.

10. You are ready to begin the measurements. Start measuring the pH by clicking Collect. Note that a new button Keep is available.

11. Click Keep to take a pH measurement. A text box will appear. You should enter the number of drops of acid you add to the water in the beaker. Since you didn't add any acid yet, type "0" in the text box and press ENTER.

12. Add 1 drop of H_2SO_4 (sulfuric acid) to the water. Stir thoroughly, until the pH is stable.

CAUTION: *Handle the sulfuric acid with care. It can cause painful burns if it comes into contact with skin.*

13. Click Keep. Enter the number of drops of acid added to the beaker. Type "1" in the text box and press ENTER.
14. Repeat Steps 12 – 13, adding 1 drop at a time, until you have added 10 drops of acid.
15. Click Stop when all measurements have been made.
16. From the table, record the pH values in Table 2.
17. Store your data by choosing Store Latest Run from the Experiment menu.

H_2SO_4 and Water from the Ocean

18. Complete these steps:
    a. Clean the pH electrode.
    b. Wash and dry the 100 mL beaker.
    c. Get a 50 mL portion of "Ocean Water" in the 100 mL beaker, lower the pH electrode into this water, and then briefly swirl the water about the electrode.
    d. Repeat Steps 10 – 17 for this sample.

H_2SO_4 and Lake Water

19. Repeat Step 18, using "Lake Water" from your teacher.

H_2SO_4 and Buffer Solution

20. Repeat Step 18, using "Buffer Solution" from your teacher.
21. Use the Text Annotation feature from the Insert menu to label each of the plots on the graph with the water type used. Adjust the position of the arrow.
22. Print a graph with the four types of water tested. Enter your name(s) and number of copies of the graph. Use your graph to answer the discussion questions at the end of this experiment.

PROCESSING THE DATA

Part I

Table 1
Adding CO_2 from your breath to water

Maximum pH	Minimum pH	ΔpH

Part II

Table 2				
Drops	pH of this water type			
	distilled	ocean	lake	buffer
0				
1				
2				
3				
4				
5				
6				
7				
8				
9				
10				
ΔpH				

QUESTIONS

1. Calculate the change in pH (ΔpH) for the water in Part I. Subtract the final pH from the initial pH. What conclusion can you make about your breath?
2. Why does the pH change rapidly at first, and remain stable after a time?
3. Calculate the change in pH (ΔpH) for each of the Part II trials. Subtract the final pH from the initial pH.
4. Compare the ΔpH values. Which test gave the largest pH change? Which test gave the smallest pH change?
5. Water from the ocean is said to be "naturally buffered." From the result of this experiment, what does this mean?
6. How does water from the ocean become buffered?
7. Many aquatic life forms can only survive in water with a narrow range of pH values. In which body of water—lakes or oceans—would living things be more threatened by acid rain? Explain.

8. There are numerous coal-burning electric power plants along the Ohio River in Southern Indiana, where the river and lake waters are naturally buffered. However, air pollution produced there is more harmful to water life in Upstate New York, where the river and lake waters are NOT buffered than in Southern Indiana. A similar situation exists in Europe where air pollutants from highly industrialized Germany are more harmful to Scandinavian water life than to water life in Germany. Use the results of this experiment to explain these situations.

9. Summarize your conclusions about this laboratory experiment. Use your data to answer the purposes of this experiment.

EXTENSIONS

1. Test hard and soft water in the same way you tested lake and ocean water. How do they compare?

2. Do library research to get more information on the effects of acid rain on streams and lakes.

3. Do library research and prepare a report on "naturally buffered" streams and lakes.

Acid Rain

1. The water in Part I should be boiled, cooled, and stored in a bottle filled to capacity. This will prevent the CO_2 from the atmosphere from affecting its pH. Alternatively, you can adjust the pH with 0.050 M NaOH to about 7. The latter option is much easier to accomplish.

2. To prepare the dilute (0.02 M) H_2SO_4 solution, dilute 1.1 mL of concentrated sulfuric acid into distilled water to make a total volume of 1 L.

3. If ocean water is not available, synthetic ocean water can easily be prepared.
   a. Dissolve the following ingredients, one at a time in the order given, in 800 mL of distilled water: 23.2 g NaCl; 1.11 g $CaCl_2$; 6.46 g $MgCl_2$.
   b. Add 0.84 g of $NaHCO_3$ slowly with rapid stirring.
   c. Add 5.40 g Na_2SO_4.
   d. Bring the volume up to 1 L with distilled water.
   e. Bring the pH of the solution to 7.8, using 0.01 M NaOH.
   f. Fill the bottle completely and stopper it tightly. This will prevent the pH from lowering due to the absorption of CO_2.

4. If lake water is not available, synthetic water can easily be prepared. Hard or soft water might be used, depending upon the local water characteristics.

5. A recipe for hard water follows:
   a. To 800 mL of distilled water, add 0.12 g $CaCO_3$; 0.10 g $CaSO_4$; 0.038 g $MgCl_2$.
   b. With rapid stirring, slowly add 0.10 g $NaHCO_3$.
   c. Bring the volume up to 1 L with distilled water.

6. A recipe for *softened* hard water follows:
   a. To 800 mL of distilled water, add 0.191 g NaCl; 0.107 g Na_2SO_4.
   b. With rapid stirring, slowly add 0.109 g $NaHCO_3$.
   c. Bring the volume up to 1 L with distilled water.

7. To prepare a buffer solution, add up to four times the recommended amount of water to a commercially available pH 7 buffer tablet. Since the amount of acid added is very small, the buffer tablet need not be made to full strength. The pH of the buffer will raise a bit after dilution—the exact value is not critical.

   Vernier Software sells a pH buffer package for preparing buffer solutions with pH values of 4, 7, and 10 (order code PHB). To prepare a standard solution, simply add the capsule contents to 100 mL of distilled water. This solution can be diluted to 500 mL for this experiment.

   You can also prepare a pH 7 buffer using the following recipe: Add 582 mL of 0.1 M NaOH to 1000 mL of 0.1 M potassium dihydrogen phosphate.

8. The stored pH calibration works well for this experiment. For more accurate pH readings, you (or your students) can do a 2-point calibration for each pH Sensor using pH-4 and pH-7 buffers.

9. If you wish to calibrate your pH Sensor, use this procedure. Using the stored pH calibration in the Logger *Pro* program, check a pH Sensor in pH-7 buffer to see if it gives a reading close to pH of 7. If it does, then one option is to have your students use the stored calibration. For even better results use buffers of pH 4.0 and 7.0 to calibrate the pH Sensor. To do this:

   First Calibration Point

   a. Choose Calibrate ▸ CH1: pH from the Experiment menu and then click Calibrate Now.
   b. For the first calibration point, rinse the pH Sensor with distilled water, then place it into a buffer of pH 4.0.
   c. Type “4” in the edit box as the pH value.
   d. Swirl the sensor, wait until the voltage for Input 1 stabilizes, then click Keep.

   Second Calibration Point

   e. Rinse the pH Sensor with distilled water, and place it into a buffer of pH 7.0.
   f. Type “7” in the edit box as the pH value for the second calibration point.
   g. Swirl the sensor and wait until the voltage for Input 1 stabilizes. Click Keep, then click Done. This completes the calibration.
   h. You are now ready to collect data, using the calibration. Or, to use the calibration values for a specific pH Sensor at a later time, you can simply choose Calibration from the pull-down menu on the Calibration tab. Choose to store the calibration with the experiment file and resave this experiment file with a unique name—the new calibration is saved along with the experiment file itself.

SAMPLE RESULTS

The following data will be different from students’ results.

Part I

Table 1 Adding CO_2 from breath to water.		
Maximum pH	Minimum pH	ΔpH
6.1	4.6	1.5

Part II

The following data were taken from the Puget Sound area.

Table 2				
Drops	pH of this water type			
	distilled	ocean	lake	buffer
0	6.5	6.4	7.2	7.1
1	4.3	6.3	6.4	7.0
2	3.9	6.2	6.1	6.6
3	3.7	6.2	5.8	6.4
4	3.5	6.1	5.6	6.3
5	3.4	6.1	5.4	6.2
6	3.4	6.0	5.3	6.1
7	3.3	6.0	5.2	6.0
8	3.2	5.9	5.1	5.9
9	3.2	5.9	5.0	5.8
10	3.1	5.8	5.0	5.7
ΔpH	3.4	0.6	2.2	1.4

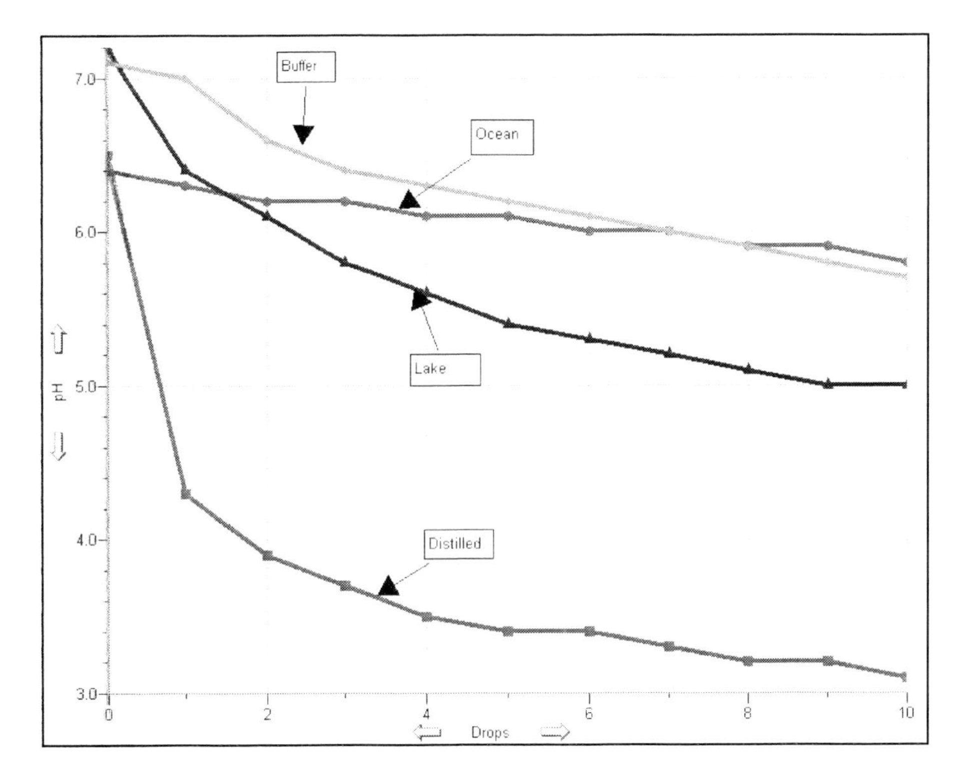

Change in pH using four different solutions

ANSWERS TO QUESTIONS

1. Exhaled breath contains carbon dioxide that causes the water to become acidic—CO_2 reacts with the water to make carbonic acid, H_2CO_3.

2. As the water becomes saturated with CO_2, no more CO_2 can dissolve in the water. When this happens, no more H_2CO_3 is formed, so the pH remains stable.

3. Answers may vary.

4. The distilled water should yield the largest change, followed by lake water, ocean water, and the buffer. The buffer should resist pH changes more than the others.

5. Buffers resist pH changes. As acid is added to the ocean water, the pH changes less than with other natural water sources.

6. The ocean water is buffered due to a mixture of salts. Some of these salts are capable of resisting changes in pH.

7. Living things would be more threatened by acid rain in lake water, as it is not as well buffered as ocean water.

8. Winds may blow polluted air vast distances. Acid rain may affect bodies of water far from the polluted air's origin, especially if the acid rain falls in unbuffered waters.

9. Student answers may vary.

Experiment
19

Dissolved Oxygen in Water

Although water is composed of oxygen and hydrogen atoms, biological life in water depends upon another form of oxygen—molecular oxygen. Oxygen is used by organisms in aerobic respiration, where energy is released by the combustion of sugar in the mitochondria. This form of oxygen can fit into the spaces between water molecules and is available to aquatic organisms.

Fish, invertebrates, and other aquatic animals depend upon the oxygen dissolved in water. Without this oxygen, they would suffocate. Some organisms, such as salmon, mayflies, and trout, require high concentrations of oxygen in their water. Other organisms, such as catfish, midge fly larvae, and carp can survive with much less oxygen. The ecological quality of the water depends largely upon the amount of oxygen the water can hold.

The following table indicates the normal tolerance of selected animals to temperature and oxygen levels. The quality of the water can be assessed with fair accuracy by observing the aquatic animal populations in a stream.

Table 1

Animal	Temperature Range (°C)	Minimum Dissolved Oxygen (mg/L)
Trout	5 – 20	6.5
Smallmouth bass	5 – 28	6.5
Caddisfly larvae	10 – 25	4.0
Mayfly larvae	10 – 25	4.0
Stonefly larvae	10 – 25	4.0
Catfish	20 – 25	2.5
Carp	10 – 25	2.0
Mosquito	10 – 25	1.0
Water boatmen	10 – 25	2.0

OBJECTIVES

In this experiment, you will

- Use a Dissolved Oxygen Probe to measure the concentration of dissolved oxygen in water.
- Study the effect of temperature on the amount of dissolved oxygen in water.
- Predict the effect of water temperature on aquatic life.

MATERIALS

computer
Vernier computer interface
Vernier Dissolved Oxygen Probe
Vernier Temperature Probe
Logger*Pro*
100 mL beaker
two 250 mL beakers
hot and cold water
1 gallon plastic milk container
Styrofoam cup

PRE-LAB PROCEDURE

Important: Prior to each use, the Dissolved Oxygen Probe must warm up for a period of 10 minutes as described below. If the probe is not warmed up properly, inaccurate readings will result. Perform the following steps to prepare the Dissolved Oxygen Probe.

1. Prepare the Dissolved Oxygen Probe for use.
   a. Remove the blue protective cap.
   b. Unscrew the membrane cap from the tip of the probe.
   c. Using a pipet, fill the membrane cap with 1 mL of DO Electrode Filling Solution.
   d. Carefully thread the membrane cap back onto the electrode.
   e. Place the probe into a container of water.

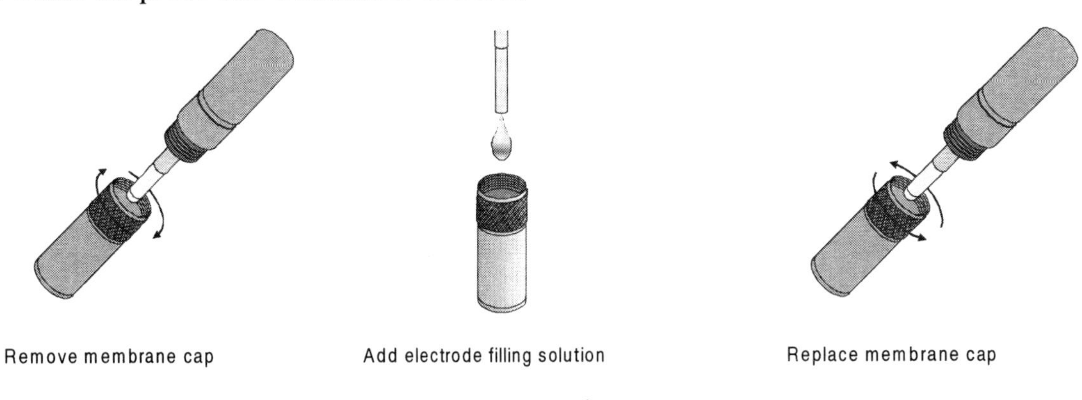

Figure 1

2. Plug the Dissolved Oxygen Probe into Channel 1 of the Vernier interface. Connect the Temperature Probe to Channel 2.
3. Prepare the computer for data collection by opening the file “19 Dissolved Oxygen” from the *Biology with Computers* folder of Logger*Pro*.
4. It is necessary to warm up the Dissolved Oxygen Probe for 10 minutes before taking readings. To warm up the probe, leave it connected to the interface, with Logger *Pro* running, for 10 minutes. The probe must stay connected at all times to keep it warmed up. If disconnected for a few minutes, it will be necessary to warm up the probe again.

5. You are now ready to calibrate the Dissolved Oxygen Probe.
   - If your instructor directs you to use the calibration stored in the experiment file, then proceed to Step 6.
   - If your instructor directs you to perform a new calibration for the Dissolved Oxygen Probe, follow this procedure.

   Zero-Oxygen Calibration Point

   a. Choose Calibrate ▸ CH1: Dissolved Oxygen (mg/L) from the Experiment menu and then click Calibrate Now.
   b. Remove the probe from the water and place the tip of the probe into the Sodium Sulfite Calibration Solution. **Important:** No air bubbles can be trapped below the tip of the probe or the probe will sense an inaccurate dissolved oxygen level. If the voltage does not rapidly decrease, tap the side of the bottle with the probe to dislodge any bubbles. The readings should be in the 0.2 to 0.5 V range.
   c. Type "0" (the known value in mg/L) in the edit box.
   d. When the displayed voltage reading for Reading 1 stabilizes, click Keep.

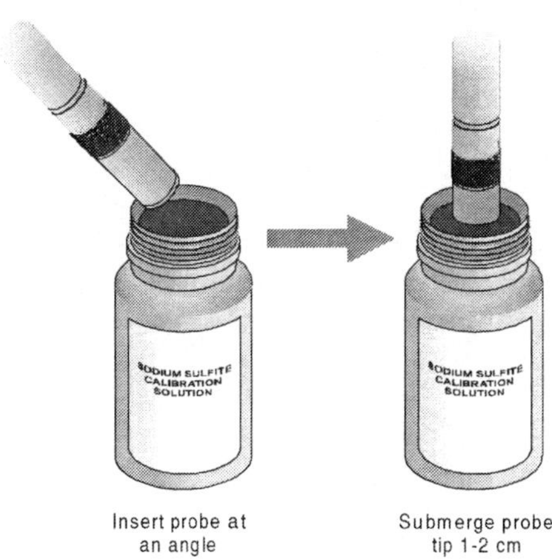

Figure 2

   Saturated DO Calibration Point

   e. Rinse the probe with distilled water and gently blot dry.
   f. Unscrew the lid of the calibration bottle provided with the probe. Slide the lid and the grommet about 2 cm onto the probe body.

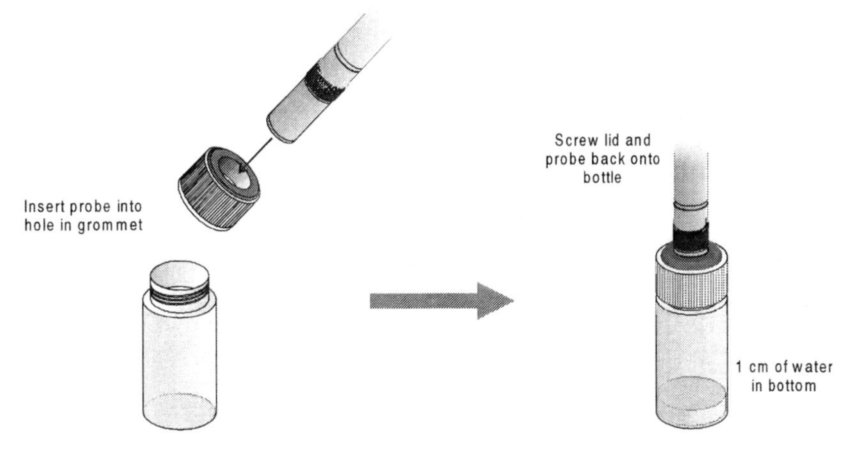

Figure 3

   g. Add water to the bottle to a depth of about 1 cm and screw the bottle into the cap, as shown. **Important:** Do not touch the membrane or get it wet during this step. Keep the probe in this position for about a minute.
   h. Type the correct saturated dissolved-oxygen value (in mg/L) from Table 3 (for example, "8.66") using the current barometric pressure and air temperature values. If you do not have the current air pressure, use Table 4 to estimate the air pressure at your altitude.
   i. When the displayed voltage reading for Reading 2 stabilizes (readings should be above 2.0 V), click Keep.

j. Choose Calibration Storage from the calibration pull-down menu. Choose Experiment File (calibration stored with the current document). Click Done.

k. From the File menu, select Save As and save the current experiment file with a new name.

PROCEDURE

6. Prepare for data collection by clicking Collect.

7. Obtain two 250 mL beakers. Fill one beaker with ice and cold water. Fill the second beaker with warm water about 40 – 50°C.

8. Place approximately 100 mL of cold water and a couple small pieces of ice, from the beaker filled with ice, into a clean plastic one-gallon milk container. Seal the container and vigorously shake the water for a period of 2 minutes. This will allow the air inside the container to dissolve into the water sample. Pour the water into the Styrofoam cup.

9. Place the Temperature Probe in the Styrofoam cup as shown in Figure 4. Place the shaft of the Dissolved Oxygen Probe into the water and gently stir. Avoid hitting the edge of the cup with the probe.

10. Monitor the dissolved oxygen readings in the meter. Give the dissolved oxygen readings ample time to stabilize (90 – 120 seconds). At colder temperatures the probe will require a greater amount of time to stabilize. When the readings have stabilized, click Keep.

11. Remove the probes from the water sample and place the Dissolved Oxygen Probe into a beaker filled with distilled water.

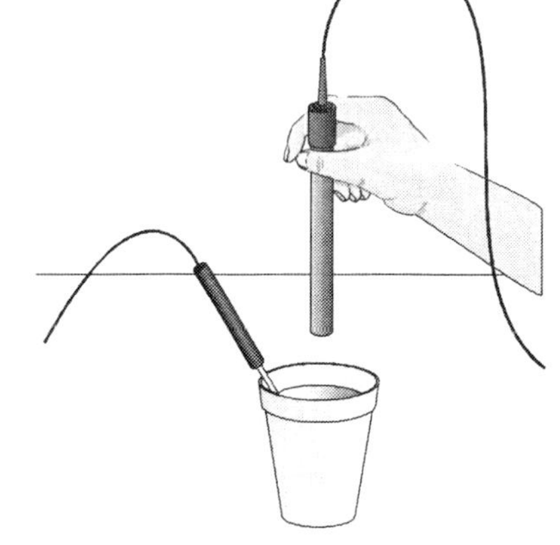

Figure 4

12. Pour the water from the Styrofoam cup back into the milk container. Seal the container and shake the water vigorously for 1 minute.

13. Repeat Steps 9 – 12 until the water sample reaches room temperature. When room temperature has been reached then begin adding about 25 mL of warm water (40°C – 50°C) prior to shaking the water sample. This will allow you to take warmer water readings. Take dissolved oxygen readings until the water temperature reaches 35°C.

14. When all readings have been taken click Stop.

15. In Table 2, record the dissolved oxygen and temperature readings from the table.

16. Print the graph of dissolved oxygen *vs.* temperature. Enter your name(s) and the number of copies of the graph.

DATA

Table 2	
Temperature (°C)	Dissolved Oxygen (mg/L)

QUESTIONS

1. At what temperature was the dissolved oxygen concentration the highest? Lowest?
2. Does your data indicate how the amount of dissolved oxygen in the water is affected by the temperature of water? Explain.
3. If you analyzed the invertebrates in a stream and found an abundant supply of caddisflies, mayflies, dragonfly larvae, and trout, what minimum concentration of dissolved oxygen would be present in the stream? What maximum temperature would you expect the stream to sustain?
4. Mosquito larvae can tolerate extremely low dissolved oxygen concentrations, yet cannot survive at temperatures above approximately 25°C. How might you account for dissolved oxygen concentrations of such a low value at a temperature of 25°C? Explain.
5. Why might trout be found in pools of water shaded by trees and shrubs more commonly than in water where the trees have been cleared?

CALIBRATION TABLES

Table 3: 100% Dissolved Oxygen Capacity (mg/L)												
	770 mm	**760 mm**	**750 mm**	**740 mm**	**730 mm**	**720 mm**	**710 mm**	**700 mm**	**690 mm**	**680 mm**	**670 mm**	**660 mm**
0°C	14.76	14.57	14.38	14.19	13.99	13.80	13.61	13.42	13.23	13.04	12.84	12.65
1°C	14.38	14.19	14.00	13.82	13.63	13.44	13.26	13.07	12.88	12.70	12.51	12.32
2°C	14.01	13.82	13.64	13.46	13.28	13.10	12.92	12.73	12.55	12.37	12.19	12.01
3°C	13.65	13.47	13.29	13.12	12.94	12.76	12.59	12.41	12.23	12.05	11.88	11.70
4°C	13.31	13.13	12.96	12.79	12.61	12.44	12.27	12.10	11.92	11.75	11.58	11.40
5°C	12.97	12.81	12.64	12.47	12.30	12.13	11.96	11.80	11.63	11.46	11.29	11.12
6°C	12.66	12.49	12.33	12.16	12.00	11.83	11.67	11.51	11.34	11.18	11.01	10.85
7°C	12.35	12.19	12.03	11.87	11.71	11.55	11.39	11.23	11.07	10.91	10.75	10.59
8°C	12.05	11.90	11.74	11.58	11.43	11.27	11.11	10.96	10.80	10.65	10.49	10.33
9°C	11.77	11.62	11.46	11.31	11.16	11.01	10.85	10.70	10.55	10.39	10.24	10.09
10°C	11.50	11.35	11.20	11.05	10.90	10.75	10.60	10.45	10.30	10.15	10.00	9.86
11°C	11.24	11.09	10.94	10.80	10.65	10.51	10.36	10.21	10.07	9.92	9.78	9.63
12°C	10.98	10.84	10.70	10.56	10.41	10.27	10.13	9.99	9.84	9.70	9.56	9.41
13°C	10.74	10.60	10.46	10.32	10.18	10.04	9.90	9.77	9.63	9.49	9.35	9.21
14°C	10.51	10.37	10.24	10.10	9.96	9.83	9.69	9.55	9.42	9.28	9.14	9.01
15°C	10.29	10.15	10.02	9.88	9.75	9.62	9.48	9.35	9.22	9.08	8.95	8.82
16°C	10.07	9.94	9.81	9.68	9.55	9.42	9.29	9.15	9.02	8.89	8.76	8.63
17°C	9.86	9.74	9.61	9.48	9.35	9.22	9.10	8.97	8.84	8.71	8.58	8.45
18°C	9.67	9.54	9.41	9.29	9.16	9.04	8.91	8.79	8.66	8.54	8.41	8.28
19°C	9.47	9.35	9.23	9.11	8.98	8.86	8.74	8.61	8.49	8.37	8.24	8.12
20°C	9.29	9.17	9.05	8.93	8.81	8.69	8.57	8.45	8.33	8.20	8.08	7.96
21°C	9.11	9.00	8.88	8.76	8.64	8.52	8.40	8.28	8.17	8.05	7.93	7.81
22°C	8.94	8.83	8.71	8.59	8.48	8.36	8.25	8.13	8.01	7.90	7.78	7.67
23°C	8.78	8.66	8.55	8.44	8.32	8.21	8.09	7.98	7.87	7.75	7.64	7.52
24°C	8.62	8.51	8.40	8.28	8.17	8.06	7.95	7.84	7.72	7.61	7.50	7.39
25°C	8.47	8.36	8.25	8.14	8.03	7.92	7.81	7.70	7.59	7.48	7.37	7.26
26°C	8.32	8.21	8.10	7.99	7.89	7.78	7.67	7.56	7.45	7.35	7.24	7.13
27°C	8.17	8.07	7.96	7.86	7.75	7.64	7.54	7.43	7.33	7.22	7.11	7.01
28°C	8.04	7.93	7.83	7.72	7.62	7.51	7.41	7.30	7.20	7.10	6.99	6.89
29°C	7.90	7.80	7.69	7.59	7.49	7.39	7.28	7.18	7.08	6.98	6.87	6.77
30°C	7.77	7.67	7.57	7.47	7.36	7.26	7.16	7.06	6.96	6.86	6.76	6.66

Table 4: Approximate Barometric Pressure at Different Elevations					
Elevation (m)	**Pressure (mm Hg)**	**Elevation (m)**	**Pressure (mm Hg)**	**Elevation (m)**	**Pressure (mm Hg)**
0	760	800	693	1600	628
100	748	900	685	1700	620
200	741	1000	676	1800	612
300	733	1100	669	1900	604
400	725	1200	661	2000	596
500	717	1300	652	2100	588
600	709	1400	643	2200	580
700	701	1500	636	2300	571

TEACHER INFORMATION

Dissolved Oxygen in Water

1. The Dissolved Oxygen Probe must be calibrated the first day of use. Follow the pre-lab procedure to prepare and calibrate the Dissolved Oxygen Probe. To save time, you may wish to calibrate the probe and record the calibration values on paper. The students can then skip the pre-lab procedure and they will have the calibration values available for manual entry in case the values stored in the program are lost.
2. At the end of class instruct the students to leave Logger *Pro* running. This will keep power going to the probes. When the next group of students come in, they can begin at Step 6 of the procedure. They can skip the pre-lab procedure because the initial group of students has completed all of the setup. Have the last group of students for the day shut everything off and put things away.
3. If you are using the Stainless Steel Temperature Probe, you can put the probe into the same solution with the Dissolved Oxygen Probe. There are no interference problems when using this probe combination.

   If you are using a Direct-Connect Temperature Probe, wrap the temperature probe in plastic wrap and secure with a rubber band. This is to keep the probe from coming into direct contact with the water sample being measured. Make sure that the probe is completely covered and there are no holes where water could leak in. If the probe does come in contact with the water sample, the dissolved oxygen and temperature readings will be negatively affected because of interference from each other probe.
4. Temperatures of 5, 10, 15, 20, 25, and 30°C are suggested.
5. Between classes store the Dissolved Oxygen Probes in a beaker of distilled water. At the end of the day be sure to empty out the electrode filling solution in the Dissolved Oxygen Probe and rinse the inside of the membrane cap with distilled water.
6. In Step 8 of the procedure warn students not to put more ice into their milk container than will melt while they are shaking the container.
7. When setting up the Dissolved Oxygen Probe, be sure to remove the blue plastic cap from the end of the probe. The cap is made of a soft plastic material and easily slides off the probe end.

SAMPLE RESULTS

The following data may be different from students' results. Note that the actual temperatures will be close to, but not necessarily the same as, the assigned temperatures.

Table 1

Temperature (°C)	Dissolved Oxygen (mg/L)
8.96	11.19
14.89	9.66
18.74	8.81
22.71	7.99
33.65	5.71

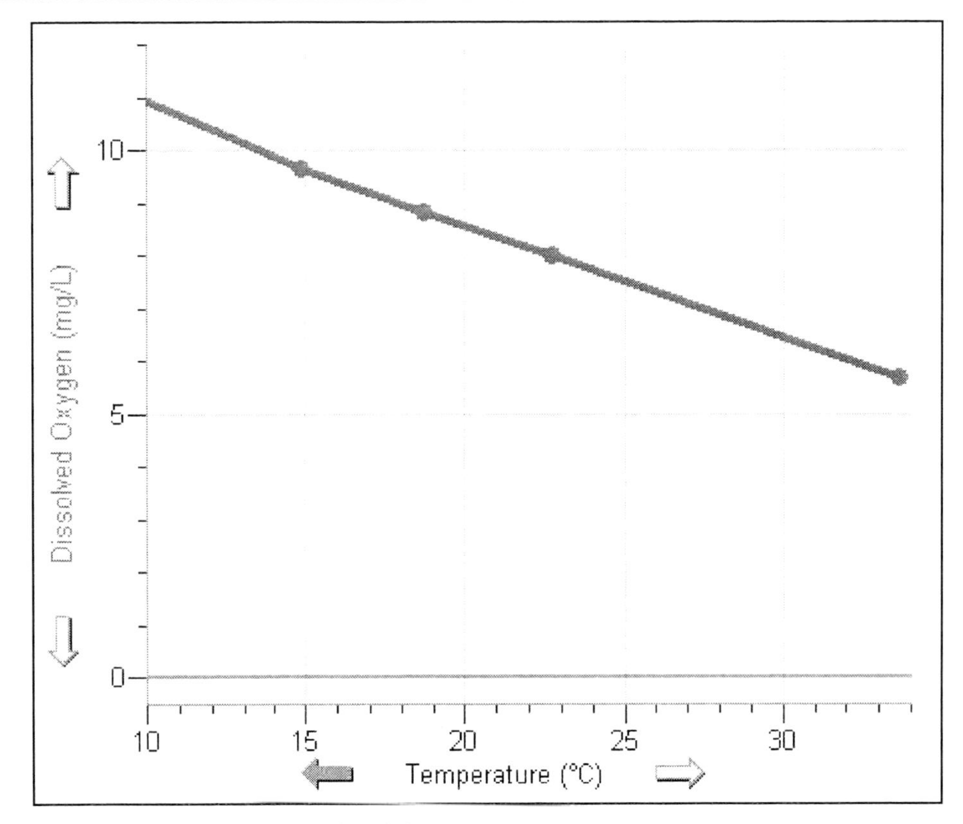

Dissolved Oxygen vs. Temperature

ANSWERS TO QUESTIONS

1. The amount of dissolved oxygen will be highest at the coldest temperature and lowest at the warmest temperature.

2. As the temperature increases, the amount of dissolved oxygen decreases. The relationship does not appear to be linear.

3. Since trout were present, the minimum amount of dissolved oxygen would be 6.5 mg/L. According to these data, fast-moving water could not be warmer than 30 – 32°C. This temperature is much higher than trout generally live in, however.

4. Other factors must have caused the low dissolved oxygen levels. Bacteria and other organisms can lower the dissolved oxygen of water when they respire aerobically.

5. Trout require high dissolved oxygen levels. Since trees shade the water from the sun, they help keep it cool. Cool water has a higher dissolved oxygen level than warm water and is preferred by trout.

Experiment

20

Watershed Testing

There are many reasons for determining water quality. You may want to compare the water quality upstream and downstream to locate a possible source of pollutants along a river or stream. Another reason may be to track the water quality of a watershed over time by making measurements periodically. When comparing the quality of a watershed at different times, it is important that measurements be taken from the same location and at the same time of day.

In 1970, the National Sanitation Foundation, in cooperation with 142 state and local environmental specialists and educators, devised a standard index for measuring water quality. This index, known as the Water Quality Index, or *WQI*, consists of nine tests to determine water quality. These nine tests are; temperature, pH, turbidity, total solids, dissolved oxygen, biochemical oxygen demand, phosphates, nitrate, and fecal coliform. A graph for each of the nine tests indicates the water quality value (or Q-value) corresponding to the data obtained. Once the Q-value for a test has been determined, it is multiplied by a weighting factor. Each of the tests is weighted based on its relative importance to a stream's overall quality. The resulting values for all nine tests are totaled and used to gauge the stream's health (excellent, good, medium, poor, or very poor).

While the WQI can be a useful tool, it is best used in light of historical data. Not all streams are the same, and without historical data it is difficult to determine if a stream is truly at risk. For example, a stream may earn a very low WQI value and appear to be in poor health. By looking at historical data, however, you may find that samples were collected just after a heavy rain with an overflow from the local city sewer system and do not accurately reflect the stream's health.

For the purpose of this exercise, you will perform only four of the WQI tests: water temperature, dissolved oxygen, pH, and total dissolved solids. A modified version of the WQI for these four tests, will allow you to determine the general quality of the stream or lake you are sampling.

OBJECTIVES

In this experiment, you will

- Use a Dissolved Oxygen, Temperature, Conductivity, and pH Probe to make on-site measurements.
- Calculate the water quality based on your findings.

MATERIALS

computer(laptop)
Vernier computer interface
Vernier Temperature Probe
Vernier Conductivity Probe
Vernier pH Sensor
Vernier Dissolved Oxygen Probe
Logger*Pro*
water sampling bottle, stoppered
small plastic cup or beaker

PRE-LAB PROCEDURE

Important: Prior to each use, the Dissolved Oxygen Probe must warm up for a period of 10 minutes as described below. If the probe is not warmed up properly, inaccurate readings will result. Perform the following steps to prepare the Dissolved Oxygen Probe.

1. Prepare the Dissolved Oxygen Probe for use.
   a. Remove the blue protective cap.
   b. Unscrew the membrane cap from the tip of the probe.
   c. Using a pipet, fill the membrane cap with 1 mL of DO Electrode Filling Solution.
   d. Carefully thread the membrane cap back onto the electrode.
   e. Place the probe into a container of water.

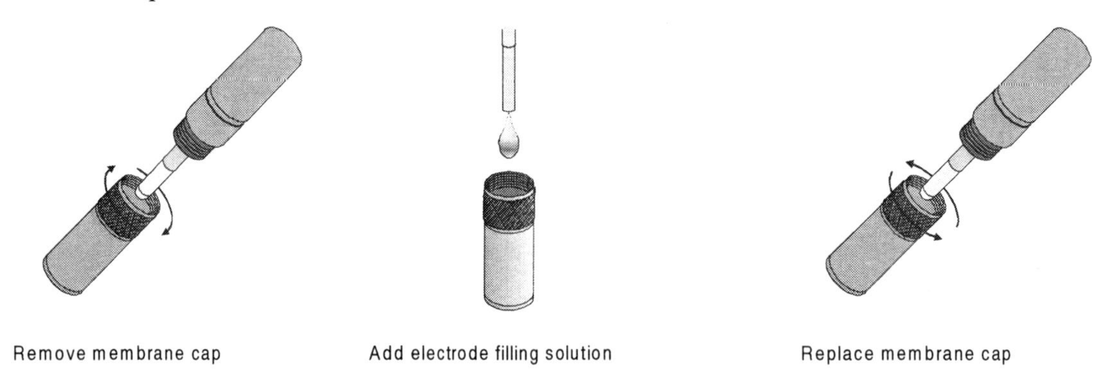

Figure 1

2. Plug the Dissolved Oxygen Probe into Channel 1 of the Vernier computer interface. Connect the Conductivity Probe to Channel 2. Set the selector switch on the side of the Conductivity Probe to the 0 – 2000 μS/cm range.

3. Prepare the computer for data collection by opening the file "20a Watershed Testing" from the *Biology with Computers* folder of Logger*Pro*. The meters display the live dissolved oxygen and TDS (Total Dissolved Solids) readings.

4. It is necessary to warm up the Dissolved Oxygen Probe for 10 minutes before taking readings. To warm up the probe, leave it connected to the interface, with Logger *Pro* running, for 10 minutes. The probe must stay connected at all times to keep it warmed up. If disconnected for a few minutes, it will be necessary to warm up the probe again.

5. You are now ready to calibrate the Dissolved Oxygen Probe.
   - If your instructor directs you to use the calibration stored in the experiment file, then proceed to Step 6.
   - If your instructor directs you to perform a new calibration for the Dissolved Oxygen Probe, follow this procedure.

Zero-Oxygen Calibration Point

a. Choose Calibrate ▸ CH1: Dissolved Oxygen (mg/L) from the Experiment menu and then click Calibrate Now.

b. Remove the probe from the water and place the tip of the probe into the Sodium Sulfite Calibration Solution. **Important:** No air bubbles can be trapped below the tip of the probe or the probe will sense an inaccurate dissolved oxygen level. If the voltage does not rapidly decrease, tap the side of the bottle with the probe to dislodge any bubbles. The readings should be in the 0.2 to 0.5 V range.

c. Type “0” (the known value in mg/L) in the edit box.

d. When the displayed voltage reading for Reading 1 stabilizes, click Keep.

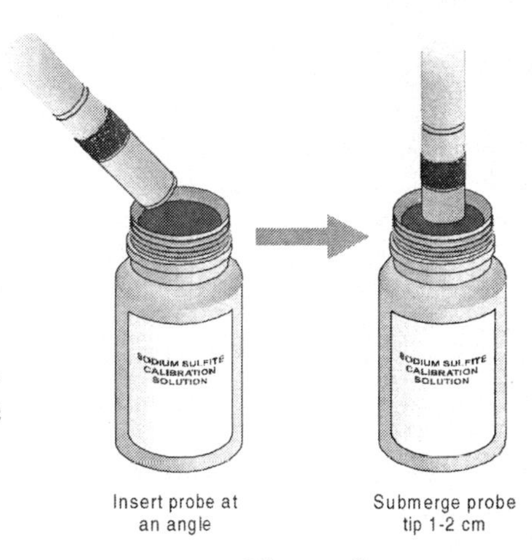

Figure 2

Saturated DO Calibration Point

e. Rinse the probe with distilled water and gently blot dry.

f. Unscrew the lid of the calibration bottle provided with the probe. Slide the lid and the grommet about 2 cm onto the probe body.

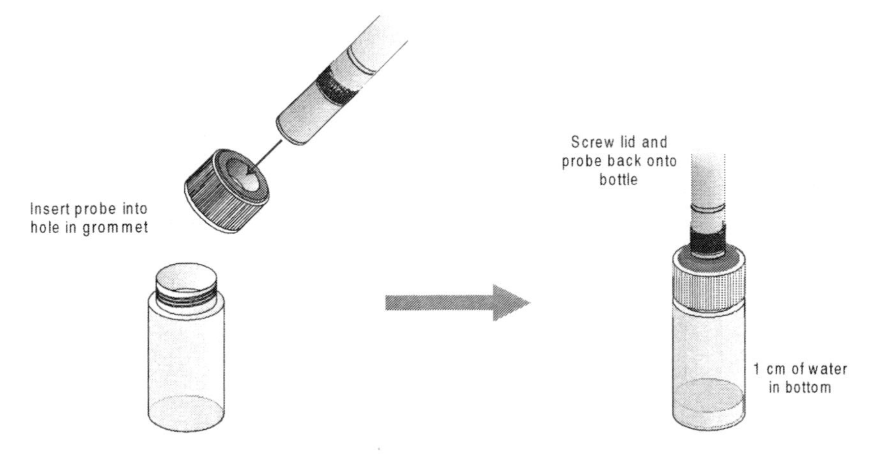

Figure 3

g. Add water to the bottle to a depth of about 1 cm and screw the bottle into the cap, as shown. **Important:** Do not touch the membrane or get it wet during this step. Keep the probe in this position for about a minute.

h. Type the correct saturated dissolved-oxygen value (in mg/L) from Table 9 (for example, “8.66”) using the current barometric pressure and air temperature values. If you do not have the current air pressure, use Table 10 to estimate the air pressure at your altitude.

i. When the displayed voltage reading for Reading 2 stabilizes (readings should be above 2.0 V), click Keep.

j. Choose Calibration Storage from the calibration pull-down menu. Choose Experiment File (calibration stored with the current document). Click Done.

k. From the File menu, select Save As and save the current experiment file with a new name.

l. Prepare the probe for transport by filling the calibration bottle half full with water. Secure the Dissolved Oxygen Probe far enough down in the bottle that the membrane is completely covered by water. Screw the calibration bottle lid completely onto the bottle so that no water will leak out.

PROCEDURE

The data collection is performed in two parts. First, you will measure dissolved oxygen and total dissolved solids concentration of the water sample from the lake or stream being studied. Then you will measure pH and temperature of the lake or stream water.

Measuring Dissolved Oxygen And Total Dissolved Solids

Because the Dissolved Oxygen Probe requires 10 minutes to polarize before it can be used, it will already be connected to the interface box, and will be in a sample of distilled water. **Important:** The 10 minute warm-up must be done in order to get accurate dissolved oxygen readings.

6. Choose a desirable location to perform your measurements. It is best to obtain samples as far from the shore edge as is safe. Your site should be representative of the watershed as a whole.
7. Rinse the sampling bottle out a few times with stream water. Fill the sampling bottle so that it is completely full and stopper the bottle under water. This should minimize the amount of atmospheric oxygen that gets into the water until the measurements have been made.
8. Position the computer and the probes safely away from the water. The computer will be damaged if it gets wet.
9. Remove the Dissolved Oxygen Probe from the storage bottle. Place the probe into the water sampling bottle. Gently and continuously swirl it to allow water to move past the probe's tip. After 30 seconds, or when the dissolved oxygen reading stabilizes, record the dissolved oxygen reading, in mg/L, in Table 1. Return the Dissolved Oxygen Probe to the storage bottle.
10. Place the Conductivity Probe into the water sampling bottle and gently swirl to allow water to move past the probe's tip. The Conductivity Probe is measuring Total Dissolved Solids (TDS). When the TDS reading stabilizes, record it in Table 1. Disconnect the Dissolved Oxygen Probe and Conductivity Probes from the interface box. Important: Handle both probes with care. The Dissolved Oxygen Probe should be reconnected to the interface box as soon as Step 12 is completed (so that it remains polarized).

Measuring Temperature And pH

11. Connect the Temperature Probe to Channel 1 and the pH Sensor to Channel 2. Prepare the computer for data collection by opening the file "20b Watershed Testing" from the *Biology with Computers* folder of Logger*Pro.*
12. Place the Temperature Probe into the center of the water sampling bottle. When the reading stabilizes, record the temperature reading in Table 1. Disconnect the Temperature Probe from the interface box and reconnect the Dissolved Oxygen Probe (so it stays polarized).
13. Using the small plastic cup, obtain some stream water to rinse the pH Sensor.
14. Remove the pH Sensor from the storage bottle. Rinse the pH electrode thoroughly with the stream water. Then place the electrode into the water sampling bottle and gently swirl. When the reading stabilizes, record the pH value in Table 1.
15. Repeat Steps 2 – 3 and 6 – 14 at one more location 6 meters from the first location.
16. Repeat Steps 2 – 3 and 6 – 14 at two locations about 1.6 km from the first location.

DATA

Table 1				
Location	Dissolved Oxygen (mg/L)	pH	Total Dissolved Solids (mg/L)	Temperature (°C)
Site 1a				
Site 1b				
Average				
Site 2a				
Site 2b				
Average				
			Temperature Difference:	

Table 2 DO (% Saturated)			
	Dissolved Oxygen (mg/L)	DO in Saturated Water	% Saturated
Site 1			
Site 2			

PROCESSING THE DATA

1. Calculate the averages for measurements at each location and record the results in Table 1.
2. Determine the % saturation of dissolved oxygen:
   a. Copy the value of dissolved oxygen measured at each site from Table 1 to Table 2.
   b. Obtain the barometric pressure, in mm Hg, using either a barometer or a table of barometric pressure values according to elevation (your instructor will provide either the barometer reading or the table of values).
   c. Note the water temperature at each site.
   d. Using the pressure and temperature values, look up the level of dissolved oxygen for air-saturated water (in mg/L) from a second table provided by your instructor. Record the results for each site in Table 2.
   e. To determine the % saturation, use this formula:

$$\% \text{ saturation} = \frac{\text{measured D.O. level}}{\text{saturated D.O. level}} \times 100$$

   f. Record the % saturation of dissolved oxygen in Table 3.

3. Using Tables 3 – 5, determine the water quality value (Q value) for each of the following measurements: dissolved oxygen, pH, and TDS. You may need to interpolate to obtain the correct Q values. Record your result in Table 7 for Site 1 and in Table 8 for Site 2.

Table 3	
Dissolved Oxygen (DO) Test Results	
DO (% saturation)	Q Value
0	0
10	5
20	12
30	20
40	30
50	45
60	57
70	75
80	85
90	95
100	100
110	95
120	90
130	85
140	80

Table 4	
pH Test Results	
pH	Q Value
2.0	0
2.5	1
3.0	3
3.5	5
4.0	8
4.5	15
5.0	25
5.5	40
6.0	54
6.5	75
7.0	88
7.5	95
8.0	85
8.5	65
9.0	48
9.5	30
10.0	20
10.5	12
11.0	8
11.5	4
12.0	2

4. Subtract the two average temperatures from the sites that are about 1.6 km apart. Record the result as the temperature difference in the blank below Table 1.

5. Using Table 6 and the value you calculated above, determine the water quality value (Q value) for the temperature difference measurement. You may need to interpolate to obtain the correct Q values. Record your result in Table 7 and Table 8. The temperature Q-value will be the same in both tables.

6. Multiply each Q-value by the weighting factor in Table 7 for Site 1 and in Table 8 for Site 2. Record the total Q-value in Tables 7 – 8.

7. Determine the overall water quality of your stream by adding the four total Q-values in Table 7 for Site 1 and in Table 8 for Site 2. Record the result in the line next to the label "Overall Quality." The closer this value is to 100, the better the water quality of the stream at this site.

   Note that this quality index is not a complete one—this value uses only four measurements. For a more complete water quality determination, you should measure fecal coliform counts, biological oxygen demand, phosphate and nitrate levels, and turbidity. It is also very valuable to do a "critter count"—that is, examine the macroinvertebrates in the stream.

Table 5	
Total Dissolved Solids (TDS) Test Results	
TDS (mg/L)	Q Value
0	80
50	90
100	85
150	78
200	72
250	65
300	60
350	52
400	46
450	40
500	30

Table 6	
Temperature Test Results	
ΔTemp (°C)	Q Value
0	95
5	75
10	45
15	30
20	20
25	15
30	10

Table 7 Site 1			
Test	Q-Value	Weight	Total Q-Value
DO		0.38	
pH		0.24	
TDS		0.16	
Temperature		0.22	

Overall Quality: ______________

Table 8 Site 2			
Test	Q-Value	Weight	Total Q-Value
DO		0.38	
pH		0.24	
TDS		0.16	
Temperature		0.22	

Overall Quality: ____________________

QUESTIONS

1. Using your measurements, what is the quality of the watershed? Explain.
2. How do you account for each of the measurements? For example, if the pH of the downstream site is very low, and you took measurements above and below an auto repair station, perhaps battery acid leaked into the stream.

   DO:

   pH:

   TDS:

   Temperature:
3. How did measurements between the two sites compare? How might you account for any differences, if any?
4. Compare the measurements you obtained with those from previous months or years. Has the water quality improved, remained about the same, or declined? Explain.
5. Why would you expect the DO in a pond to be less than in a rapidly moving stream? If applicable, did your measurements confirm this assumption? Explain.
6. What could be done to improve the quality of the watershed?

CALIBRATION TABLES

Table 9: 100% Dissolved Oxygen Capacity (mg/L)

	770 mm	760 mm	750 mm	740 mm	730 mm	720 mm	710 mm	700 mm	690 mm	680 mm	670 mm	660 mm
0°C	14.76	14.57	14.38	14.19	13.99	13.80	13.61	13.42	13.23	13.04	12.84	12.65
1°C	14.38	14.19	14.00	13.82	13.63	13.44	13.26	13.07	12.88	12.70	12.51	12.32
2°C	14.01	13.82	13.64	13.46	13.28	13.10	12.92	12.73	12.55	12.37	12.19	12.01
3°C	13.65	13.47	13.29	13.12	12.94	12.76	12.59	12.41	12.23	12.05	11.88	11.70
4°C	13.31	13.13	12.96	12.79	12.61	12.44	12.27	12.10	11.92	11.75	11.58	11.40
5°C	12.97	12.81	12.64	12.47	12.30	12.13	11.96	11.80	11.63	11.46	11.29	11.12
6°C	12.66	12.49	12.33	12.16	12.00	11.83	11.67	11.51	11.34	11.18	11.01	10.85
7°C	12.35	12.19	12.03	11.87	11.71	11.55	11.39	11.23	11.07	10.91	10.75	10.59
8°C	12.05	11.90	11.74	11.58	11.43	11.27	11.11	10.96	10.80	10.65	10.49	10.33
9°C	11.77	11.62	11.46	11.31	11.16	11.01	10.85	10.70	10.55	10.39	10.24	10.09
10°C	11.50	11.35	11.20	11.05	10.90	10.75	10.60	10.45	10.30	10.15	10.00	9.86
11°C	11.24	11.09	10.94	10.80	10.65	10.51	10.36	10.21	10.07	9.92	9.78	9.63
12°C	10.98	10.84	10.70	10.56	10.41	10.27	10.13	9.99	9.84	9.70	9.56	9.41
13°C	10.74	10.60	10.46	10.32	10.18	10.04	9.90	9.77	9.63	9.49	9.35	9.21
14°C	10.51	10.37	10.24	10.10	9.96	9.83	9.69	9.55	9.42	9.28	9.14	9.01
15°C	10.29	10.15	10.02	9.88	9.75	9.62	9.48	9.35	9.22	9.08	8.95	8.82
16°C	10.07	9.94	9.81	9.68	9.55	9.42	9.29	9.15	9.02	8.89	8.76	8.63
17°C	9.86	9.74	9.61	9.48	9.35	9.22	9.10	8.97	8.84	8.71	8.58	8.45
18°C	9.67	9.54	9.41	9.29	9.16	9.04	8.91	8.79	8.66	8.54	8.41	8.28
19°C	9.47	9.35	9.23	9.11	8.98	8.86	8.74	8.61	8.49	8.37	8.24	8.12
20°C	9.29	9.17	9.05	8.93	8.81	8.69	8.57	8.45	8.33	8.20	8.08	7.96
21°C	9.11	9.00	8.88	8.76	8.64	8.52	8.40	8.28	8.17	8.05	7.93	7.81
22°C	8.94	8.83	8.71	8.59	8.48	8.36	8.25	8.13	8.01	7.90	7.78	7.67
23°C	8.78	8.66	8.55	8.44	8.32	8.21	8.09	7.98	7.87	7.75	7.64	7.52
24°C	8.62	8.51	8.40	8.28	8.17	8.06	7.95	7.84	7.72	7.61	7.50	7.39
25°C	8.47	8.36	8.25	8.14	8.03	7.92	7.81	7.70	7.59	7.48	7.37	7.26
26°C	8.32	8.21	8.10	7.99	7.89	7.78	7.67	7.56	7.45	7.35	7.24	7.13
27°C	8.17	8.07	7.96	7.86	7.75	7.64	7.54	7.43	7.33	7.22	7.11	7.01
28°C	8.04	7.93	7.83	7.72	7.62	7.51	7.41	7.30	7.20	7.10	6.99	6.89
29°C	7.90	7.80	7.69	7.59	7.49	7.39	7.28	7.18	7.08	6.98	6.87	6.77
30°C	7.77	7.67	7.57	7.47	7.36	7.26	7.16	7.06	6.96	6.86	6.76	6.66

Table 10: Approximate Barometric Pressure at Different Elevations

Elevation (m)	Pressure (mm Hg)	Elevation (m)	Pressure (mm Hg)	Elevation (m)	Pressure (mm Hg)
0	760	800	693	1600	628
100	748	900	685	1700	620
200	741	1000	676	1800	612
300	733	1100	669	1900	604
400	725	1200	661	2000	596
500	717	1300	652	2100	588
600	709	1400	643	2200	580
700	701	1500	636	2300	571

TEACHER INFORMATION

Watershed Testing

1. When setting up the Dissolved Oxygen Probe, be sure to remove the blue plastic cap from the end of the probe. The cap is made of a soft plastic material and easily slides off the probe end.
2. The Dissolved Oxygen Probe must be connected to an active interface for at least 10 minutes prior to use. This time period allows polarization of the electrode to take place. If the 10 minute period is skipped, inaccurate readings will result. If the probe has polarized once and is unplugged for a very short period of time (under 5 minutes) it will require approximately the same period of time to polarize.
3. The Dissolved Oxygen Probe must be calibrated the day of use. Follow the pre-lab procedure to prepare and calibrate the Dissolved Oxygen Probe. To save time, you may wish to calibrate the probe and record the calibration values on paper. The students can then skip the pre-lab procedure and they will have the calibration values available for manual entry in case the values stored in the program are lost.
4. To ensure the most accurate measurements of pH, the pH System should be calibrated prior to use. Refer to the teacher's section of experiment 18 for additional calibration information.
5. When transporting the Dissolved Oxygen Probe to the field site, you should store it in the plastic calibration bottle filled with distilled water. This plastic bottle is shipped with the Dissolved Oxygen Probe. It is important that the students understand the fragile nature of the electrode membrane and proper handling procedures.
6. A glass-stoppered water sampling bottle is recommended for collecting samples. Filling this bottle to the brim, followed by stoppering, will prevent additional oxygen from dissolving after water is collected.
7. Two sites 1.6 km apart should be selected for comparison. Have students take samples at two points for each site. Each of the sample points should be approximately 6 m (20 feet) apart.
8. To determine the D.O. concentration for a solution saturated with dissolved oxygen, refer to Table 9 and Table 10. **Important:** Be sure to bring a copy of these tables on the day you collect and test water samples! Use Table 10 to estimate barometric pressure using your approximate elevation above sea level. Temperature and barometric pressure values can then be used in Table 9 to determine the saturated level of dissolved oxygen, in mg/L. Use this formula to calculate % saturation of dissolved oxygen:

$$\% \text{ saturation} = \frac{\text{measured D.O. level}}{\text{saturated D.O. level}} \times 100$$

9. When measuring total dissolved solids, you may wish to have students use the 0-200 µS/cm (equal to 100 mg/L TDS) range to improve accuracy. This should only be done if TDS levels are below 100 mg/L.

10. A more complete water quality index can be obtained by measuring fecal coliform counts; biological oxygen demand, phosphate and nitrate levels, and turbidity. It is also very valuable to do a "critter count"—that is, examine the macroinvertebrates in the stream.

    For more information on the Water Quality Index, you may be interested in the Vernier book *Water Quality with Computers.*

SAMPLE DATA

Table 1				
Location	Dissolved Oxygen (mg/L)	pH	Total Dissolved Solids (mg/L)	Temperature (°C)
Site 1 Average	10.2	7.4	88.4	11.0
Site 2 Average	8.1	7.4	94.0	8.0

Table 2 DO (% Saturated)			
	Dissolved Oxygen (mg/L)	DO in Saturated Water	% Saturated
Site 1	10.2	11.1	91.9
Site 2	8.1	11.9	68.0

Table 7 Site 1			
Test	Q-Value	Weight	Total Q-Value
DO	97	0.38	36.9
pH	95	0.24	22.8
TDS	84	0.16	13.4
Temperature	85	0.22	18.7
		Overall Quality:	91.8

Table 8 Site 2			
Test	Q-Value	Weight	Total Q-Value
DO	70	0.38	26.6
pH	95	0.24	22.8
TDS	83	0.16	13.3
Temperature	85	0.22	18.7
		Overall Quality:	81.4

ANSWERS TO QUESTIONS

1. The water quality indices for the above sites are 91.8 and 81.4. These are very high indices, considering that they were obtained in an urban Seattle watershed. The first site was from a small, rapidly moving stream (~3.4 m^3/s), and the second from a pond 1.6 km upstream. Other measurements corroborated these measurements—the water quality was very high.

2. Answers will vary.

3. Answers will vary. The two sites compared equally except for the DO value. Since water at the second site was hardly moving, it had less dissolved oxygen than in rapidly moving, highly aerated water.

4. Answers will vary.

5. Water in rapidly moving stream is aerated as it flows through riffles, and may have more dissolved oxygen than in slowly moving water.

6. Answers will vary.

Experiment

21

Physical Profile of a Lake

Lakes are different from streams and rivers because the water they contain is not quickly replaced by fresh water. In a lake, the flushing and changing of water can take anywhere from a year to 100 years, depending on the size of the lake and the watershed that flows into it. This makes lakes very susceptible to damage by pollution. Acid deposition is common in lakes and can result in acid shock if a lake has a low alkaline content or if the soils surrounding it have very little acid-neutralizing capacity. Acid shock can damage or kill aquatic life in the lake.

Lakes can be characterized in three ways. Lakes with large or excessive supplies of nutrients are called *Eutrophic* (well nourished). This type of lake is typically shallow and murky. Lakes with a small supply of nutrients are called *Oligotrophic* (poorly nourished). This type of lake is typically deep and clear with a blue or green color. Most lakes are somewhere in between, and are called *Mesotrophic*.

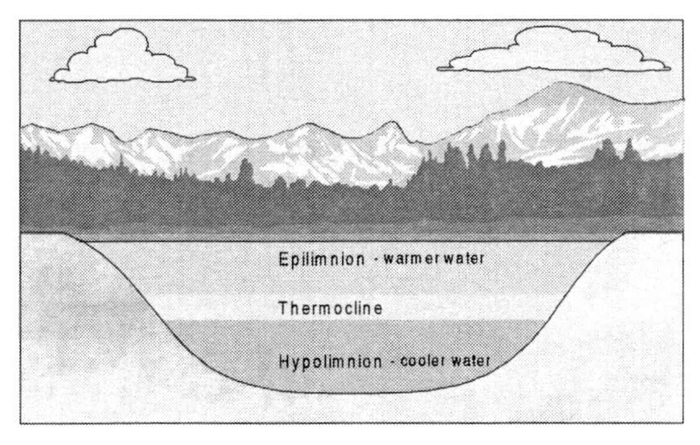

Summer

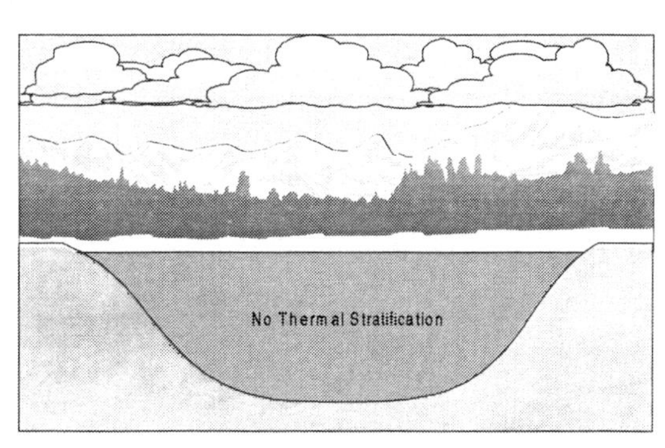

Winter

The density of water increases as the temperature decreases. When water reaches 4°C its density begins to decrease until it freezes. Because the density of water differs with temperature, lakcs undergo a process known as *thermal stratification*. In summer, thermal stratification separates a lake into different regions at different depths. This prevents mixing of water and nutrients between the lake surface and the lake bottom. In winter, the water temperature decreases at the surface, and the cooler water sinks to the lake bottom. Because the water at the bottom of the lake is warmer than the sinking surface water, it begins to rise to the surface. This causes a mixing of the water which brings nutrients from the bottom to the surface, and dissolved oxygen in the surface waters to the bottom.

In this experiment, you will investigate thermal stratification and how it affects the placement of nutrients and dissolved oxygen. You will be taking water samples at various depths throughout the lake. The water will then be analyzed for dissolved oxygen (DO), pH, and total dissolved solids (TDS). An extended Temperature Probe will then be used to measure water temperature at the same depths the water samples were taken from.

OBJECTIVES

In this experiment, you will

- Use a Water Depth Sampler to collect water samples at different depths in the lake.
- Measure DO, pH, and TDS of the collected water samples.
- Use a Temperature Probe to measure water temperature at various depths.

MATERIALS

computer (laptop)
Vernier computer interface
Vernier Conductivity Probe
Vernier pH Sensor
Logger *Pro*
Vernier Dissolved Oxygen Probe
Vernier Extra Long Temperature Probe
Water Depth Sampler
500 mL water sampling bottles
D.O. calibration bottle

PRE-LAB PROCEDURE

Important: Prior to each use, the Dissolved Oxygen Probe must warm up for a period of 10 minutes as described below. If the probe is not warmed up properly, inaccurate readings will result. Perform the following steps to prepare the Dissolved Oxygen Probe.

1. Prepare the Dissolved Oxygen Probe for use.
   a. Remove the blue protective cap.
   b. Unscrew the membrane cap from the tip of the probe.
   c. Using a pipet, fill the membrane cap with 1 mL of DO Electrode Filling Solution.
   d. Carefully thread the membrane cap back onto the electrode.
   e. Place the probe into a container of water.

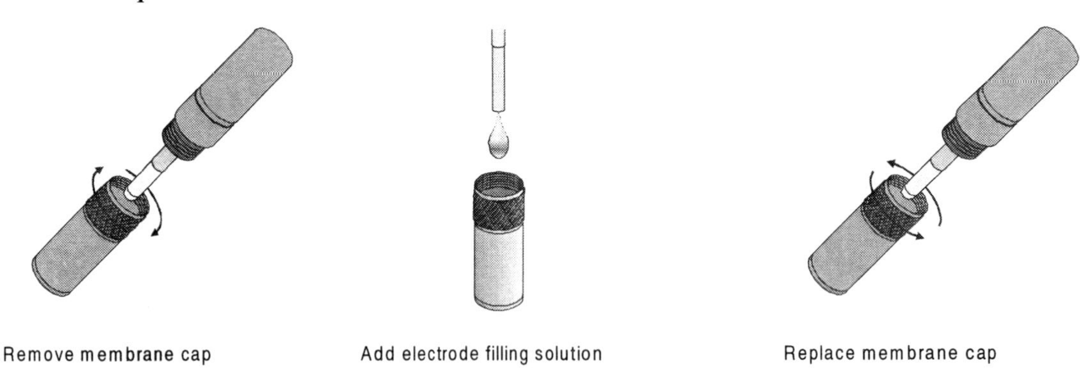

Figure 1

2. Plug the Dissolved Oxygen Probe into Channel 1 of the Vernier computer interface. Connect the pH Sensor to Channel 2.

3. Prepare the computer for data collection by opening the file "21a Phys Profile Lakes" from the *Biology with Computers* folder of Logger*Pro*.

4. It is necessary to warm up the Dissolved Oxygen Probe for 10 minutes before taking readings. To warm up the probe, leave it connected to the interface, with Logger *Pro* running, for 10 minutes. The probe must stay connected at all times to keep it warmed up. If disconnected for a few minutes, it will be necessary to warm up the probe again.

5. You are now ready to calibrate the Dissolved Oxygen Probe.
   - If your instructor directs you to use the calibration stored in the experiment file, then proceed to Step 6.
   - If your instructor directs you to perform a new calibration for the Dissolved Oxygen Probe, follow this procedure.

Zero-Oxygen Calibration Point

a. Choose Calibrate ▸ CH1: Dissolved Oxygen (mg/L) from the Experiment menu and then click Calibrate Now.

b. Remove the probe from the water and place the tip of the probe into the Sodium Sulfite Calibration Solution. **Important:** No air bubbles can be trapped below the tip of the probe or the probe will sense an inaccurate dissolved oxygen level. If the voltage does not rapidly decrease, tap the side of the bottle with the probe to dislodge any bubbles. The readings should be in the 0.2 to 0.5 V range.

c. Type "0" (the known value in mg/L) in the edit box.

d. When the displayed voltage reading for Reading 1 stabilizes, click Keep.

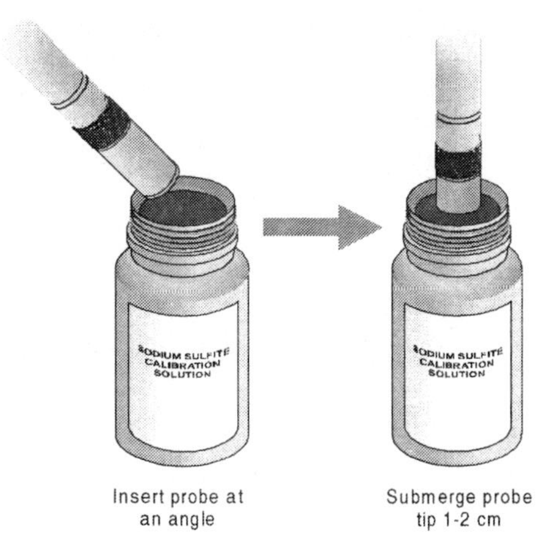

Figure 2

Saturated DO Calibration Point

e. Rinse the probe with distilled water and gently blot dry.

f. Unscrew the lid of the calibration bottle provided with the probe. Slide the lid and the grommet about 2 cm onto the probe body.

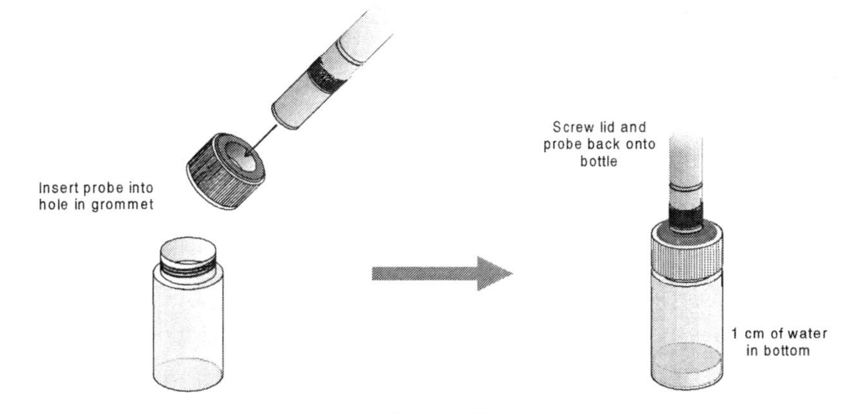

Figure 3

g. Add water to the bottle to a depth of about 1 cm and screw the bottle into the cap, as shown. **Important:** Do not touch the membrane or get it wet during this step. Keep the probe in this position for about a minute.

h. Type the correct saturated dissolved-oxygen value (in mg/L) from Table 2 (for example, "8.66") using the current barometric pressure and air temperature values. If you do not have the current air pressure, use Table 3 to estimate the air pressure at your altitude.

i. When the displayed voltage reading for Reading 2 stabilizes (readings should be above 2.0 V), click Keep.

j. Choose Calibration Storage from the calibration pull-down menu. Choose Experiment File (calibration stored with the current document). Click Done.

k. From the File menu, select Save As and save the current experiment file with a new name.

l. Prepare the probe for transport by filling the calibration bottle half full with water. Secure the Dissolved Oxygen Probe far enough down in the bottle that the membrane is completely covered by water. Screw the calibration bottle lid completely onto the bottle so that no water will leak out.

PROCEDURE

When collecting samples at different depths in a lake or pond, it is best to choose a sampling site as far from shore as possible. This will generally require a boat or other form of floating vessel to reach the site. Once at the site proceed to Step 6.

Part I: Measuring Dissolved Oxygen and pH

6. Rinse the sampling bottle a few times with lake water.

7. Arm the water sampler by doing the following:
    a. At each end of the water sampler, there is a small metal tube with two holes cut out. Take hold of the metal tubes and pull the balls at each end of the tube outward at the same time.
    b. Slip the two metal tubes together and align the holes.
    c. Insert the metal trigger pin attached to the rope through the aligned holes of the two metal tubes. The water sampler is now armed. You can now let go of the metal tubes and they will stay in place due to the trigger pin.

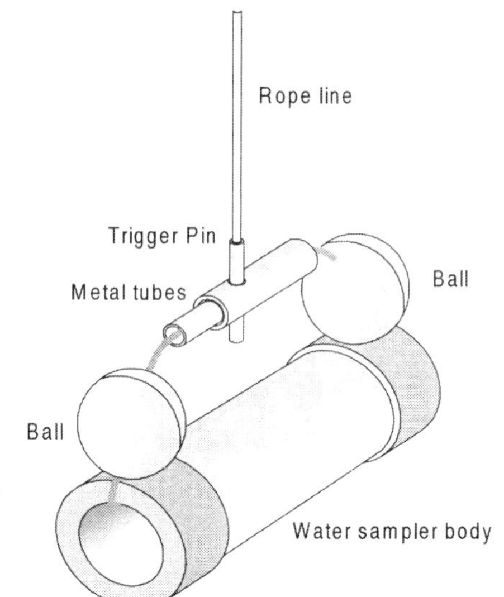

Figure 4

8. Test the water sampler.
    a. Take hold of the rope 1.5 meters up from the sampler.
    b. Place the sampler and slack rope in the water. Lower the armed water sampler to a depth of 1.5 meters. **Important:** Provide plenty of slack rope when lowering the sampler in the water. It is best to take hold of the rope at the depth you will be sampling and place the slack rope in the water with the water sampler.
    c. Give the rope line two quick tugs upward. At a 1.5 meter depth you should be able to see the sampler trigger and the balls pop into place at each end of the sampler tube.
    d. If the water sampler does not function properly, notify your instructor. Otherwise, empty the water from the sampler and rearm it.

9. Lower the water sampler to the bottom depth you are going to be measuring. Remember to provide plenty of slack rope when lowering the water sampler.

10. Allow the water sampler to remain at that depth for 1 minute. Trigger the sampler with two quick tugs upward on the rope line. Pull the sampler up into the boat.

11. Empty the water from the sampler using the clear plastic tubing connected to the bottom of the sampler. Hold the sampler at an angle of 30° with the clear plastic tubing pushed down into your sampling bottle. Minimize the introduction of oxygen into the sample by tipping the sampling bottle at an angle and letting the water pour down the inside wall of the bottle. Water will flow from the tubing when the white plastic stop valve is opened (see Figure 5).

Push forward to open

Figure 5

12. Measure the dissolved oxygen and pH of your water sample.
    a. Open the experiment file that was saved in Step 5k.
    b. Remove the Dissolved Oxygen Probe from its storage bottle. Place the probe into the water sample and gently swirl to allow water to move past the probe's tip.
    c. Monitor the dissolved oxygen reading displayed on the screen. When the reading is stable record the dissolved oxygen concentration in Table 1. Remove the probe from the water sample and place it back into the storage bottle.
    d. Remove the pH Sensor from its storage bottle. Rinse the probe tip with lake water. Place the probe into the water sample.
    e. Monitor the pH reading displayed on the screen. Record the pH in Table 1 when the reading has stabilized. Remove the probe from the water sample and place it back into the storage bottle.
    f. Seal the water sampling bottle and mark the container with the depth the sample was taken. Place the bottle aside to measure total dissolved solids with the Conductivity Probe back in the classroom.
13. Reset the sampler and repeat Steps 9 – 12 every 1.5 meters up from the first sample taken. The final measurement should be taken just below the surface.

Part II: Measuring Water Temperature

14. Disconnect the pH Sensor and move the Dissolved Oxygen Probe to Channel 2 of the Vernier computer interface. Connect the Extra Long Temperature Probe to Channel 1. Prepare the computer for data collection by opening the file "21b Phys Profile Lakes" in the *Biology with Computers* folder.
15. Start measuring water temperature by clicking [Collect].
16. Measure the water temperature at various depths.
    a. Lower the Temperature Probe to the bottom depth you will be measuring. Be sure to always maintain a solid grip on the grey cable. Do not support the cable by holding the amplifier box. Depending on how much weight has been added to the end of the Temperature Probe cable, the probe could disconnect from the amplifier box and sink to the bottom of the lake.
    b. Monitor the temperature reading displayed on the screen. When the reading has stabilized, click [Keep].
    c. In the text box, enter the depth of the Temperature Probe and press ENTER.
    d. Pull the probe up to the next depth to be measured.
17. Repeat Step 16 at the same depths samples were taken in Part I. When all measurements have been made click [Stop].
18. Obtain the temperature values for each depth from the table and record them in Table 1.

Part III: Measuring Total Dissolved Solids in the classroom

19. Connect the Conductivity Probe to Channel 1 of the Vernier computer interface. Set the selector switch on the Conductivity Probe to the middle setting of 0-200 μS/cm range (equivalent to 0-100 mg/L). Prepare the computer for data collection by opening the file "21c Phys Profile Lakes" in *Biology with Computers* folder.

20. Measure the Total Dissolved Solids of each water sample collected
    a. Place the Conductivity Probe into a water sample bottle. The hole near the probe end must be completely submerged in the water sample.
    b. Once the total dissolved solid reading has stabilized, record the reading in the Table 1.
    c. To avoid contaminating the water samples, rinse the probe with clean, distilled water after each test. Blot the outside of the probe tip dry with a tissue or paper towel. It is *not* necessary to dry the *inside* of the hole near the probe end.
    d. Repeat for each water sample collected.

DATA

Table 1				
Depth (m)	Dissolved oxygen (mg/L)	pH	Total dissolved solids (mg/L)	Temperature (°C)
0				
1.5				
3.0				
4.5				
6.0				
7.5				
9.0				
10.5				
12.0				
13.5				
15.0				

PROCESSING THE DATA

1. Remove all sensors from the interface. Open the file "Exp 21d Phys Profile Lakes" in the *Biology with Computers* folder.

2. Enter the data recorded in Table 1 into the appropriate column in the table.

3. Click once on the displayed graph with the mouse pointer to make it active. Print a copy of your graph. Enter your name(s) and the number of copies of the graph. Use your graph to answer the following questions.

QUESTIONS

1. Using your measurements, is there any evidence of thermal stratification in the lake you investigated? Explain.
2. A thermocline is the region where water temperature changes rapidly with depth. Was there evidence of a thermocline? If so, at what depth was it apparent?
3. How would you expect the temperature measurements to change if you repeated the measurements during a different season?
4. How did the dissolved oxygen levels change as the depth changed? How does this relate to thermal stratification?
5. Based on your results, at what depth would you expect to find most of the aquatic life?
6. Was there evidence of a thermocline? If so, at what depth was it apparent?
7. Would you classify this lake as oligotrophic, eutrophic, or mesotrophic?
8. Describe what happened to the pH measurements as depth changed? Did you notice any kind of pattern? Explain.
9. Describe what happened to the concentration of total dissolved solids as depth changed? Is there any pattern? Explain.

CALIBRATION TABLES

Table 2: 100% Dissolved Oxygen Capacity (mg/L)												
	770 mm	760 mm	750 mm	740 mm	730 mm	720 mm	710 mm	700 mm	690 mm	680 mm	670 mm	660 mm
0°C	14.76	14.57	14.38	14.19	13.99	13.80	13.61	13.42	13.23	13.04	12.84	12.65
1°C	14.38	14.19	14.00	13.82	13.63	13.44	13.26	13.07	12.88	12.70	12.51	12.32
2°C	14.01	13.82	13.64	13.46	13.28	13.10	12.92	12.73	12.55	12.37	12.19	12.01
3°C	13.65	13.47	13.29	13.12	12.94	12.76	12.59	12.41	12.23	12.05	11.88	11.70
4°C	13.31	13.13	12.96	12.79	12.61	12.44	12.27	12.10	11.92	11.75	11.58	11.40
5°C	12.97	12.81	12.64	12.47	12.30	12.13	11.96	11.80	11.63	11.46	11.29	11.12
6°C	12.66	12.49	12.33	12.16	12.00	11.83	11.67	11.51	11.34	11.18	11.01	10.85
7°C	12.35	12.19	12.03	11.87	11.71	11.55	11.39	11.23	11.07	10.91	10.75	10.59
8°C	12.05	11.90	11.74	11.58	11.43	11.27	11.11	10.96	10.80	10.65	10.49	10.33
9°C	11.77	11.62	11.46	11.31	11.16	11.01	10.85	10.70	10.55	10.39	10.24	10.09
10°C	11.50	11.35	11.20	11.05	10.90	10.75	10.60	10.45	10.30	10.15	10.00	9.86
11°C	11.24	11.09	10.94	10.80	10.65	10.51	10.36	10.21	10.07	9.92	9.78	9.63
12°C	10.98	10.84	10.70	10.56	10.41	10.27	10.13	9.99	9.84	9.70	9.56	9.41
13°C	10.74	10.60	10.46	10.32	10.18	10.04	9.90	9.77	9.63	9.49	9.35	9.21
14°C	10.51	10.37	10.24	10.10	9.96	9.83	9.69	9.55	9.42	9.28	9.14	9.01
15°C	10.29	10.15	10.02	9.88	9.75	9.62	9.48	9.35	9.22	9.08	8.95	8.82
16°C	10.07	9.94	9.81	9.68	9.55	9.42	9.29	9.15	9.02	8.89	8.76	8.63
17°C	9.86	9.74	9.61	9.48	9.35	9.22	9.10	8.97	8.84	8.71	8.58	8.45
18°C	9.67	9.54	9.41	9.29	9.16	9.04	8.91	8.79	8.66	8.54	8.41	8.28
19°C	9.47	9.35	9.23	9.11	8.98	8.86	8.74	8.61	8.49	8.37	8.24	8.12
20°C	9.29	9.17	9.05	8.93	8.81	8.69	8.57	8.45	8.33	8.20	8.08	7.96
21°C	9.11	9.00	8.88	8.76	8.64	8.52	8.40	8.28	8.17	8.05	7.93	7.81
22°C	8.94	8.83	8.71	8.59	8.48	8.36	8.25	8.13	8.01	7.90	7.78	7.67
23°C	8.78	8.66	8.55	8.44	8.32	8.21	8.09	7.98	7.87	7.75	7.64	7.52
24°C	8.62	8.51	8.40	8.28	8.17	8.06	7.95	7.84	7.72	7.61	7.50	7.39
25°C	8.47	8.36	8.25	8.14	8.03	7.92	7.81	7.70	7.59	7.48	7.37	7.26
26°C	8.32	8.21	8.10	7.99	7.89	7.78	7.67	7.56	7.45	7.35	7.24	7.13
27°C	8.17	8.07	7.96	7.86	7.75	7.64	7.54	7.43	7.33	7.22	7.11	7.01
28°C	8.04	7.93	7.83	7.72	7.62	7.51	7.41	7.30	7.20	7.10	6.99	6.89
29°C	7.90	7.80	7.69	7.59	7.49	7.39	7.28	7.18	7.08	6.98	6.87	6.77
30°C	7.77	7.67	7.57	7.47	7.36	7.26	7.16	7.06	6.96	6.86	6.76	6.66

Table 3: Approximate Barometric Pressure at Different Elevations					
Elevation	Pressure	Elevation	Pressure	Elevation	Pressure
0	760	800	693	1600	628
100	748	900	685	1700	620
200	741	1000	676	1800	612
300	733	1100	669	1900	604
400	725	1200	661	2000	596
500	717	1300	652	2100	588
600	709	1400	643	2200	580
700	701	1500	636	2300	571

Experiment
21

TEACHER INFORMATION

Physical Profile of a Lake

1. When setting up the Dissolved Oxygen Probe, be sure to remove the blue plastic cap from the end of the probe. The cap is made of a soft plastic material and easily slides off the probe.
2. The Dissolved Oxygen Probe must be connected to an active interface for at least 10 minutes prior to use. This time period allows polarization of the electrode to take place. If the 10 minute period is skipped, inaccurate readings will result. If the probe has polarized once and is unplugged for a very short period of time (under 5 minutes) it will require approximately the same period of time to polarize.
3. The Dissolved Oxygen Probe must be calibrated the day of use. Follow the pre-lab procedure to prepare and calibrate the Dissolved Oxygen Probe. To save time, you may wish to calibrate the probe and record the calibration values on paper. The students can then skip the pre-lab procedure and they will have the calibration values available for manual entry in case the values stored in the program are lost.
4. To ensure the most accurate measurements of pH, the pH System should be calibrated prior to use. Refer to the teacher's section of experiment 18 for additional calibration information.
5. When transporting the Dissolved Oxygen Probe to the field site, you should store it in the plastic calibration bottle filled with distilled water. This plastic bottle is shipped with the Dissolved Oxygen Probe. It is important that the students understand the fragile nature of the electrode membrane and proper handling procedures.
6. We suggest that you take your students to a swimming pool to practice using the Water Depth Sampler. Swimming pools are great for testing because the clear water enables the students to see what is necessary to trigger the sampler and how to prevent triggering before the sampler reaches the designated depth.
7. When the Water Depth Sampler is being lowered, it is important that plenty of slack rope be placed in the water with the sampler. This will prevent the rope from becoming taut before it reaches the correct depth. If the rope does become taut before reaching the measuring depth, it could possibly trigger the sampler prematurely.
8. The rope of the Water Depth Sampler and the cable of the Extra Long Temperature Probe should be sectioned off in 1.5 meter increments with water proof tape and marked accordingly.
9. It will be necessary to weigh down the end of the Extra Long Temperature Probe so it does not float. This can be done with duct tape and a rock or weight. Another option would be to use a large fishing weight with an eyeloop at the top. Run the end of the temperature probe through the eyeloop and tie the cable into a square knot. This should provide adequate weight so the probe will hang straight down when lowered into the lake. When the cable of the temperature probe has a weight attached, make sure that the students hold onto the probe cable at all times and not the amplifier box.
10. Students should always wear proper flotation equipment when in or around water.
11. To prevent water damage, it is best to store all probes and electronic equipment in plastic bags or containers when not in use.

SAMPLE DATA

Table 1				
Depth (m)	Dissolved Oxygen (mg/L)	pH	Total Dissolved Solids (mg/L)	Temperature (°C)
0	9.38			17.7
1.5	9.54	7.18	26.1	17.7
3.0				17.7
4.5	9.68	6.85	26.1	17.0
6.0				17.0
7.5	9.01	7.33	25.6	16.7
9.0				16.4
10.5	7.2	6.67	25.9	16.1
12.0				14.8
13.5	2.53	6.28	24.1	10.7
15.0		6.15	24.1	9.8

Water sample data collected from Hagg Lake, Oregon on October 8, 1996. Hagg Lake is a man-made reservoir classified as a eutrophic lake. Dissolved oxygen, pH, and total dissolved solids readings were taken from raised samples using the Water Depth Sampler. Temperature data was taken using the Vernier Extra Long Temperature Probe.

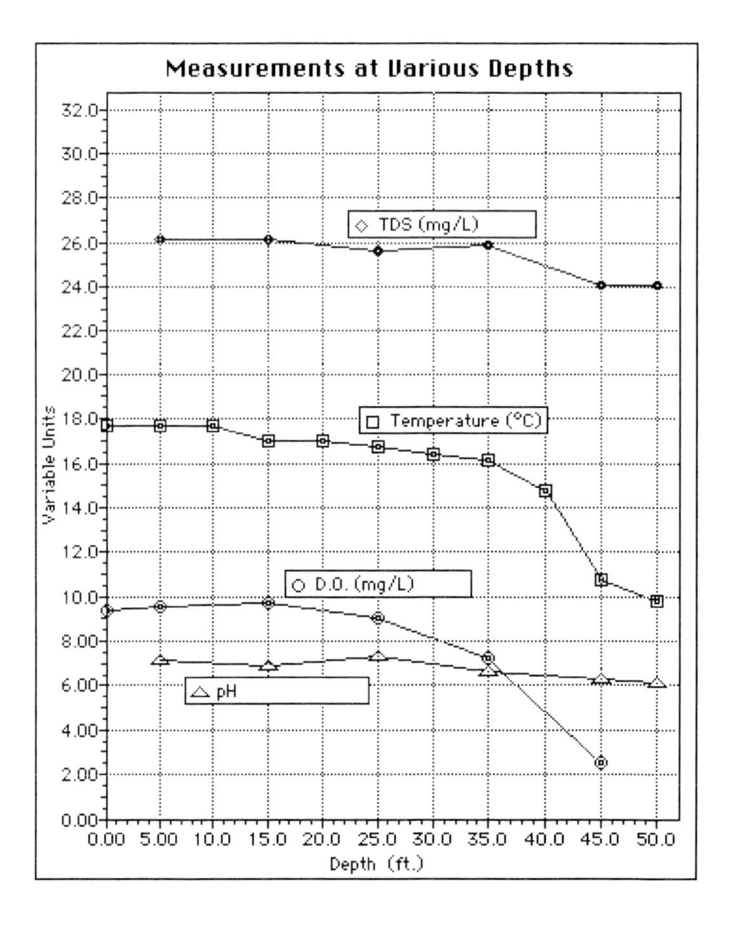

ANSWERS TO QUESTIONS

1. Answers will vary.

2. Answers will vary.

3. If original measurements are made in the summer, there should be thermal stratification apparent, consisting of warmer temperatures at the surface and colder temperatures at the bottom. If measurements are made in the winter, there should be little temperature difference throughout the lake due to fall turnover and winter temperatures.

4. Answers will vary. Generally dissolved oxygen levels will be higher near the lake surface and lower near the lake bottom. This is primarily due to dissolving of oxygen into the water at the lake surface and the presence of photosynthetic producers in the water near the surface. Dissolved oxygen levels will drop off quickly below the thermocline. Thermal stratification prevents mixing of nutrients and dissolved oxygen between the surface waters and the lake bottom.

5. Most aquatic life will be found in the littoral and limnetic zone in the top 10 meters of the lake.

6. Answers will vary.

7. Answers will vary.

8. Answers will vary.

9. Answers will vary.

Experiment

22

Osmosis

In order to survive, all organisms need to move molecules in and out of their cells. Molecules such as gases (e.g., O_2, CO_2), water, food, and wastes pass across the cell membrane. There are two ways that the molecules move through the membrane: *passive transport* and *active transport*. While active transport requires that the cell uses chemical energy to move substances through the cell membrane, passive transport does not require such energy expenditures. Passive transport occurs spontaneously, using heat energy from the cell's environment.

Diffusion is the movement of molecules by passive transport from a region in which they are highly concentrated to a region in which they are less concentrated. Diffusion continues until the molecules are randomly distributed throughout the system. Osmosis, the movement of water across a membrane, is a special case of diffusion. Water molecules are small and can easily pass through the membrane. Other molecules, such as proteins, DNA, RNA, and sugars are too large to pass through the cell membrane, so they cannot diffuse from one side of the membrane to the other. The membrane is said to be semipermeable, since it allows some molecules to pass through but not others.

If the concentration of water on one side of the membrane is different than on the other side, water will attempt to move through the membrane until the concentration of water is the same on both sides. If the concentration of water outside of a cell is greater, the water moves into the cell faster than it leaves, and the cell swells. The cell membrane acts somewhat like a balloon. If too much water enters the cell, the cell can burst, killing the cell. Cells usually have some mechanism for preventing too much water from entering, such as pumping excess water out of the cell or making a tough outer coat that will not rupture. If the concentration of water inside of a cell is greater, the water moves out of the cell faster than it enters, and the cell shrinks. If a cell becomes too dehydrated, it may not be able to survive.

In this experiment, you will use a Gas Pressure Sensor to measure the rate of pressure change as water moves in or out of potato wells filled with various concentrations of sugar solution.

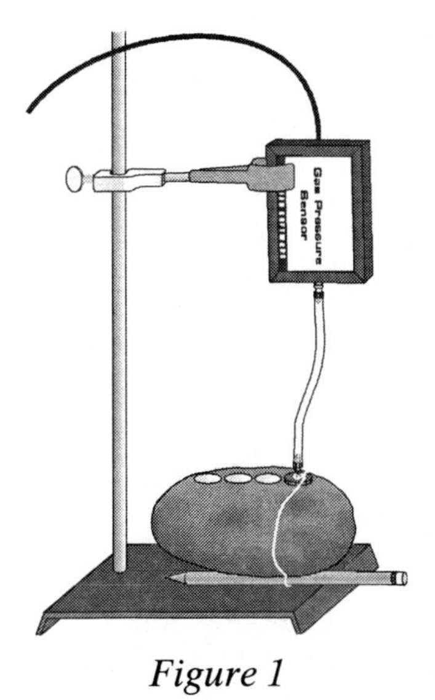

Figure 1

OBJECTIVES

In this experiment, you will

- Use a Gas Pressure Sensor to investigate the relationship between water movement and solute concentration.
- Determine the molar concentration of potato cells.
- Determine the water potential of potato cells.

MATERIALS

computer
Vernier computer interface
Logger *Pro*
Vernier Gas Pressure Sensor
1-hole rubber stopper assembly
0 M, 0.33 M, 0.67 M, and 1.0 M sugar solution
waste beaker
utility clamp
ring stand
paper towels
potato with four wells
pencil
string

PROCEDURE

1. Connect the plastic tubing to the valve on the Gas Pressure Sensor.
2. Connect the Temperature Probe to the computer interface. Prepare the computer for data collection by opening the file "22 Osmosis" from the *Biology with Computers* folder of Logger*Pro*.
3. Obtain four sugar solutions. They should be labeled 0 M, 0.33 M, 0.67 M, and 1.0 M.
4. Obtain a tray, if available, and a potato with four wells bored into it. Rinse the wells out with tap water and blot the well dry.
5. Using a ring stand and clamp, mount the pressure sensor above the potato, as in Figure 1. **Note:** Be careful not to get any solution on the sensor.
6. Fill the first well with the sugar solution labeled "0 M Sugar." The zero indicates that the solution has no sugar in it, only distilled water. Allow the solution to remain in the well for five minutes. At the end of five minutes, proceed to Step 7.
7. Empty the 0 M Sugar solution from the well and refill the well with fresh 0 M Sugar solution.
8. Place the stopper assembly firmly into the well. You will need to tie the stopper down with a small piece of string, as it may pop out of the well unless it is held in place.
   a. Loop the string around the plastic stem that is inserted into the one-hole stopper.

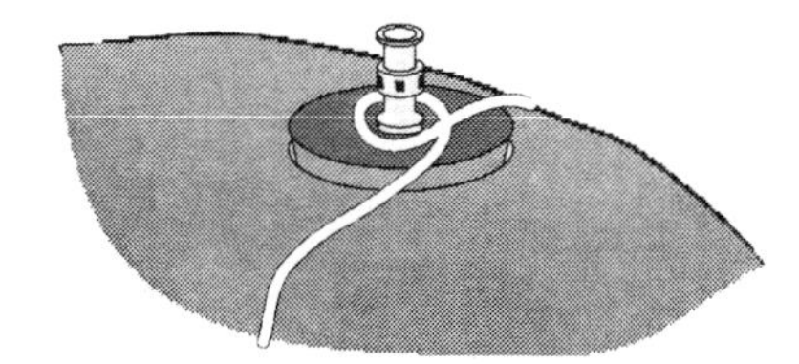

   b. Loosely wrap the string around the potato.

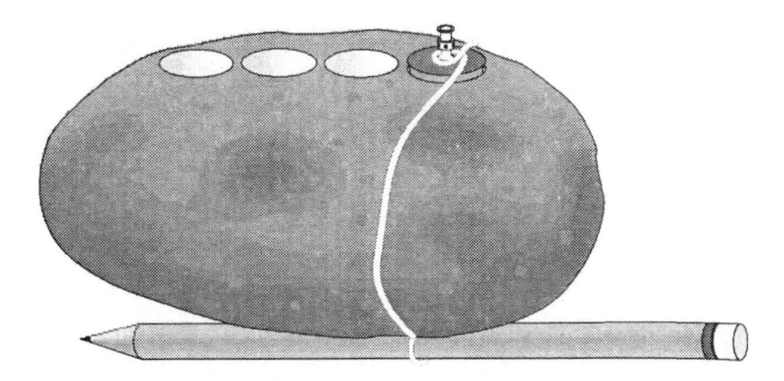

c. Tie the string securely around the potato so that the knot will not slip. Leave enough slack to insert a pencil as shown above.

d. To cinch the string tightly, place a pencil between the string and the potato. Twist the pencil until the stopper is firmly held in place.

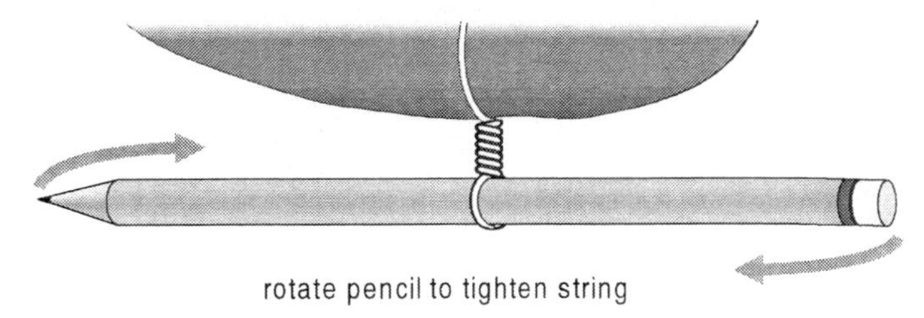

9. Allow the potato to rest for three minutes, making sure the string stays tight and that the pencil does not unwind.

10. After three minutes, connect the free-end of the plastic tubing to the connector in the rubber stopper as shown in Figure 3. Click Collect to begin data collection.

11. Fill the next well with the next higher concentration of sugar, 0.33 M. Allow it to sit in the well during the five minutes that data is being collected.

12. When data collection has finished, disconnect the plastic tubing connector from the rubber stopper. Remove the pencil, string, and rubber stopper, and empty the sugar solution from the potato into a waste beaker.

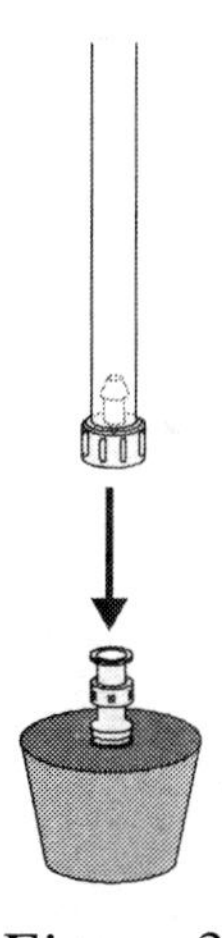

Figure 3

13. Determine the rate of pressure change for the curve of pressure *vs*. time. Perform a linear regression to calculate the rate of change.

    a. Move the mouse pointer to the point where the pressure values begin to change. Hold down the mouse button. Drag the mouse pointer to the end of the data and release the mouse button.

    b. Click the Linear Fit button, , to perform a linear regression. A floating box will appear with the formula for a best fit line.

    c. Record the slope of the line, m, as the rate of pressure change in Table 1.

    d. Close the linear regression floating box.

14. Fill the second well with fresh 0.33 M sugar solution and repeat Steps 8 – 13.

15. Fill the third well with fresh 0.67 M sugar solution and repeat Steps 8 – 13.

16. Fill the last well with fresh 1.0 M sugar solution and repeat Steps 8 – 13. Skip Step 11 since this is the final sugar solution to be tested.

DATA

Table 1	
Concentration	Rate of pressure change (kPa/min)
0.00 M	
0.33 M	
0.67 M	
1.0 M	

PROCESSING THE DATA

1. On Page 2 of the experiment file, create a graph with the rate of pressure change (slope) on the y-axis and the concentration of sugar on the x-axis.

QUESTIONS

1. Which solutions, if any, produced a positive slope? Was water moving in or out of the potato cells under these circumstances? Explain.
2. Which solutions, if any, produced a negative slope? Was water moving in or out of the potato cells under these circumstances? Explain.
3. Does sugar move in or out of the cells? Explain.
4. Examine the plot of the rate of pressure change *vs.* the sugar concentration. Describe any pattern you discern from the data.
5. Using the plot, estimate the concentration of sugar that would yield no change in pressure. Why is this biologically significant?
6. When wilted plants are watered, they tend to become rigid. Explain how this might happen.

EXTENSION – WATER POTENTIAL

Water potential is a term used when predicting the movement of water into or out of plant cells. Water always moves from an area of higher water potential to an area of lower water potential. The symbol for water potential is the Greek letter Psi, Ψ. Water potential consists of a physical pressure component called pressure potential Ψ_p, and the effects of solutes called solute potential, Ψ_s.

$$\Psi = \Psi_p + \Psi_S$$

Water potential	=	Pressure potential	+	Solute potential

Distilled water in an open beaker has a water potential of zero. The addition of solute decreases water potential while the addition of pressure increases water potential. A water potential value can be positive, negative, or zero. Water potential is usually measured in bars, a metric measure of pressure. (1 kPa = .1 bar)

In this experiment, you will measure the percent change in mass of potato cores after they have soaked in various concentrations of sugar solutions for a 24 hour period. You will use this data to calculate the water potential of the potato cells.

MATERIALS

computer
Logger *Pro*
four potato cores
0 M, 0.33 M, 0.67 M, or 1.0 M sugar solution
250 mL beaker
plastic wrap
balance
paper towels

PROCEDURE

You will be assigned one or more of the sugar solutions in which to soak your potato cores.

1. Pour 100 mL of the assigned sugar solution into a 250 mL beaker.
2. Measure and record the mass of the four potato cores together.
3. Put the four cores into the beaker of sugar solution.
4. Cover the beaker with plastic wrap and allow it to stand for a 24-hour period.
5. Remove the cores from the beaker, blot with a paper towel, and determine the mass of the four cores together after soaking.
6. Calculate the percent change in mass and record your data for the sugar concentration tested in Table 2 as well as on the class data sheet.
7. Repeat Steps 1 – 6 for any additional assigned sugar solutions.

DATA

Table 2			
Sugar solution concentration	Initial mass (g)	Final mass (g)	Percent change in mass
0.00 M			
0.33 M			
0.67 M			
1.0 M			

Table 3 Class data of percent change in mass										
Sugar solution concentration	Group 1	Group 2	Group 3	Group 4	Group 5	Group 6	Group 7	Group 8	Total	Class average
0.00 M										
0.33 M										
0.67 M										
1.00 M										

Table 4	
Sugar molar concentration	=
Solute potential	=
Water potential	=

PROCESSING THE DATA

1. On Page 3 of the experiment file, create a graph with the percent change in mass on the y-axis and concentration of sugar on the x-axis.

2. Perform a linear regression to determine the molar concentration of sugar solution at which the mass of the potato cores does not change.

   a. Move the mouse pointer to the first data point. Press the left mouse button. Drag the pointer to the last data point.

b. Click the Linear Fit button, , to perform a linear regression. A floating box will appear with the formula for a best fit line.

c. Choose Interpolate from the Analyze menu. Move the mouse pointer along the regression line to the point where the line crosses the x axis. This point represents the molar concentration of sugar with a water potential equal to the potato core water potential. Record this concentration in Table 4.

3. Use the sugar molar concentration from the previous step to calculate the solute potential of the sugar solution.

   a. Calculate the solute potential of the sugar solution using the equation

$$\Psi_s = -iCRT$$

where i = ionization constant (1.0 for sugar since it doesn't ionize in water)

C = sugar molar concentration (determined from the graph)

R = pressure constant (R = 0.0831 liter bars/mol-K)

T = temperature (K)

   b. Record this value with units in Table 4.

4. Calculate the water potential of the solution using the equation

$$\Psi = \Psi_p + \Psi_s$$

The pressure potential of the solution in this case is zero because the solution is at equilibrium. Record the water potential value with units in Table 4.

QUESTIONS

1. What may happen to an animal cell if water moves into it? How does this differ from what would happen in a plant cell?

2. What factors affect water potential?

3. If a plant cell has a higher water potential than its surrounding environment and the pressure is equal to zero, will water move in or out of the cell? Explain why.

Experiment
22

TEACHER INFORMATION

Osmosis

1. The Extension on water potential has been added to this experiment to further address the concept of water movement in and out of cells.

2. Select one large potato per lab team.

3. You will need to bore four wells into each potato prior to class. To do so,

   a. Choose a cork borer to match the size of the rubber stopper being used. A number one stopper works well, with a 14 mm (9/16 inch) borer. The stopper should fit snugly when inserted halfway into the well.
   b. Use an ink marker or length of tape to place a mark three centimeters from the cutting edge of the cork borer. The wells should all be of the same depth.
   c. Insert the borer three centimeters into the potato, up to the mark. Do not remove the borer.
   d. Insert a knife or other object with a flat blade down the center of the borer and into the potato. A thin pair of scissors will also work well.
   e. Twist the borer and knife one-half turn or so. You will feel the potato snap off at the base of the well. Remove the borer and knife. Discard the piece of potato from the center of the cork borer.
   f. Repeat this until four wells are formed.
   g. If the potato must be stored for a time before use, fill each well with a 0.5 M sucrose solution. Prior to use, rinse the sugar solution out of the wells. The wells should not be allowed to dry out. The sugar solution is isotonic to the potato cells, and will not adversely affect the experiment.
   h. Save the potato cores for the Extension.

4. If the surface of the potato near a well cracks, it will leak air. If the crack is not too deep, you can cut off a part of the potato. Make the cut perpendicular to the axis of the well.

5. The rubber stopper assembly may slowly pop out of the well, especially if a positive pressure develops. It is necessary to tie the assembly down. One easy way to do this is given in the student's procedures. You may want to demonstrate how it is done. Figure 1 illustrates how this might look.

Figure 1

6. Between classes, fill each well with a 0.5 M sucrose solution. This isotonic solution should allow the potato to be used repeatedly.

7. To make a 1.0 M solution of sucrose, dissolve 342.2 g of table sugar per liter of solution.

8. To prepare a 0.33 M sucrose solution, add enough distilled water to 333 mL of 1.0 M sucrose solution to make 1 L of solution. To prepare a 0.67 M sucrose solution, add 667 mL of 1.0 M sucrose solution to make 1 L of solution.

9. Each lab team will require about 10 mL of each sucrose solution for the main lab and 100 mL of one or more of the solutions for the Extension.

SAMPLE RESULTS

Table 1	
Concentration	Rate of pressure change (kPa/min)
0.00 M	–0.564
0.33 M	–0.035
0.67 M	0.175
1.0 M	0.508

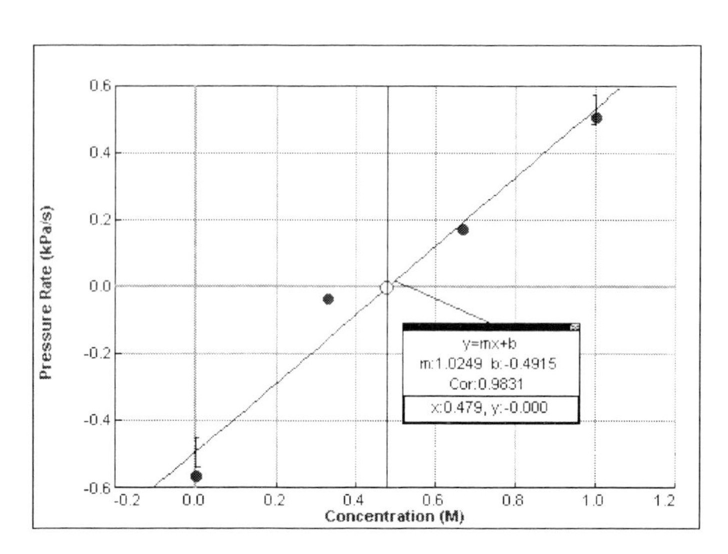

EXTENSION SAMPLE DATA

Average percent change in mass - class data	
0.00 M	18.5
0.33 M	1.6
0.67 M	–15.0
1.0 M	–18.0

Sugar molar concentration	0.41 M
Solute potential	–10.0 bars
Water potential	–10.0 bars

ANSWERS TO QUESTIONS

1. The 0.67 M and 1.0 M sugar solutions produced a positive slope. Water moved out of the cell and into the sugar solution, since there was more water in the cell than in the sugar solution. This produced a pressure increase, causing the slope to be positive.

2. The 0 M and 0.33 M sugar solutions produced a negative slope. Water moved into the cell and out of the sugar solution, since there was more water in the sugar solution than in the cell. This produced a pressure decrease, causing the slope to be negative.

3. Sugar will not move in or out of the cells, since it is too large to pass through the cell membrane.

4. As the sugar concentration increases, the rate of pressure change also increases. The increase should be linear. Since the pressure changes are very small, consistency is very important and may affect the results. While inexperienced students may not produce quantitative results, they should easily obtain qualitatively valid data. More experienced students will find a linear relationship between the sugar concentration and the rate of pressure change in this experiment.

5. Students should find that a 0.5 M sugar solution yields no change in pressure. The significance of this is that the water concentration in the cell is the same as in a 0.5 M sugar solution.

6. If plants are watered, the amount of water inside the cell is less than outside of the cells. This causes water to move into the cells—this makes the plant turgid and stiff. Since plants have a strong cell wall, they will not burst.

ANSWERS TO EXTENSION QUESTIONS

1. An animal cell will swell and maybe burst when water moves into it. A plant cell has a cell wall that prevents the cell from bursting.

2. The two factors that affect water potential are solute potential and pressure potential.

3. Water will move out of the plant cell because water always moves from an area of higher water potential to an area of lower water potential.

FURTHER EXTENSIONS

Some other possible extensions include:

1. Design and perform an experiment to determine how table salt (NaCl) affects the rate of osmosis.

2. When Hannibal conquered Carthage, his soldiers salted the fields. Using your data, what effect do you think this had on plants? Explain.

3. When one makes beef jerky, strips of fresh beef are covered with salt for many hours. What effect would this have in the process?

Experiment

23A

Effect of Temperature on Cold-Blooded Organisms

In cold-blooded organisms, *poikilotherms*, there is a link between the temperature of the environment and the organism's metabolic rate. Reptiles are a common example of a cold-blooded organism with which most people are familiar. If you have ever seen a lizard or snake in the early morning when the air and ground are cool, you may have noticed how slowly they move. They move slow when the environment is cold because they require heat from their surroundings to increase their internal temperature and metabolism. Once their internal body temperature has warmed, they can metabolize foods more quickly and produce the energy they need. Oxidative respiration is the process of metabolism where sugars are broken down. Under aerobic conditions, respiration yields chemical energy, carbon dioxide, and water.

$$C_6H_{12}O_6 + 6\ O_2 \rightarrow 6\ CO_2 + 6\ H_2O + \text{energy}$$

glucose oxygen carbon dioxide water

Crickets will be used to study the effect of temperature on the metabolism of cold-blooded organisms. You will determine how temperature affects the respiration rate of crickets by monitoring oxygen gas consumption with an O_2 Gas Sensor.

OBJECTIVES

In this experiment, you will

- Use an O_2 Gas Sensor to measure concentrations of oxygen gas.
- Determine the rate of respiration by crickets at different temperatures.
- Determine the effect of temperature on metabolism of crickets.

Figure 1

MATERIALS

computer
Vernier computer interface
Logger*Pro*
Vernier O_2 Gas Sensor
10 adult crickets
balance
250 mL respiration chamber
600 mL beaker
1 L beaker
thermometer
ice
basting bulb

PROCEDURE

1. Connect the O_2 Gas Sensor to the computer interface.

2. Prepare the computer for data collection by opening the file "23A Temp Cold-Blood (O2)" from the *Biology with Computers* folder of Logger*Pro*.

3. Obtain and weigh ten adult crickets in a 600 mL beaker and record the mass at the bottom of Table 1.

4. Data will be collected at three different temperatures according to your assigned group number (I, II, or III). You will set up a water bath at a different temperature prior to each data collection run until you have collected data at all three assigned temperatures.

Group I: Cold Temperatures

a. Your group will collect respiration data at 5 - 10°C, 10 - 15°C, and 15 - 20°C. Set up a water bath for the desired temperature. A water bath is simply a large beaker of water at a certain temperature. This ensures that the crickets will remain at a constant and controlled temperature. To prepare the water bath, obtain some cool water and some ice from your teacher. Combine the cool water and ice into the 1 liter beaker until it reaches the desired temperature range. The beaker should be filled with about 600 – 700 mL water. Leave the thermometer in the water bath during the experiment. It may be necessary for one group member to hold the respiration chamber down in the water bath during the course of the experiment.
b. Place the 250 mL respiration chamber in the water bath. Be sure to keep the temperature of the water bath constant while you are collecting data. If you need to add more hot or cold water, first remove about as much water as you will be adding or the beaker may overflow. Use a basting bulb or Beral pipet to remove excess water.
c. Record the water bath temperature in Table 1. Perform Steps 5 – 12 for each of the three temperature ranges.

Group II: Warm Temperatures

a. Your group will collect respiration data at 20 - 25°C, 25 - 30°C, and 30 - 35°C. Set up a water bath for the desired temperature. A water bath is simply a large beaker of water at a certain temperature. This ensures that the crickets will remain at a constant and controlled temperature. To prepare the water bath, obtain some hot and cold water from your teacher. Combine the hot and cold water into the 1 liter beaker until it reaches the desired temperature range. The beaker should be filled with about 600 – 700 mL water. Leave the thermometer in the water bath during the experiment. One group member can hold the respiration chamber down in the water bath during the course of the experiment.
b. Place the 250 mL respiration chamber in the water bath. Be sure to keep the temperature of the water bath constant while you are collecting data. If you need to add more hot or

cold water, first remove about as much water as you will be adding or the beaker may overflow. Use a basting bulb or Beral pipet to remove excess water.

c. Record the water bath temperature in Table 1. Perform Steps 5 – 12 for each of the three temperature ranges.

Group III: Hot Temperatures

a. Your group will collect respiration data at 35 - 40°C, 40 - 45°C, and 45 - 50°C. Set up a water bath for the desired temperature. To prepare the water bath, obtain some hot and cold water from your teacher. Combine the hot and cold water into the 1-liter beaker until it reaches the desired temperature range. The beaker should be filled with about 600 – 700 mL water. Leave the thermometer in the water bath during the experiment. It may be necessary for one group member to hold the respiration chamber down in the water bath during the course of the experiment.

b. Place the 250 mL respiration chamber in the water bath. Be sure to keep the temperature of the water bath constant while you are collecting data. If you need to add more hot or cold water, first remove about as much water as you will be adding or the beaker may overflow. Use a basting bulb or Beral pipet to remove excess water.

c. Record the water bath temperature in Table 1. Perform Steps 5 – 12 for each of the three temperature ranges.

5. Place the crickets into the respiration chamber.

6. Place the shaft of the O_2 Gas Sensor in the opening of the respiration chamber.

7. Wait one minute, then begin measuring oxygen concentration by clicking Collect. Data will be collected for 10 minutes.

8. Remove the O_2 Gas Sensor from the respiration chamber. Place the crickets in a 600 mL beaker.

9. Use a notebook or notepad to fan air across the openings in the probe shaft of the O_2 Gas Sensor for 1 minute.

10. Fill the respiration chamber with water and then empty it. Thoroughly dry the inside of the respiration chamber with a paper towel.

11. Determine the rate of respiration:

    a. Move the mouse pointer to the point where the data values begin to decrease. Hold down the left mouse button. Drag the mouse pointer to the end of the data and release the mouse button.

    b. Click the Linear Fit button, , to perform a linear regression. A floating box will appear with the formula for a best fit line.

    c. Record the absolute value of the slope of the line, *m*, in the slope column of Table 1.

    d. Close the linear regression floating box.

12. Move your data to a stored run by choosing Store Latest Run from the Experiment menu.

13. Repeat Steps 4 – 12 for the other assigned temperatures.

14. To print a graph of oxygen concentration *vs.* time showing all data runs:

a. Label each curve by choosing Text Annotation from the Insert menu, and entering the temperature tested in the edit box. Repeat for each temperature tested. Then drag each box to a position near its respective curve. Adjust the alignment of the arrow heads.

b. Print a copy of the graph, with all data sets displayed. Enter your name(s) and the number of copies of the graph you want.

DATA

Table 1			
Temperature (°C)	Actual Temperature (°C)	Slope (%/min)	Respiration Rate (%/min/g)
5 – 10°C			
10 – 15°C			
15 – 20°C			
20 – 25°C			
25 – 30°C			
30 – 35°C			
35 – 40°C			
40 – 45°C			
45 – 50°C			

Mass of crickets	_______ g

Table 2	
Temperature (°C)	Respiration Rate (%/min/g)
5 – 10°C	
10 – 15°C	
15 – 20°C	
20 – 25°C	
25 – 30°C	
30 – 35°C	
35 – 40°C	
40 – 45°C	
45 – 50°C	

PROCESSING THE DATA

1. For each temperature you tested, divide the slope of the regression line by the mass of the crickets. Record this value as the rate of respiration in Table 1.

2. Record the temperatures your group tested along with the respiration rates on the classroom board. When all other groups have posted their results, calculate the average for each temperature range. Record the average rate values in Table 2.

3. On Page 2 of the experiment file, create a graph of respiration rate *vs.* temperature using the data recorded in Table 2. The respiration rate values should be plotted on the y-axis, and the temperature on the x-axis.

QUESTIONS

1. At what temperature was the rate of oxygen consumption highest? How does this relate to the internal body temperature of warm-blooded organisms?

2. How does temperature affect the rate of respiration in crickets? How does this compare to your prediction?

3. What errors might affect the results of this experiment? How could you help reduce those errors?

4. Predict the rate of respiration for crickets at 60°C. Explain.

EXTENSIONS

1. Perform the same experiment with different species of insects.

2. Investigate the difference in metabolic rate of an insect at different life stages.

3. Compare cellular respiration between insects and plants.

TEACHER INFORMATION

Effect of Temperature on Cold-Blooded Organisms

1. Assign one of the following sets of temperatures to each lab group:

Group I	Group II	Group III
5 – 10°C	20 – 25°C	35 – 40°C
10 – 15°C	25 – 30°C	40 – 45°C
15 – 20°C	30 – 35°C	45 – 50°C

   These are the temperatures that groups will be responsible for testing. Try to test each temperature an equal number of times. It is important to be in the temperature range, but not critical to be at a specific temperature.

2. At temperatures 5 - 10°C and 45 - 50°C, instruct students to closely monitor their crickets. If the crickets appear to be in danger of dying, remove them from the water bath. It is best if the students have their water baths closer to 10°C when in the range of 5 - 10°C and closer to 45°C in the 45 - 50°C range.
3. The O_2 Gas Sensor must be kept dry at all times. Students should be very careful when using the sensor around water baths.
4. To extend the life of the O_2 Gas Sensor, always store the sensor upright in the box it was shipped in.
5. The calibration stored in this experiment file works well for this experiment. The calibration is for the O_2 Gas Sensor (%).
6. The morning of the experiment fill a 1 L beaker with ice and water so students will have cold water for Step 4 of the procedure.

SAMPLE RESULTS

Table 2

Temperature (°C)	Respiration rate (%/min/g)	Temperature (°C)	Respiration rate (%/min/g)
5 – 10°C	0.007	30 – 35°C	0.051
10 – 15°C	0.013	35 – 40°C	0.062
15 – 20°C	0.016	40 – 45°C	0.059
20 – 25°C	0.019	45 – 50°C	
25 – 30°C	0.034		

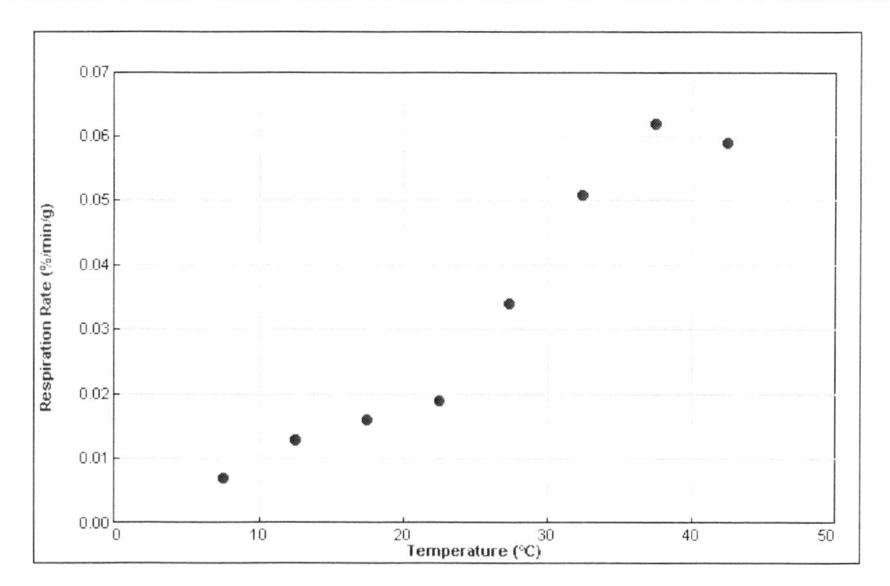

Respiration rates for eight different temperatures

ANSWERS TO QUESTIONS

1. The highest levels of oxygen consumption will occur at temperatures between 30°C and 40°C. This corresponds with the internal temperatures of many warm-blooded organisms. This temperature range is optimal for many enzyme-catalyzed reactions.

2. As temperature increases, the rate of respiration also increases. At temperatures above 45°C the respiration rate begins to decline. This is primarily due to death of many of the organisms.

3. The rate of respiration for all crickets is assumed to be the same, but this may not be so. Since different students used different crickets for each measurement, the rates of respiration may not be comparable.

4. The rate of respiration for crickets at 60°C should be very low. Crickets cannot survive at these high temperatures.

Experiment

23B

Effect of Temperature on Cold-Blooded Organisms

In cold-blooded organisms, *poikilotherms*, there is a link between the temperature of the environment and the organism's metabolic rate. Reptiles are a common example of a cold-blooded organism with which most people are familiar. If you have ever seen a lizard or snake in the early morning when the air and ground are cool, you may have noticed how slowly they move. They move slow when the environment is cold because they require heat from their surroundings to increase their internal temperature and metabolism. Once their internal body temperature has warmed, they can metabolize foods more quickly and produce the energy they need. Oxidative respiration is the process of metabolism where sugars are broken down. Under aerobic conditions, respiration yields chemical energy, carbon dioxide, and water.

$$\underset{\text{glucose}}{C_6H_{12}O_6} + \underset{\text{oxygen}}{6\ O_2} \rightarrow \underset{\text{carbon dioxide}}{6\ CO_2} + \underset{\text{water}}{6\ H_2O} + \text{energy}$$

Crickets will be used to study the effect of temperature on the metabolism of cold-blooded organisms. You will determine how temperature affects the respiration rate of crickets by monitoring carbon dioxide production with a CO_2 Gas Sensor.

OBJECTIVES

In this experiment, you will

- Use a CO_2 Gas Sensor to measure concentrations of carbon dioxide.
- Determine the rate of respiration by crickets at different temperatures.
- Determine the effect of temperature on metabolism of crickets.

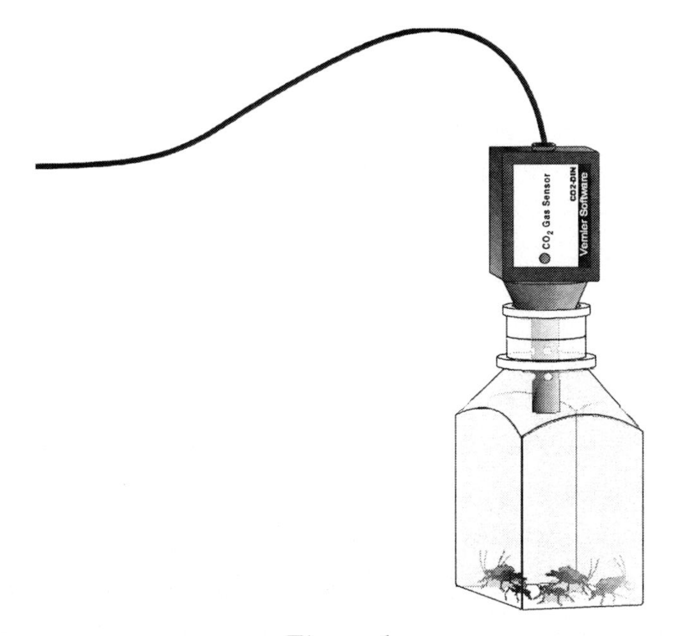

Figure 1

MATERIALS

computer
Vernier computer interface
Logger*Pro*
Vernier CO_2 Gas Sensor
10 adult crickets
balance
250 mL respiration chamber
600 mL beaker
1 L beaker
thermometer
ice
basting bulb

PROCEDURE

1. Connect the CO_2 Gas Sensor to the Vernier computer interface.
2. Prepare the computer for data collection by opening the file "23B Temp Cold-Blood (CO2)" from the *Biology with Computers* folder of Logger*Pro*.
3. Obtain and weigh ten adult crickets in a 600 mL beaker and record the mass at the bottom of Table 1.
4. Data will be collected at three different temperatures according to your assigned group number (I, II, or III). You will set up a water bath at a different temperature prior to each data collection run until you have collected data at all three assigned temperatures.

Group I: Cold Temperatures

a. Your group will collect respiration data at 5 - 10°C, 10 - 15°C, and 15 - 20°C. Set up a water bath for the desired temperature. A water bath is simply a large beaker of water at a certain temperature. This ensures that the crickets will remain at a constant and controlled temperature. To prepare the water bath, obtain some cool water and some ice from your teacher. Combine the cool water and ice into the 1 liter beaker until it reaches the desired temperature range. The beaker should be filled with about 600 – 700 mL water. Leave the thermometer in the water bath during the experiment. It may be necessary for one group member to hold the respiration chamber down in the water bath during the course of the experiment.
b. Place the 250 mL respiration chamber in the water bath. Be sure to keep the temperature of the water bath constant while you are collecting data. If you need to add more hot or cold water, first remove about as much water as you will be adding or the beaker may overflow. Use a basting bulb or Beral pipet to remove excess water.
c. Record the water bath temperature in Table 1. Perform Steps 5 – 12 for each of the three temperature ranges.

Group II: Warm Temperatures

a. Your group will collect respiration data at 20 - 25°C, 25 - 30°C, and 30 - 35°C. Set up a water bath for the desired temperature. A water bath is simply a large beaker of water at a certain temperature. This ensures that the crickets will remain at a constant and controlled temperature. To prepare the water bath, obtain some hot and cold water from your teacher. Combine the hot and cold water into the 1 liter beaker until it reaches the desired temperature range. The beaker should be filled with about 600 – 700 mL water. Leave the thermometer in the water bath during the experiment. It may be necessary for one group member to hold the respiration chamber down in the water bath during the course of the experiment.
b. Place the 250 mL respiration chamber in the water bath. Be sure to keep the temperature of the water bath constant while you are collecting data. If you need to add more hot or

cold water, first remove about as much water as you will be adding or the beaker may overflow. Use a basting bulb or Beral pipet to remove excess water.

c. Record the water bath temperature in Table 1. Perform Steps 5 – 12 for each of the three temperature ranges.

Group III: Hot Temperatures

a. Your group will collect respiration data at 35 - 40°C, 40 - 45°C, and 45 - 50°C. Set up a water bath for the desired temperature. To prepare the water bath, obtain some hot and cold water from your teacher. Combine the hot and cold water into the 1-liter beaker until it reaches the desired temperature range. The beaker should be filled with about 600 – 700 mL of water. Leave the thermometer in the water bath during the experiment. It may be necessary for one group member to hold the respiration chamber down in the water bath during the course of the experiment.

b. Place the 250 mL respiration chamber in the water bath. Be sure to keep the temperature of the water bath constant while you are collecting data. If you need to add more hot or cold water, first remove about as much water as you will be adding or the beaker may overflow. Use a basting bulb or Beral pipet to remove excess water.

c. Record the water bath temperature in Table 1. Perform Steps 5 – 12 for each of the three temperature ranges.

5. Place the crickets into the respiration chamber.

6. Place the shaft of the CO_2 Gas Sensor in the opening of the respiration chamber. Gently twist the stopper on the shaft of the CO_2 Gas Sensor into the chamber opening. Do not twist the shaft of the sensor or you may damage it.

7. Wait one minute, then begin measuring carbon dioxide concentration by clicking Collect. Data will be collected for 5 minutes.

8. Remove the CO_2 Gas Sensor from the respiration chamber. Place the crickets in a 600 mL beaker.

9. Use a notebook or notepad to fan air across the openings in the probe shaft of the CO_2 Gas Sensor for 1 minute.

10. Fill the respiration chamber with water and then empty it. Thoroughly dry the inside of the respiration chamber with a paper towel.

11. Determine the rate of respiration:

    a. Move the mouse pointer to the point where the data values begin to increase. Hold down the left mouse button. Drag the mouse pointer to the end of the data and release the mouse button.

    b. Click the Linear Fit button, , to perform a linear regression. A floating box will appear with the formula for a best fit line.

    c. Record the slope of the line, *m*, in the slope column of Table 1.

    d. Close the linear regression floating box.

12. Move your data to a stored run by choosing Store Latest Run from the Experiment menu.

13. Repeat Steps 4 – 12 for the other assigned temperatures.

14. To print a graph of carbon dioxide *vs.* time showing all data runs:

    a. Label each curve by choosing Text Annotation from the Insert menu, and entering the temperature tested in the edit box. Repeat for each temperature tested. Then drag each box to a position near its respective curve. Adjust the position of the arrow heads.

b. Print a copy of the graph, with all data sets displayed. Enter your name(s) and the number of copies of the graph you want.

DATA

Table 1			
Temperature (°C)	Actual temperature (°C)	Slope (ppm/min)	Respiration rate (ppm/min/g)
5 – 10°C			
10 – 15°C			
15 – 20°C			
20 – 25°C			
25 – 30°C			
30 – 35°C			
35 – 40°C			
40 – 45°C			
45 – 50°C			

Mass of crickets	_______ g

Table 2	
Temperature (°C)	Respiration rate (ppm/min/g)
5 – 10°C	
10 – 15°C	
15 – 20°C	
20 – 25°C	
25 – 30°C	
30 – 35°C	
35 – 40°C	
40 – 45°C	
45 – 50°C	

PROCESSING THE DATA

1. For each temperature you tested, divide the slope of the regression line by the mass of the crickets. Record this value as the rate of respiration in Table 1.

2. Record the temperatures your group tested along with the respiration rates on the classroom board. When all other groups have posted their results, calculate the average for each temperature range. Record the average rate values in Table 2.

3. On Page 2 of the experiment file, create a graph of respiration rate *vs.* temperature using the data recorded in Table 2. The respiration rate values should be plotted on the y-axis, and the temperature on the x-axis.

QUESTIONS

1. At what temperature was the rate of carbon dioxide production highest? How does this relate to the internal body temperature of warm blooded organisms?

2. How does temperature affect the rate of respiration in crickets? How does this compare to your prediction?

3. What errors might affect the results of this experiment? How could you help reduce those errors?

4. Predict the rate of respiration for crickets at 60°C. Explain.

EXTENSIONS

1. Perform the same experiment with different species of insects.

2. Investigate the difference in metabolic rate of an insect at different life stages.

3. Compare cellular respiration between insects and plants.

Effect of Temperature on Cold-Blooded Organisms

1. Assign one of the following sets of temperatures to each lab group:

Group I	Group II	Group III
5 – 10°C	20 – 25°C	35 – 40°C
10 – 15°C	25 – 30°C	40 – 45°C
15 – 20°C	30 – 35°C	45 – 50°C

   These are the temperatures that they will be responsible for testing. Try to get each temperature tested an equal number of times. It is important to be in the temperature range, but not critical to be at a specific temperature.

2. At temperatures 5 - 10°C and 45 - 50°C instruct students to closely monitor their crickets. If the crickets appear to be in danger of dying, remove them from the water bath. It is best if the students have their water baths closer to 10°C when in the range of 5 - 10°C and closer to 45°C in the 45 - 50°C range.

3. The CO_2 Gas Sensor must be kept dry at all times. Students should be very careful when using the sensor around water baths.

4. The stopper included with the CO_2 Gas Sensor is slit to allow it to be easily added or removed from the probe. When students are placing the probe in the respiration chamber, they should gently twist the stopper into the chamber opening. Warn the students not to twist the probe shaft or they may damage the sensing unit.

5. The CO_2 Gas Sensor is dependent on the diffusion of gases into the probe shaft. Students should allow a couple of minutes between trials so that gases from the previous trial will have diffused out of the probe shaft. Alternatively, the students can use a firm object such as a book or notepad to fan air through the probe shaft. This method is used in Step 9 of the student procedure.

6. The morning of the experiment fill a 1 L beaker with ice and water so students will have cold water for Step 4 of the procedure.

7. The stored calibration works well for this experiment. This calibration is for the CO_2 Gas Sensor (ppm). The CO_2 Gas Sensor was calibrated at the time of manufacturing. You only need to recalibrate the sensor if it appears to be giving unusual readings. Refer to the CO_2 Gas Sensor probe booklet if you need to recalibrate the sensor.

SAMPLE RESULTS

Table 2: Class Averages			
Temperature (°C)	Respiration Rate (ppm/min/g)	Temperature (°C)	Respiration Rate (ppm/min/g)
5 – 10°C	16.31	30 – 35°C	214.57
10 – 15°C	44.78	35 – 40°C	292.94
15 – 20°C	83.97	40 – 45°C	
20 – 25°C	97.86	45 – 50°C	
25 – 30°C	139.12		

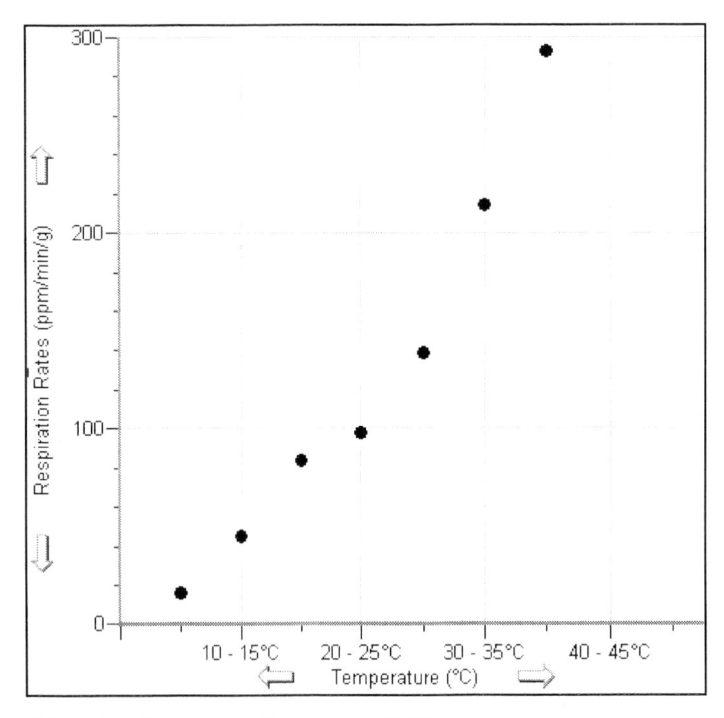

Respiration rates for seven different temperatures

ANSWERS TO QUESTIONS

1. The highest levels of oxygen consumption will occur at temperatures between 30°C and 40°C. This corresponds with the internal temperatures of many warm blooded organisms. This temperature range is optimal for many enzyme-catalyzed reactions.

2. As temperature increases, the rate of respiration also increases. At temperatures above 45°C the respiration rate begins to decline. This is primarily due to death of many of the specimens.

3. The rate of respiration for all crickets is assumed to be the same, but this may not be so. Since different students used different crickets for each measurement, the rates of respiration may not be comparable.

4. The rate of respiration for crickets at 60°C should be very low. Crickets cannot survive at these high temperatures.

Experiment

24A

Lactase Action

Intestinal gas occurs normally in humans. It is composed primarily of carbon dioxide, nitrogen and oxygen gases. Gas is produced when indigestible sugars are metabolized by certain bacteria in the intestines. Nearly one out of four individuals lack the genetic ability to digest these sugars.

Lactase is an enzyme responsible for breaking bonds in lactose, a disaccharide sugar. When lactose cannot be metabolized within the cells of a person, bacteria digest lactose in the intestines. This bacterial digestion produces carbon dioxide gas as a by-product. When this gas builds up in the intestines, flatulence is the result. Recently, researchers have discovered a way to mass produce the enzyme lactase. Lactase converts lactose into glucose and galactose, both easily digestible monosaccharides. Using biotechnological processes, they have formulated the product lactase as a way to lower the production and expulsion of rectal gas. The enzyme lactase is used to break the bond in lactose before it is ingested.

In this lab, you will test the function of lactase. One way to test its activity is to determine if the enzymes are capable of converting the disaccharide into glucose and galactose. It is easy to test for glucose. You can use a special test paper originally made for diabetics to help detect blood sugar in urine. If glucose is present in a liquid, the test paper turns a different color.

An alternative way to test the activity of lactase is to determine whether the sugar can support life. Presumably, yeast are unable digest lactose. They can, however, digest glucose and use it as an energy source. Yeast can burn sugar aerobically during respiration, according to the equation:

$$\underset{\text{glucose}}{C_6H_{12}O_6} + \underset{\text{oxygen}}{6\ O_2} \rightarrow \underset{\text{carbon dioxide}}{6\ CO_2} + \underset{\text{water}}{6\ H_2O} + \text{energy}$$

Carbon dioxide is released when yeast metabolize sugar. By monitoring the production of CO_2, we can use yeast to indicate the activity of lactase.

OBJECTIVES

In this experiment, you will

- Test the action of lactase.
- Use glucose test paper to monitor the presence of glucose.
- Determine if yeast can metabolize glucose, lactose, or galactose.

MATERIALS

Power Macintosh or Windows PC
Vernier computer interface
Logger*Pro*
Vernier CO_2 Gas Sensor
Lactase (droplet form)
5% galactose solution
5% glucose solution
5% lactose solution
10 mL graduated cylinder
600 mL beaker (for water bath)
graduated Beral pipet
forceps
five 18 × 150 mm test tubes
hot and cold water
stopwatch
Tes-Tape or other glucose test paper
test-tube rack
thermometer
yeast suspension

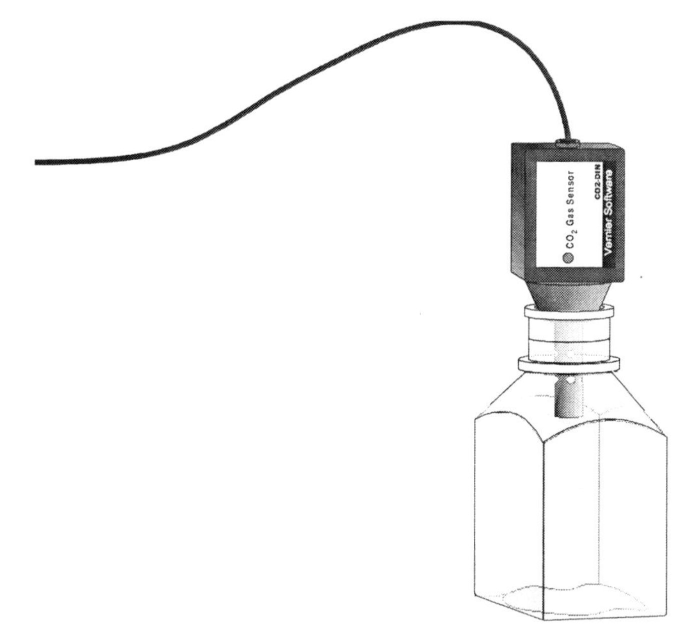

Figure 1

PROCEDURE

Testing for the Production of Glucose

1. You will determine if lactase can produce glucose from the conversion of lactose.

2. Obtain two test tubes and label them 1 and 2.

Table 1		
Test tube	Sugar solution (2.5 mL)	Lactase
1	lactose	2 drops
2	lactose	none

3. Obtain the lactose sugar solution. Add 2.5 mL of the sugar solution to both test tubes, as shown in Table 1.

4. Prepare a water bath for the sugar solutions. A water bath is simply a large beaker of water at a constant and controlled temperature. To prepare the water bath, obtain some warm and cool water from your teacher. Combine the warm and cool water into the 600 mL beaker until it reaches 35°C. The beaker should be filled with about 400 mL of water.

   If you need to add more hot or cold water to maintain a constant temperature, first remove about as much water as you will be adding, or the beaker may overflow. Use a basting bulb or a Beral pipet to remove excess water.

5. Measure the glucose concentration. To do this:
   a. If the test paper is supplied in a continuous strip, tear off a small piece (0.5 cm) of glucose test paper. Otherwise, obtain one test strip.
   b. Using a dropper pipet, withdraw a drop or two of sugar solution from test tube 1.

c. Place one drop of sugar solution onto the glucose test paper.
d. Follow the instructions on the glucose test paper package to develop the test paper. This usually requires a 30 or 60 second wait before you compare the color of the tape to the supplied color chart.
e. Record the approximate concentration of glucose in Table 3 in the *Time 0* column.
f. Discard any sugar solution remaining in the dropper. Rinse the dropper by taking up clean water and expelling it into a waste beaker.

6. Repeat Step 5 for the second test tube.
7. Add lactase to test tube 1:

    a. Place 2 drops of lactase into test tube 1.
    b. Gently mix the contents of the tube.

8. Set both tubes in the water bath. Start the stopwatch. Be sure to keep the temperature of the water bath close to 35°C.
9. Incubate the test tubes for 10 minutes, taking a glucose test once a minute for 10 minutes. Repeat Step 5 and record the concentrations of glucose in Table 3 once every minute.

Testing for the Ability of Yeast to Metabolize Sugars

10. Plug the Vernier CO_2 Gas Sensor into the Vernier computer interface.
11. Prepare the computer for data collection by opening the file "24A Lactase Act (CO2)" from the *Biology with Computers* folder of Logger*Pro*.
12. Check the water bath so it remains at a constant temperature range between 33°C and 37°C.
13. You will perform five tests outlined in Table 2. Obtain three additional test tubes and label them 3 through 5.

Table 2

Tube	Sugar	Volume (mL)	Enzyme
1	Lactose	2.5	Lactase
2	Lactose	2.5	none
3	Glucose	2.5	none
4	Galactose	2.5	none
5	None (water only)	2.5	none

14. Prepare the test tubes:

    a. You have already prepared the solutions for test tubes 1 – 2. Use the same test tubes as in Table 1.
    b. Place 2.5 mL of glucose solution into test tube 3.
    c. Place 2.5 mL of galactose solution into test tube 4.
    d. Place 2.5 mL of water into test tube 5, as in Table 2.

15. Set the test tubes in the water bath.

16. Obtain the yeast suspension. Gently swirl the yeast suspension to mix the yeast that settles to the bottom. Put 2.5 mL of yeast into each test tube and mix the solutions.

17. Incubate the test tubes for 10 minutes in the water bath. Be sure to keep the temperature of the water bath constant.

18. When incubation is finished, use a Beral pipet to place 1 mL of the sugar/yeast solution of test tube 1 into the 250 mL respiration chamber.

19. Quickly place the shaft of the CO_2 Gas Sensor in the opening of the respiration chamber. Gently twist the stopper on the shaft of the CO_2 Gas Sensor into the chamber opening. Do not twist the shaft of the CO_2 Gas Sensor or you may damage it.

20. Begin collecting data by clicking Collect.

21. Data collection will end after 4 minutes. Remove the CO_2 Gas Sensor from the respiration chamber.

22. Determine the rate of respiration:

    a. Move the mouse pointer to the point where the data values begin to increase. Hold down the left mouse button. Drag the pointer to the end of the data and release the mouse button.
    b. Click the Linear Fit button, , to perform a linear regression. A floating box will appear with the formula for a best fit line.
    c. Record the slope of the line, *m*, as the rate of respiration in Table 4.
    d. Close the linear regression floating box.

23. Move your data to a stored run. To do this, choose Store Latest Run from the Experiment menu.

24. Fill the respiration chamber with water and then empty it. Make sure that all yeast have been removed from the respiration chamber. Thoroughly dry the inside of the respiration chamber with a paper towel.

25. Use a notebook or notepad to fan air across the openings in the probe shaft of the CO_2 Gas Sensor for 1 minute.

26. Repeat Steps 18 – 25 for each of the sugar solutions.

27. On Page 2 of the experiment file, make a graph of respiration rate *vs.* sugar/enzyme combination.

DATA

Table 3: Glucose Concentrations		
Time (minutes)	Lactose + Lactase	Lactose only
0		
1		
2		
3		
4		
5		
6		
7		
8		
9		
10		

Table 4		
Test	Type of Sugar / Enzyme	Respiration rate (ppm/min)
1	Lactose + Lactase	
2	Lactose only	
3	Glucose	
4	Galactose	
5	None (water only)	

QUESTIONS

1. From the results of this experiment, how does lactase function? What is your evidence?
2. Considering the results of this experiment, can yeast utilize all of the sugars equally well? Explain.
3. Hypothesize why some sugars are not utilized by yeast while other sugars are metabolized.
4. How did the results of testing lactase's activity using glucose test paper compare with the results of using yeast as an indicator of activity? What is your evidence?
5. Which test tube served as a control in this experiment? What did you conclude from the control? How did this affect the interpretation of data in this experiment?

EXTENSIONS

1. Design an experiment to test the activity of Beano on the sugar melibiose. Does Beano have any effect on the sugar lactose? From the results of this experiment, how does Beano function? What is your evidence?

2. Design an experiment to test whether Beano has any effect on the sugar lactose.

Experiment
24A

TEACHER INFORMATION

Lactase Action

1. Lactase comes in two forms, caplet and droplet. For use in this experiment, it is necessary to purchase the droplet form rather than the caplet form. Lactaid no longer manufactures the enzyme in droplet form. Lacteeze, in droplet form, can be ordered from digestmilk.com or fitmart.com.

2. In this experiment, students will test the activity of lactase in two ways:
   - The appearance of glucose using Tes-Tape or other glucose test paper.
   - The ability of yeast to respire aerobically using the glucose that is enzymatically released. Yeast will not utilize lactose, or galactose. They will, however, utilize glucose. Lactose releases glucose into solution when the disaccharides are converted into monosaccharides enzymatically.

3. The stored calibration works well for this experiment. This calibration is for the CO_2 Gas Sensor (ppm). The CO_2 Gas Sensor was calibrated at the time of manufacturing. You only need to recalibrate the sensor if it appears to be giving unusual readings. Refer to the CO_2 Gas Sensor sensor booklet to calibrate the sensor.

4. Glucose test paper, used by diabetics, may be obtained at most drug stores. The resolution of the glucose test paper is not adequate for students to use for enzyme kinetic studies. Since the wet dye on the glucose test paper may stain desks, supply students with a container to discard used strips.

5. Use a glass Pasteur or a plastic Beral pipette for the dropper pipette. It should be long enough to easily withdraw a few drops of sugar solution from the test tube.

6. To prepare the necessary sugar solutions, add 5 g of sugar per 100 mL of solution.

7. To prepare the yeast solution, dissolve 7 grams (1 package) of dried yeast per 100 mL of water. Incubate the suspension in 37 – 40°C water for at least 10 minutes.

8. After the 10 minute incubation period, transfer the yeast to dispensing tubes. Each group will need about 3 mL of yeast.

9. The CO_2 Gas Sensor is dependent on the diffusion of gases into the probe shaft. Students should allow a couple of minutes between trials so that gases from the previous trial will have diffused out of the probe shaft. Alternatively, the students can use a firm object such as a book or notepad to fan air through the probe shaft. This method is used in Step 25 of the student procedure.

10. The stopper included with the CO_2 Gas Sensor is slit to allow it to be easily added or removed from the probe. When students are placing the probe in the respiration chamber, they should gently twist the stopper into the chamber opening. Warn the students not to twist the probe shaft or they may damage the sensing unit.

SAMPLE RESULTS

The following data may be different from students' results. The actual values depend upon the viability and concentration of the yeast, among other factors.

Table 4		
Test	Type of Sugar / Enzyme	Rate of Respiration (ppm/min)
1	Lactose + Lactase	500.99
2	Lactose	88.32
3	Glucose	538.56
4	Galactose	92.76
5	None (water only)	95.68

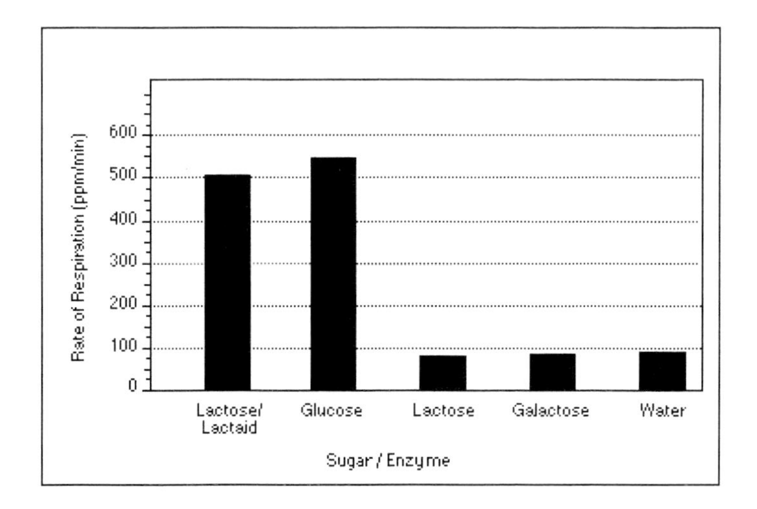

ANSWERS TO QUESTIONS

1. Lactase acts by converting lactose into glucose and, presumably, galactose. Yeast are not able to metabolize lactose, but can metabolize glucose. Yeast are able to use the lactose solution only *after* it has been acted upon by lactase. Since lactose is a disaccharide that can hydrolyze into glucose and galactose, lactase must have caused that hydrolysis.

2. Yeast cannot utilize all of the sugars equally well. Glucose was metabolized much more efficiently than lactose or galactose, as shown by their rates of respiration..

3. Yeast may not have the proper enzymes to transport certain sugars across its cell membrane, or it may not have the enzyme needed to convert sugar from disaccharides to monosaccharides.

4. The results of the two experiments should agree. The glucose test tape experiment indicated the presence of glucose with lactase + lactose. The yeast were able to respire with the products of lactase + lactose. Since glucose was the only sugar utilized by yeast in this experiment, one of the products of enzymatic activity presumably was glucose.

5. Test 5 served as the control. It contained only water and yeast in order to verify that the yeast need sugar to respire. Without the control, a person could conclude that the yeast can respire without any sugar.

Lactase Action

Intestinal gas occurs normally in humans. It is composed primarily of carbon dioxide, nitrogen and oxygen gases. Gas is produced when indigestible sugars are metabolized by certain bacteria in the intestines. Nearly one out of four individuals lack the genetic ability to digest these sugars.

Lactase is an enzyme responsible for breaking bonds in lactose, a disaccharide sugar. When lactose cannot be metabolized within the cells of a person, bacteria digest lactose in the intestines. This bacterial digestion produces carbon dioxide gas as a by-product. When this gas builds up in the intestines, flatulence is the result. Recently, researchers have discovered a way to mass produce the enzyme lactase. Lactase converts lactose into glucose and galactose, both easily digestible monosaccharides. Using biotechnological processes, they have formulated the product *Lactaid* as a way to lower the production and expulsion of rectal gas. The enzyme lactase is used to break the bond in lactose *before* it is ingested.

In this lab, you will test the function of lactase. One way to test its activity is to determine if the enzymes are capable of converting the disaccharide into glucose and galactose. It is easy to test for glucose. You can use a special test paper originally made for diabetics to help detect blood sugar in urine. If glucose is present in a liquid, the test paper turns a different color.

An alternative way to test the activity of lactase is to determine whether the sugar can support life. Presumably, yeast are unable digest lactose. They can, however, digest glucose and use it as an energy source. Yeast can burn sugar anaerobically during fermentation, according to the equation:

$$C_6H_{12}O_6 \rightarrow 2\ CH_3CH_2OH + 2\ CO_2 + \text{energy}$$

Yeast release CO_2 when they metabolize sugar. By monitoring a pressure change caused by the production of CO_2, we can use yeast to indicate the activity of lactase.

OBJECTIVES

In this experiment, you will

- Test the action of lactase.
- Use glucose test paper to monitor the presence of glucose.
- Determine if yeast can metabolize glucose, lactose, or galactose.

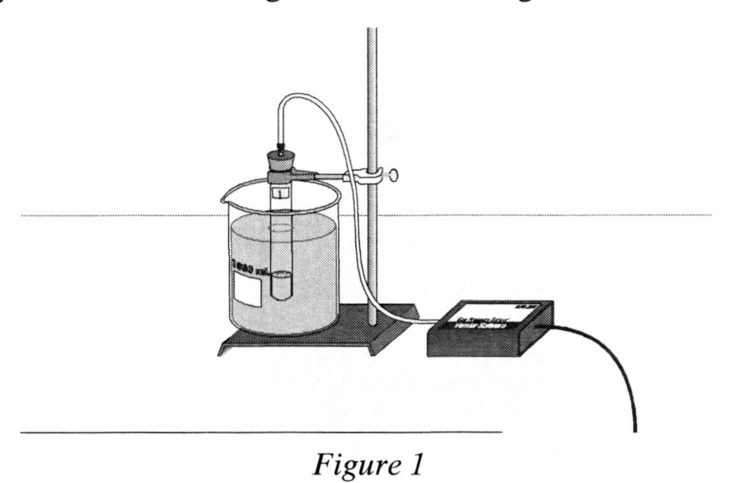

Figure 1

MATERIALS

computer
Vernier computer interface
Logger*Pro*
Vernier Gas Pressure Sensor
1-hole rubber stopper assembly
Lactase (droplet form)
5% galactose solution
5% glucose solution
5% lactose solution
10 mL graduated cylinder
600 mL beaker (for water bath)
dropper or Beral pipet
forceps
four 18 × 150 mm test tubes
hot and cold water
stopwatch
Tes-Tape or other glucose test paper
test-tube rack
thermometer
yeast suspension

PROCEDURE

Testing for the Production of Glucose

1. You will determine if lactase can produce glucose from the conversion of lactose.
2. Obtain two test tubes and label them test tube 1 and test tube 2.

Table 1		
Test Tube	Sugar Solution (2.5 mL)	Lactase
1	lactose	2 drops
2	lactose	none

3. Obtain the lactose sugar solution. Add 2.5 mL of the sugar solution to both test tubes, as shown in Table 1.
4. Prepare a water bath for the sugar solutions. A water bath is simply a large beaker of water at a constant and controlled temperature. To prepare the water bath, obtain some warm and cool water from your teacher. Combine the warm and cool water in the 600 mL beaker until it reaches 35°C. The beaker should be filled with about 400 mL of water.

   If you need to add more hot or cold water to maintain a constant temperature, first remove about as much water as you will be adding, or the beaker may overflow. Use a basting bulb or a Beral pipette to remove excess water.
5. Measure the glucose concentration. To do this:
   a. If the test paper is supplied in a continuous strip, tear off a small piece (0.5 cm) of glucose test paper. Otherwise, obtain one test strip.
   b. Using a dropper pipette, withdraw a drop or two of sugar solution from test tube 1.
   c. Place one drop of sugar solution onto the glucose test paper.
   d. Follow the instructions on the glucose test paper package to develop the test paper. This usually requires a 30 or 60 second wait before you compare the color of the tape to the supplied color chart.
   e. Record the approximate concentration of glucose in Table 3 in the "Time 0" column.
   f. Discard any sugar solution remaining in the dropper. Rinse the dropper by taking up clean water and expelling it into a waste beaker.

6. Repeat Step 5 for the second test tube.

7. Add lactase to the test tube 1:

   a. Place 2 drops of lactase into test tube 1.

   b. *Gently* mix the contents of the tube.

8. Set both tubes in the water bath. Start the stopwatch. Be sure to keep the temperature of the water bath close to 35°C.

9. Incubate the test tubes for 10 minutes, taking a glucose test once every minute for 10 minutes. Repeat Step 5 and record the concentrations of glucose in Table 3 once every minute.

Testing for the Ability of Yeast to Ferment Sugars

10. Connect the Gas Pressure Sensor to the computer interface. Prepare the computer for data collection by opening the file "24B Lactase Act (Press)" from the *Biology with Computers* folder of Logger*Pro*.

11. Connect the plastic tubing to the valve on the Gas Pressure Sensor.

12. Check the water bath so it remains at a constant temperature range between 33°C and 37°C.

13. You will perform one of the four tests outlined in Table 2 and obtain the results of the other tests from your classmates. Ask your instructor which test you will be performing and record the test number in Table 4.

Table 2

Tube	Sugar	Volume (mL)	Enzyme
1	Lactose	2.5	Lactase
2	Lactose	2.5	none
3	Glucose	2.5	none
4	Galactose	2.5	none
5	None (water only)	2.5	none

14. Prepare the sugar solution:

   - If you are assigned tests 1 or 2, you have already made the solution. Use the same test tube as in Table 1.
   - If you are assigned tests 3, 4, or 5, obtain the appropriate solution. Place 2.5 mL of the solution into the test tube, as in Table 2.

15. Set the test tube in the water bath.
16. Obtain the yeast suspension. Gently swirl the yeast suspension to mix the yeast that settles to the bottom. Put 2.5 mL of yeast into the test tube and mix the solution.
17. In the test tube, place enough vegetable oil to completely cover the surface of the yeast/sugar mixture as shown in Figure 3. Be careful to not get oil on the inside wall of the test tube. Set the test tube in the water bath.
18. Insert the single-holed rubber-stopper into the test tube. **Note:** *Firmly* twist the stopper for an *airtight* fit. Secure the test tube with a utility clamp and ring-stand as shown in Figure 1.
19. Incubate the test tube for 10 minutes in the water bath. Be sure to keep the temperature of the water bath constant. If you need to add more hot or cold water, first remove about as much water as you will be adding, or the beaker may overflow. Use a basting bulb to remove excess water.

    Note: Be sure that most of the test tube is completely covered by the water in the water bath. The temperature of the air in the tube must be constant for this experiment to work well.

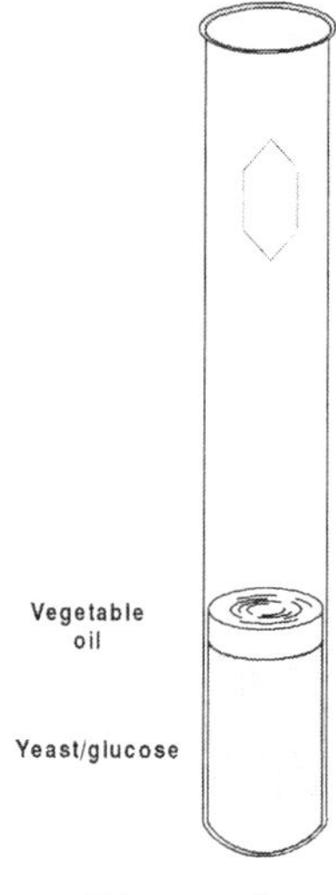

Figure 2

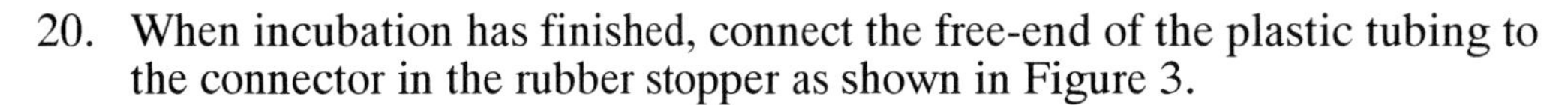

20. When incubation has finished, connect the free-end of the plastic tubing to the connector in the rubber stopper as shown in Figure 3.

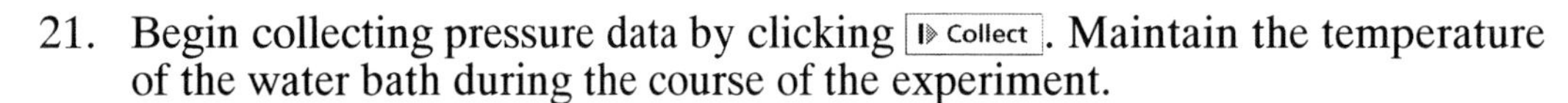

21. Begin collecting pressure data by clicking Collect. Maintain the temperature of the water bath during the course of the experiment.
22. Data collection will end after 15 minutes. Monitor the pressure readings displayed on the computer screen. If the pressure exceeds 130 kilopascals, the pressure inside the tube will be too great and the rubber stopper is likely to pop off. Disconnect the plastic tubing from the Gas Pressure Sensor if the pressure exceeds 130 kilopascals.
23. When data collection has finished, disconnect the plastic tubing connector from the rubber stopper. Remove the rubber stopper from the test tube and discard the contents in a waste beaker.

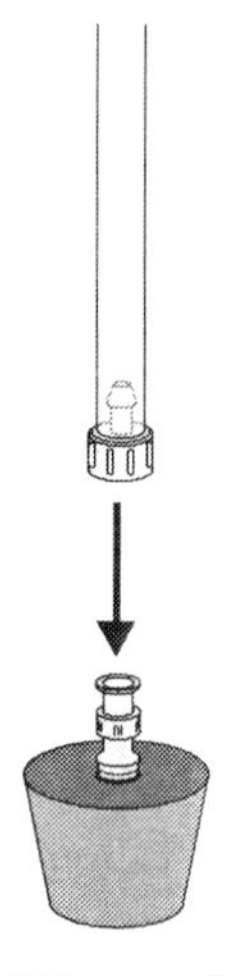

Figure 3

PROCESSING THE DATA

You will now determine the *rate* of fermentation. The rate of fermentation can be measured by examining the *slope* of the pressure *vs.* time plot for each test.

1. Find the rate of fermentation:
   a. Move the mouse pointer to the point where the pressure values begin to increase. Hold down the mouse button. Drag the pointer to the end of the data and release the mouse button.
   b. Click the Linear Fit button, , to perform a linear regression. A floating box will appear with the formula for a best fit line.
   c. Record the slope of the line, *m*, as the rate of fermentation in Table 4.
   d. Close the linear regression floating box.
   e. Share your data with other classmates by recording the test tube number and the rate of fermentation on the board.

2. Print the graph of pressure *vs.* time if directed to do so by your instructor.
3. Average the rate values for each of the four tests performed by the class and record them in Table 5.
4. On Page 2 of the experiment file, make a graph of rate of fermentation *vs.* sugar/enzyme combination.

DATA

Table 3: Glucose Concentrations		
Time (minutes)	Lactose + Lactase	Lactose only
0		
1		
2		
3		
4		
5		
6		
7		
8		
9		
10		

Table 4: Your results	
Test	Rate (kPa/min)

Table 5		
Test	Type of Sugar / Enzyme	Fermentation Rate (kPa/min)
1	Lactose + Lactase	
2	Lactose only	
3	Glucose	
4	Galactose	
5	None (water only)	

QUESTIONS

1. From the results of this experiment, how does lactase function? What is your evidence?
2. Considering the results of this experiment, can yeast utilize all of the sugars equally well? Explain.
3. Hypothesize why some sugars are not utilized by yeast while other sugars are metabolized.
4. How did the results of testing lactase's activity using glucose test paper compare with the results of using yeast as an indicator of activity? What is your evidence?
5. Which test tube served as a control in this experiment? What did you conclude from the control? How did this affect the interpretation of data in this experiment?

EXTENSIONS

1. Design an experiment to test the activity of Beano on the sugar melibiose. Does Beano have any effect on the sugar lactose? From the results of this experiment, how does Beano function? What is your evidence?
2. Design an experiment to test whether Beano has any effect on the sugar lactose.

TEACHER INFORMATION

Experiment
24B

Lactase Action

1. Lactase comes in two forms, caplet and droplet. For use in this experiment, it is necessary to purchase the droplet form rather than the caplet form. Lactaid no longer manufactures the enzyme in droplet form. Lacteeze, in droplet form, can be ordered from digestmilk.com or fitmart.com.

2. In this experiment, students will test the activity of lactase in two ways:
   - The appearance of glucose using Tes-Tape or other glucose test paper.
   - The ability of yeast to respire anaerobically using the glucose that is enzymatically released. Yeast will not utilize lactose, or galactose. They will, however, utilize glucose. Lactose releases glucose into solution when the disaccharides are converted into monosaccharides enzymatically.

3. If time allows, you may consider having students do a second trial. This would increase the chances that they have at least one trial in which there is a significant pressure change.

4. Glucose test paper, used by diabetics, may be obtained at most drug stores. The resolution of the glucose test paper is not adequate for students to use for enzyme kinetic studies. Since the wet dye on the glucose test paper may stain desks, supply students with a container to discard used strips.

5. Use a glass Pasteur or a plastic Beral pipette for the dropper pipette. It should be long enough to easily withdraw a few drops of sugar solution from the test tube.

6. You will need to assign each group one of the five tests detailed in Table 2. If two or more groups perform the same test and record their values on the board, the class should use the average of these values. If there are sufficient measurements, take the average of the data after the highest and lowest values are discarded.

7. Students should record their rate value on the board for other groups to use.

8. To prepare the yeast solution, dissolve 7 g (1 package) of dried yeast per 100 mL of water. Incubate the suspension in 37 – 40°C water for at least 10 minutes.

9. To prepare the 5% sugar solutions, add 5 g of sugar per 100 mL of solution.

10. After the 10 minute incubation period, transfer the yeast to dispensing tubes. Each group will need about 3 mL of yeast.

11. All of the pressure valves, tubing, and connectors used in this experiment are included with Vernier Gas Pressure Sensors shipped after February 15, 1998. These accessories are also helpful when performing respiration/fermentation experiments such as Experiments 11C, and 12B in this manual.

If you purchased your Gas Pressure Sensor at an earlier date, Vernier has a Pressure Sensor Accessories Kit (PS-ACC) that includes all of the parts shown here for doing pressure-related experiments. Using this kit allows for easy assembly of a completely airtight system. The kit includes the following parts:

- two ribbed, tapered valve connectors inserted into a No. 5 rubber stopper
- one ribbed, tapered valve connectors inserted into a No. 1 rubber stopper
- two Luer-lock connectors connected to either end of a piece of plastic tubing
- one two-way valve
- one 20 mL syringe
- two tubing clamps for transpiration experiments

12. The accessory items used in this experiment are the #1 single hole stopper fitted with a tapered valve connector and the section of plastic tubing fitted with Luer-lock connectors.

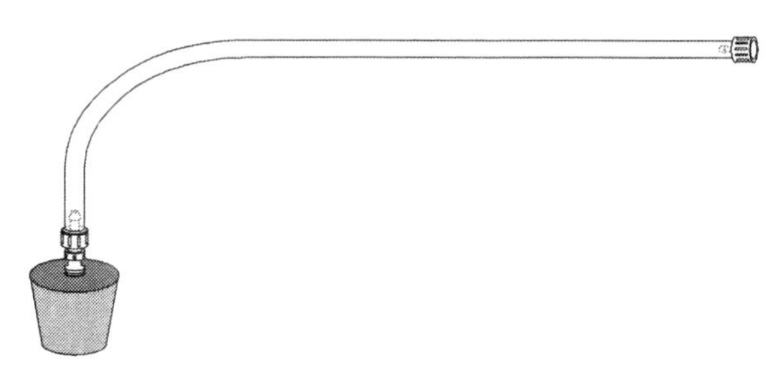

13. The length of plastic tubing connecting the rubber stopper assemblies to each gas pressure sensor must be the same for all groups. It is best to keep the length of tubing reasonably small to keep the volume of gas in the test tube low. **Note:** If pressure changes during data collection are too small, you may need to decrease the total gas volume in the system. Shortening the length of tubing used will help to decrease the volume.

SAMPLE RESULTS

The following data may be different from students' results. The actual values depend upon the viability and concentration of the yeast, among other factors.

Table 5

Test	Type of sugar/enzyme	Rate of fermentation (kPa/min)
1	Lactose + Lactase	1.3
2	Lactose	0
3	Glucose	1.6
4	Galactose	0
5	None (water only)	0

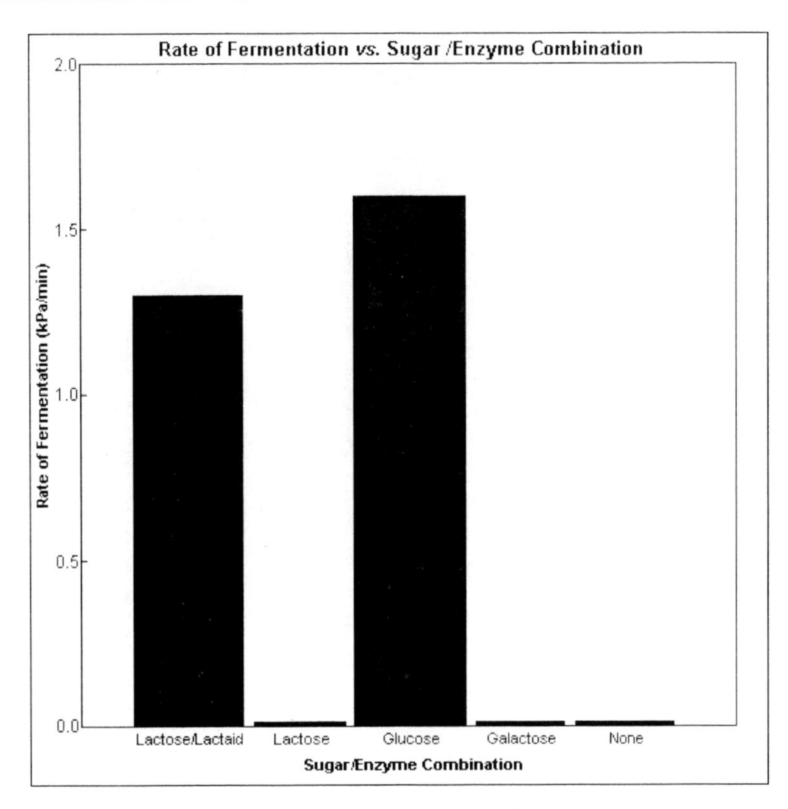

Rate of fermentation values for each test

ANSWERS TO QUESTIONS

1. Lactase acts by converting lactose into glucose and, presumably, galactose. Yeast are not able to metabolize lactose, but can metabolize glucose. Yeast are able to use the lactose solution only *after* it has been acted upon by lactase. Since lactose is a disaccharide that can hydrolyze into glucose and galactose, lactase must have caused that hydrolysis.
2. Yeast cannot utilize all of the sugars equally well. Of all the sugars tested, yeast can only metabolize glucose—not lactose or galactose. The rates of the latter three tests were zero.
3. Yeast may not have the proper enzymes to transport lactose across its cell membrane, or it may not have the enzyme needed to convert it from a disaccharide to a monosaccharide.
4. The results of the two experiments should agree. The glucose test tape experiment indicated the presence of glucose with lactase + lactose. The yeast were able to respire with the products of lactase + lactose. Since glucose was the only sugar utilized by yeast in this experiment, one of the products of enzymatic activity presumably was glucose.
5. Test 5 served as the control. It contained only water and yeast in order to verify that the yeast need sugar to respire. Without the control, a person could conclude that the yeast can respire without any sugar.

Experiment
25

Primary Productivity

Oxygen is vital to life. In the atmosphere, oxygen comprises over 20% of the available gases. In aquatic ecosystems, however, oxygen is scarce. To be useful to aquatic organisms, oxygen must be in the form of molecular oxygen, O_2. The concentration of oxygen in water can be affected by many physical and biological factors. Respiration by plants and animals reduces oxygen concentrations, while the photosynthetic activity of plants increases it. In photosynthesis, carbon is assimilated into the biosphere and oxygen is made available, as follows:

$$6\ H_2O + 6\ CO_2(g) + \text{energy} \rightarrow C_6H_{12}O_6 + 6O_2(g)$$

The rate of assimilation of carbon in water depends on the type and quantity of plants within the water. Primary productivity is the measure of this rate of carbon assimilation. As the above equation indicates, the production of oxygen can be used to monitor the primary productivity of an aquatic ecosystem. A measure of oxygen production over time provides a means of calculating the amount of carbon that has been bound in organic compounds during that period of time. Primary productivity can also be measured by determining the rate of carbon dioxide utilization or the rate of formation of organic compounds.

One method of measuring the production of oxygen is the *light and dark bottle* method. In this method, a sample of water is placed into two bottles. One bottle is stored in the dark and the other in a lighted area. Only respiration can occur in the bottle stored in the dark. The decrease in dissolved oxygen (DO) in the dark bottle over time is a measure of the rate of respiration. Both photosynthesis and respiration can occur in the bottle exposed to light, however. The difference between the amount of oxygen produced through photosynthesis and that consumed through aerobic respiration is the *net productivity*. The difference in dissolved oxygen over time between the bottles stored in the light and in the dark is a measure of the total amount of oxygen produced by photosynthesis. The total amount of oxygen produced is called the *gross productivity*.

The measurement of the DO concentration of a body of water is often used to determine whether the biological activities requiring oxygen are occurring and is an important indicator of pollution.

OBJECTIVES

In this experiment, you will

- Measure the rate of respiration in an aquatic environment using a Dissolved Oxygen Probe.
- Determine the net and gross productivity in an aquatic environment.

MATERIALS

computer
Vernier computer interface
Vernier Dissolved Oxygen Probe
Logger*Pro*
17 pieces of 12 cm × 12 cm (5″ × 5″) plastic window screen
250 mL beaker
aluminum foil
seven 25 × 150 mm screw top test tubes
shallow pan
scissors
siphon tube
500 mL pond, lake, seawater or algal culture
distilled water
rubber bands

PRE-LAB PROCEDURE

Important: Prior to each use, the Dissolved Oxygen Probe must warm up for a period of 10 minutes as described below. If the probe is not warmed up properly, inaccurate readings will result. Perform the following steps to prepare the Dissolved Oxygen Probe.

1. Prepare the Dissolved Oxygen Probe for use.
   a. Remove the blue protective cap.
   b. Unscrew the membrane cap from the tip of the probe.
   c. Using a pipet, fill the membrane cap with 1 mL of DO Electrode Filling Solution.
   d. Carefully thread the membrane cap back onto the electrode.
   e. Place the probe into a 250 mL beaker containing distilled water.

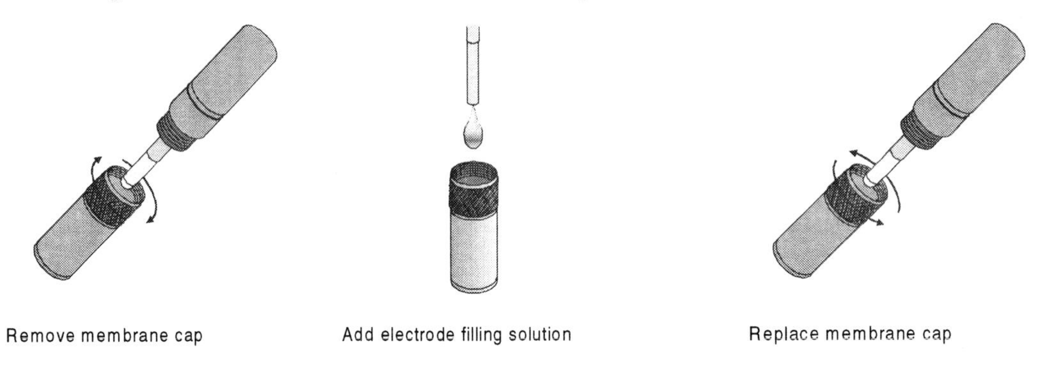

Figure 1

2. Plug the Dissolved Oxygen Probe into Channel 1 of the Vernier interface.
3. Prepare the computer for data collection by opening the file "25 Primary Productivity" from the *Biology with Computers* folder of Logger*Pro*.
4. It is necessary to warm up the Dissolved Oxygen Probe for 10 minutes before taking readings. To warm up the probe, leave it connected to the interface, with Logger *Pro* running, for 10 minutes. The probe must stay connected at all times to keep it warmed up. If disconnected for a few minutes, it will be necessary to warm up the probe again.
5. You are now ready to calibrate the Dissolved Oxygen Probe.
   - If your instructor directs you to use the calibration stored in the experiment file, then proceed to Step 6.
   - If your instructor directs you to perform a new calibration for the Dissolved Oxygen Probe, follow this procedure.

   Zero-Oxygen Calibration Point

   a. Choose Calibrate ▸ CH1: Dissolved Oxygen (mg/L) from the Experiment menu and then click Calibrate Now.
   b. Remove the probe from the water and place the tip of the probe into the Sodium Sulfite Calibration Solution. **Important:** No air bubbles can be trapped below the tip of the probe or the probe will sense an inaccurate dissolved oxygen level. If the voltage does not rapidly decrease, tap the side of the bottle with the probe to dislodge any bubbles. The readings should be in the 0.2 to 0.5 V range.

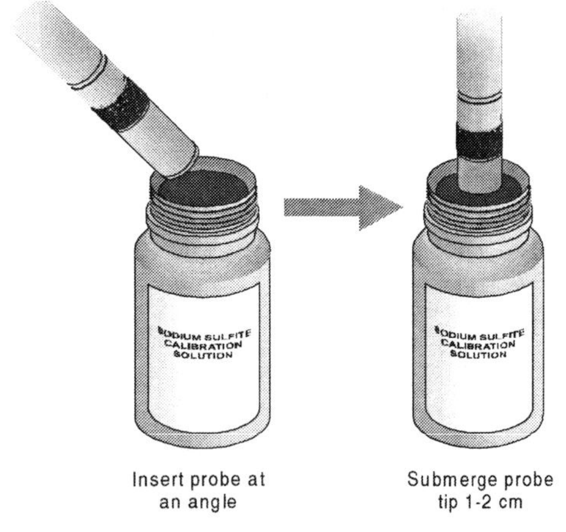

Figure 2

c. Type "0" (the known value in mg/L) in the edit box.
d. When the displayed voltage reading for Reading 1 stabilizes, click Keep.

Saturated DO Calibration Point

e. Rinse the probe with distilled water and gently blot dry.
f. Unscrew the lid of the calibration bottle provided with the probe. Slide the lid and the grommet about 2 cm onto the probe body.

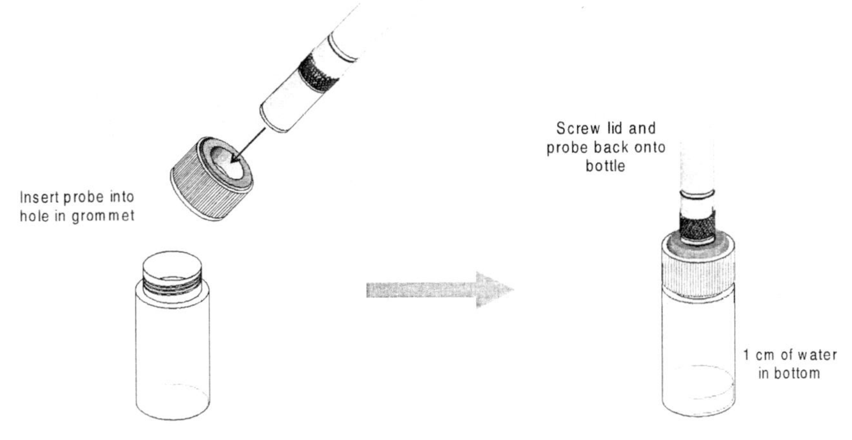

Figure 3

g. Add water to the bottle to a depth of about 1 cm and screw the bottle into the cap, as shown. **Important:** Do not touch the membrane or get it wet during this step. Keep the probe in this position for about a minute.
h. Type the correct saturated dissolved-oxygen value (in mg/L) from Table 5 (for example, "8.66") using the current barometric pressure and air temperature values. If you do not have the current air pressure, use Table 6 to estimate the air pressure at your altitude.
i. When the displayed voltage reading for Reading 2 stabilizes (readings should be above 2.0 V), click Keep.
j. Choose Calibration Storage from the pull-down menu and choose to store the calibration with the experiment file. Click Done.
k. Return the Dissolved Oxygen Probe to the distilled water beaker.
l. From the File menu, select Save As and save the current experiment file with a new name.

PROCEDURE

Day 1

6. Obtain seven water sampling test tubes.

7. Fill each of the test tubes with the water sample. To fill a test tube:
   a. Obtain a siphon tube.
   b. Insert the tube into the water sample and fill the tube completely with water.
   c. Pinch the tube (or use a tube clamp) to close off the siphon tube.
   d. Place one end of the tube in the bottom of a test tube. Keep the other end in the water sample, well below the surface. Position the test tube lower than the water sample Place a shallow pan under the test tube to collect any water that spills over.
   e. Siphon the water into the test tube. Fill the test tube until it overflows approximately

1/3 of the volume of the test tube. Fill the test tube completely to the top of the rim.

f. Tighten the cap on the test tube securely. Be sure no air is in the test tube.

8. The percentage of available natural light for each water sample is listed in Table 1 below:

Table 1		
Test tube	Number of Screens	% Light
1	0	Initial
2	0	100%
3	1	65%
4	3	25%
5	5	10%
6	8	2%
7	Aluminum Foil	Dark

9. Use masking tape to label the cap of each tube. Mark the labels as follows: dark, initial, 2%, 10%, 25%, 65%, and 100%
10. Wrap test tube 7, the dark tube, with aluminum foil so that it is light-proof. This water sample will remain in the dark.
11. Mix the contents of each tube. Be sure that there are no air bubbles present in any of the tubes. Fill a tube with more sample water if necessary.
12. Wrap screen layers around the tubes according to Table 1. Trim the screens so each one only wraps around the tube once. Hold the screens in place with rubber bands or clothes pins. The bottles will stand on end so the bottoms of the tubes need not be covered. The layers of screen will simulate the amount of natural light available for photosynthesis at different depths in a body of water.
13. Remove the Dissolved Oxygen Probe from the beaker. Place the probe into test tube 1, the initial tube, so that it is submerged half the depth of the water. Gently and continuously move the probe up and down a distance of about 1 cm in the tube. This allows water to move past the probe's tip. Note: Do not agitate the water, or oxygen from the atmosphere will mix into the water and cause erroneous readings.
14. After 60 seconds, or when the dissolved oxygen reading stabilizes, record the reading in Table 2. Discard the contents of the initial tube and clean the test tube. Rinse the Dissolved Oxygen Probe with distilled water and place it back in the distilled water beaker. The probe should remain in the beaker overnight, so that measurements can be made the following day.
15. Place test tubes 2 – 7 near the light source, as directed by your instructor.

Day 2

16. Prepare the computer for data collection by opening the file saved on Day 1 of the experiment. Allow the probe 10 – 15 minutes to warm up. Keep it in the beaker of distilled water during this time. The calibration from the previous day should have been saved with the experiment file, so no calibration is needed for Day 2 measurements.

17. Place the probe into the 100% light tube. Gently and continuously move the probe up and down a distance of about 1 cm in the tube. After 60 seconds, or when the dissolved oxygen reading stabilizes, record the reading in Table 2.

18. Repeat Step 17 for the remaining test tubes.

19. Clean your test tubes as directed by your instructor. Rinse the Dissolved Oxygen Probe with distilled water and place it back in the distilled water beaker.

DATA

Table 2		
Test tube	% Light	DO (mg/L)
1	Initial	
2	100%	
3	65%	
4	25%	
5	10%	
6	2%	
7	Dark	

PROCESSING THE DATA

1. Determine the number of hours that have passed since the onset of this experiment. Subtract the DO value in the test tube 1 (the initial DO value) from that of test tube 7 (the dark test tube's DO value). Divide the DO value by the time in hours. Record the resulting value as the respiration rate in Table 3.

   Respiration rate = (dark DO – initial DO) / time

2. Determine the gross productivity in each test tube. To do this, subtract the DO in test tube 7 (the dark test tube's DO value) from that of test tubes 2–6 (the light test tubes' DO value). Divide each DO value by the length of the experiment in hours. Record each resulting value as the gross productivity in Table 4.

   Gross productivity = (DO of test tube – dark DO) / time

3. Determine the net productivity in each test tube. To do this, subtract the DO in test tubes 2–6 (the light test tube's DO value) from that of test tube 1 (the initial DO value). Divide the result by the length of the experiment in hours. Record each resulting value as the net productivity in Table 4.

Net Productivity = (DO of test tube – initial DO) / time

Table 3	
Respiration (mg O_2/L/hr)	

Table 4			
Test Tube	% Light	Gross productivity (mg/L/hr)	Net productivity (mg/L/hr)
2	100%		
3	65%		
4	25%		
5	10%		
6	2%		

4. Prepare a graph of gross productivity and net productivity as a function of light intensity. Graph both types of productivity on the same piece of graph paper.

QUESTIONS

1. Is there evidence that photosynthetic activity added oxygen to the water? Explain.
2. Is there evidence that aerobic respiration occurred in the water? If so, what kind of organisms might be responsible for this—autotrophs? Heterotrophs? Explain.
3. What effect did light have on the primary productivity? Explain.
4. Refer to your graph of productivity and light intensity. At what light intensity do you expect there to be no net productivity? no gross productivity?
5. How would turbidity affect the primary productivity of a pond?

EXTENSIONS

1. Determine the effects of nitrogen and phosphates on primary productivity. Why would the presence of phosphates and nitrogen, in the form of nitrates and ammonium ions, be important to an aquatic ecosystem during the spring season? How do they accumulate in the watershed? What is eutrophication?
2. Measure the dissolved oxygen of a pond at different temperatures. What is the effect of temperature on the primary productivity of a pond?
3. Calculate the amount of carbon that was fixed in each of the tubes. Use the following conversion factors to do the calculations.

$$\text{mg } O_2/\text{L} \times 0.698 = \text{mL } O_2/\text{L}$$

$$\text{mL } O_2/\text{L} \times .536 = \text{mg carbon fixed/L}$$

CALIBRATION TABLES

Table 5: 100% Dissolved Oxygen Capacity (mg/L)												
	770 mm	**760 mm**	**750 mm**	**740 mm**	**730 mm**	**720 mm**	**710 mm**	**700 mm**	**690 mm**	**680 mm**	**670 mm**	**660 mm**
0°C	14.76	14.57	14.38	14.19	13.99	13.80	13.61	13.42	13.23	13.04	12.84	12.65
1°C	14.38	14.19	14.00	13.82	13.63	13.44	13.26	13.07	12.88	12.70	12.51	12.32
2°C	14.01	13.82	13.64	13.46	13.28	13.10	12.92	12.73	12.55	12.37	12.19	12.01
3°C	13.65	13.47	13.29	13.12	12.94	12.76	12.59	12.41	12.23	12.05	11.88	11.70
4°C	13.31	13.13	12.96	12.79	12.61	12.44	12.27	12.10	11.92	11.75	11.58	11.40
5°C	12.97	12.81	12.64	12.47	12.30	12.13	11.96	11.80	11.63	11.46	11.29	11.12
6°C	12.66	12.49	12.33	12.16	12.00	11.83	11.67	11.51	11.34	11.18	11.01	10.85
7°C	12.35	12.19	12.03	11.87	11.71	11.55	11.39	11.23	11.07	10.91	10.75	10.59
8°C	12.05	11.90	11.74	11.58	11.43	11.27	11.11	10.96	10.80	10.65	10.49	10.33
9°C	11.77	11.62	11.46	11.31	11.16	11.01	10.85	10.70	10.55	10.39	10.24	10.09
10°C	11.50	11.35	11.20	11.05	10.90	10.75	10.60	10.45	10.30	10.15	10.00	9.86
11°C	11.24	11.09	10.94	10.80	10.65	10.51	10.36	10.21	10.07	9.92	9.78	9.63
12°C	10.98	10.84	10.70	10.56	10.41	10.27	10.13	9.99	9.84	9.70	9.56	9.41
13°C	10.74	10.60	10.46	10.32	10.18	10.04	9.90	9.77	9.63	9.49	9.35	9.21
14°C	10.51	10.37	10.24	10.10	9.96	9.83	9.69	9.55	9.42	9.28	9.14	9.01
15°C	10.29	10.15	10.02	9.88	9.75	9.62	9.48	9.35	9.22	9.08	8.95	8.82
16°C	10.07	9.94	9.81	9.68	9.55	9.42	9.29	9.15	9.02	8.89	8.76	8.63
17°C	9.86	9.74	9.61	9.48	9.35	9.22	9.10	8.97	8.84	8.71	8.58	8.45
18°C	9.67	9.54	9.41	9.29	9.16	9.04	8.91	8.79	8.66	8.54	8.41	8.28
19°C	9.47	9.35	9.23	9.11	8.98	8.86	8.74	8.61	8.49	8.37	8.24	8.12
20°C	9.29	9.17	9.05	8.93	8.81	8.69	8.57	8.45	8.33	8.20	8.08	7.96
21°C	9.11	9.00	8.88	8.76	8.64	8.52	8.40	8.28	8.17	8.05	7.93	7.81
22°C	8.94	8.83	8.71	8.59	8.48	8.36	8.25	8.13	8.01	7.90	7.78	7.67
23°C	8.78	8.66	8.55	8.44	8.32	8.21	8.09	7.98	7.87	7.75	7.64	7.52
24°C	8.62	8.51	8.40	8.28	8.17	8.06	7.95	7.84	7.72	7.61	7.50	7.39
25°C	8.47	8.36	8.25	8.14	8.03	7.92	7.81	7.70	7.59	7.48	7.37	7.26
26°C	8.32	8.21	8.10	7.99	7.89	7.78	7.67	7.56	7.45	7.35	7.24	7.13
27°C	8.17	8.07	7.96	7.86	7.75	7.64	7.54	7.43	7.33	7.22	7.11	7.01
28°C	8.04	7.93	7.83	7.72	7.62	7.51	7.41	7.30	7.20	7.10	6.99	6.89
29°C	7.90	7.80	7.69	7.59	7.49	7.39	7.28	7.18	7.08	6.98	6.87	6.77
30°C	7.77	7.67	7.57	7.47	7.36	7.26	7.16	7.06	6.96	6.86	6.76	6.66

Table 6: Approximate Barometric Pressure at Different Elevations					
Elevation (m)	**Pressure (mm Hg)**	**Elevation (m)**	**Pressure (mm Hg)**	**Elevation (m)**	**Pressure (mm Hg)**
0	760	800	693	1600	628
100	748	900	685	1700	620
200	741	1000	676	1800	612
300	733	1100	669	1900	604
400	725	1200	661	2000	596
500	717	1300	652	2100	588
600	709	1400	643	2200	580
700	701	1500	636	2300	571

TEACHER INFORMATION

Experiment
25

Primary Productivity

1. This experiment requires two 45 minute periods to complete. There will be time during the second period to discuss the experiment and to begin the follow-up activities.
2. If time permits, the first part of this lab can be completed during a field trip.
3. The Dissolved Oxygen Probe must be calibrated the first day of use. Follow the pre-lab procedure to prepare and calibrate the Dissolved Oxygen Probe. To save time, you may wish to calibrate the probe and record the calibration values on paper. The students can then skip the pre-lab procedure and they will have the calibration values available for manual entry in case the values stored in the program are lost.
4. At the end of class instruct the students to leave Logger *Pro* running. This will keep power going to the probes. When the next group of students comes in, they can begin at Step 6 of the procedure. They can skip the pre-lab procedure because the initial group of students has completed all of the setup. Have the last group of students place the probe back in the beaker of distilled water to be stored overnight.
5. Bottles or test tubes that can form an airtight seal must be used. Air should not be in any of the tubes while they are stored overnight.
6. When setting up the Dissolved Oxygen Probe, be sure to remove the blue plastic cap from the end of the probe. The cap is made of a soft plastic material and easily slides off the probe end.
7. Either pond water (or some other fresh water source) or a *Chlorella* culture should be used as a water source. You may want to place about 5 mL of *Chlorella* culture into 1 L of sample water.
8. Plastic window screen is available from most hardware stores. The plastic screen is much easier to use than metal screen.
9. An excellent way to dispense the water sample is to use a carboy with a dispensing spout near its bottom. Attach a short length of tubing to the dispensing spout and place a shallow pan below the tube to catch any spill-over. Nalgene makes a variety of plastic carboys that work particularly well.
10. The test tubes should be water-tight when they are stored overnight. Lay the tubes horizontally on a table and lower a fluorescent light about 3 – 6 cm above the tubes. This provides ample light without heating the tubes. If the tubes are stored vertically, non-motile algae may sink to the bottom of the tube and be shielded from the light.

SAMPLE RESULTS

The following data may be different from students' results. This water sample was taken over a 25 hour period.

Table 2		
Test Tube	% Light	DO (mg/L)
1	Initial	7.5
2	100%	7.8
3	65%	7.6
4	25%	6.8
5	10%	6.7
6	2%	6.5
7	Dark	6.5

Table 3	
Respiration (mg/L/hr)	–0.04

Table 4			
Test tube	% Light	Gross productivity (mg/L/hr)	Net productivity (mg/L/hr)
2	100%	0.052	–0.012
3	65%	0.044	–0.004
4	25%	0.012	0.028
5	10%	0.008	0.032
6	2%	0	0.04

ANSWERS TO QUESTIONS

1. There is evidence that photosynthetic activity added oxygen to the water. The gross primary productivity was positive and greater than zero. The gross primary productivity is a measurement of the total amount of oxygen produced in the water sample, regardless of the amount of aerobic respiration that occurs. The net primary productivity may or may not be a positive value, depending on the BOD due to heterotrophs. If so, this does not mean that photosynthesis did not occur—just that the rate of photosynthesis was less than the rate of aerobic respiration.

2. There is evidence that aerobic respiration occurred in the water, as the respiration rate was negative and non-zero. The dark bottle had less oxygen after one day than the bottle had

initially. Both autotrophs and heterotrophs may be responsible for removing oxygen, depending upon the types of organisms in the water sample.

3. Light was necessary for the assimilation of carbon into the ecosystem within the water sample, since light is necessary for photosynthesis.

4. Student answers will vary. According to the sample data, there would be no net productivity at 70% light and no gross productivity at 0% light.

5. The higher the turbidity of the water, the less clear it is. Less light reaches the plants going through photosynthesis, and less oxygen is produced. This results in a decrease in primary productivity.

Experiment
26

Control of Human Respiration

Your respiratory system allows you to obtain oxygen, eliminate carbon dioxide, and regulate the blood's pH level. The process of taking in air is known as *inspiration*, while the process of blowing out air is called *expiration*. A respiratory cycle consists of one inspiration and one expiration. The rate at which your body performs a respiratory cycle is dependent upon the levels of oxygen and carbon dioxide in your blood.

You will monitor the respiratory patterns of one member of your group under different conditions. A respiration belt will be strapped around the test subject and connected to a computer-interfaced Gas Pressure Sensor. Each respiratory cycle will be recorded by the computer, allowing you to calculate a respiratory rate for comparison at different conditions.

OBJECTIVES

In this experiment, you will

- Use a computer to monitor the respiratory rate of an individual.
- Evaluate the effect of holding of breath on the respiratory cycle.
- Evaluate the effect of rebreathing of air on the respiratory cycle.

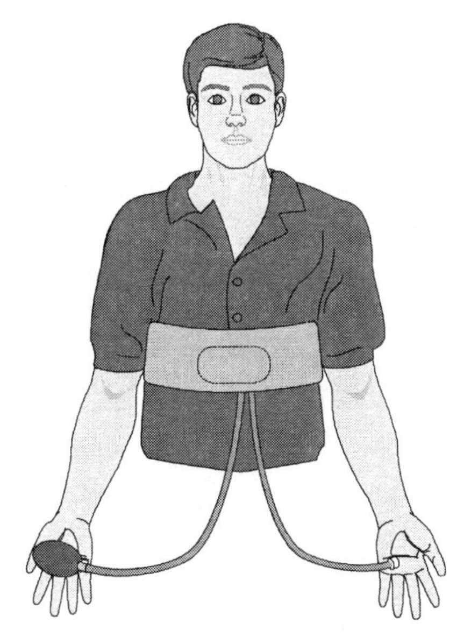

Figure 1

MATERIALS

computer
Vernier computer interface
Logger*Pro*
Vernier Gas Pressure Sensor
Vernier Respiration Monitor Belt
plastic produce bag 30 × 40 cm (12"×16")
small paper grocery bag

PROCEDURE

1. Connect the Gas Pressure Sensor to the computer interface.
2. Prepare the computer for data collection by opening the file "26a Human Respiration" from the *Biology with Computers* folder of Logger*Pro*.
3. Select one member of the group as the test subject. Wrap the Respiration Monitor Belt snugly around the test subject's chest. Press the Velcro strips together at the back. Position the belt on the test subject so that the belt's air bladder is resting over the base of the rib cage and in alignment with the elbows as shown in Figure 3.

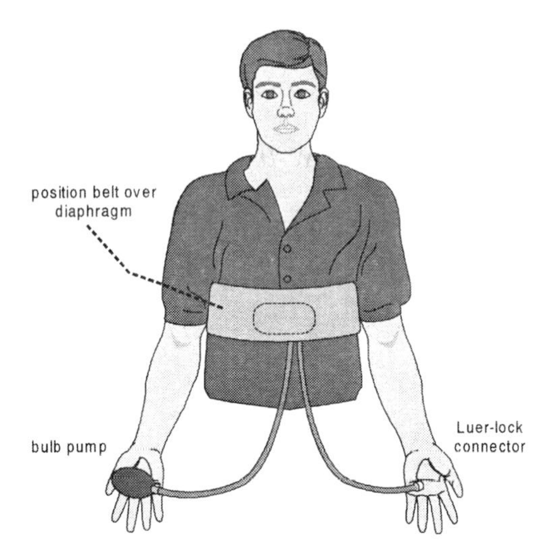

Figure 3

4. Attach the Respiration Monitor Belt to the Gas Pressure Sensor. There are two rubber tubes connected to the bladder. One tube has a white Luer-lock connector at the end and the other tube has a bulb pump attached. Connect the Luer-lock connector to the stem on the Gas Pressure Sensor with a gentle half turn.
5. Have the test subject sit upright in a chair. Close the shut-off screw of the bulb pump by turning it clockwise as far as it will go. Pump air into the bladder by squeezing on the bulb pump. Fill the bladder as full as possible without being uncomfortable for the test subject.
6. The pressure reading displayed in the meter should increase about 6 kPa above the initial pressure reading (e.g., at sea level, the pressure would increase from about 100 to 106 kPa). At this pressure, the belt and bladder should press firmly against the test subject's diaphragm. Pressures will vary, depending upon how tightly the belt was initially wrapped around the test subject.
7. As the test subject breathes in and out normally, the displayed pressure alternately increases and decreases over a range of about 2 – 3 kPa. If the range is less than 1 kPa, it may be necessary to pump more air into the bladder. Note: If you still do not have an adequate range, you may need to tighten the belt.

Part I Holding of Breath

8. Instruct the test subject to breathe normally. Start collecting data by clicking Collect. When data has been collected for 60 seconds, have the test subject hold his or her breath for 30 to 45 seconds. The test subject should breathe normally for the remainder of the data collection once breath has been released.

9. Examine the respiration rates recorded in the bottom graph by clicking the Examine button, [Examine icon]. As you move the mouse pointer from point to point on the graph the data values are displayed in the examine window. Determine the respiration rate before and after the test subject's breath was held and record the values in Table 1.

Part II Rebreathing of Air

10. Prepare the computer for data collection by opening the file "26b Human Respiration" from the *Biology with Computers* folder of Logger*Pro*.
11. Place a small paper bag into a plastic produce bag. Have the test subject cover his or her mouth with the bags, tight enough to create an air-tight seal. The test subject should breathe normally into the bags throughout the course of the data collection process.
12. Click [Collect] to begin data collection. Again, the test subject should be sitting and facing away from the computer screen. Collect respiration data for the full 300 seconds while breathing into the sack. **Important:** Anyone prone to dizziness or nausea should not be tested in this section of the experiment. If the test subject experiences dizziness, nausea, or a headache during data collection, testing should be stopped immediately.
13. Once you have finished collecting data in Step 12, calculate the maximum height of the respiration waveforms for the intervals of 0 to 30 seconds, 120 to 150 seconds, and 240 to 270 seconds:
    a. Move the mouse pointer to the beginning of the section you are examining. Hold down the mouse button. Drag the pointer to the end of the section and release the mouse button.
    b. Click the Statistics button, [Statistics icon], to determine the statistics for the selected data.
    c. Subtract the minimum pressure value from the maximum value (in kPa).
    d. Record this value for each section as the wave amplitude in Table 2.

DATA

Table 1	
Holding of Breath	
Before holding breath	After holding breath
_______ breaths / minute	_______ breaths / minute

Table 2		
Rebreathing of Air: Amplitudes of Respiration Waves		
0 to 30 seconds	120 to 150 seconds	240 to 270 seconds
________ kPa	________ kPa	________ kPa

QUESTIONS

1. Did the respiratory rate of the test subject change after holding his or her breath? If so, describe how it changed.
2. What is different about the size (amplitude) or shape (frequency) of the respiratory waveforms following the release of the test subject's breath? Explain.
3. What would be the significance of an increase in the amplitude and frequency of the waveform while the test subject was breathing into the bag?
4. How did the respiratory waveforms change while the test subject was breathing into the bag? How would you interpret this result?
5. Explain how you think carbon dioxide affects your breathing.

TEACHER INFORMATION

Control of Human Respiration

1. When putting on the Respiration Monitor Belt, try to get as tight a fit as is comfortable.
2. Always have the test subject sit in a position in which they are unable to see the computer screen. This will help ensure that they do not consciously alter their respiration rate.
3. Anyone prone to dizziness or nausea should not be tested in the section of the experiment involving the rebreathing of their own air. If the person being tested experiences dizziness, nausea, or a headache during testing, stop data collection. Students who are sensitive to hyperventilation or are nervous by nature should not be tested.

SAMPLE RESULTS

Table 1	
Holding of Breath	
Before holding breath	After holding breath
11 breaths / minute	19 breaths / minute

Table 2		
Rebreathing of Air: Amplitudes of Respiration Waves		
0 to 30 seconds	120 to 150 seconds	240 to 270 seconds
5.1 kPa	6.6 kPa	8.4 kPa

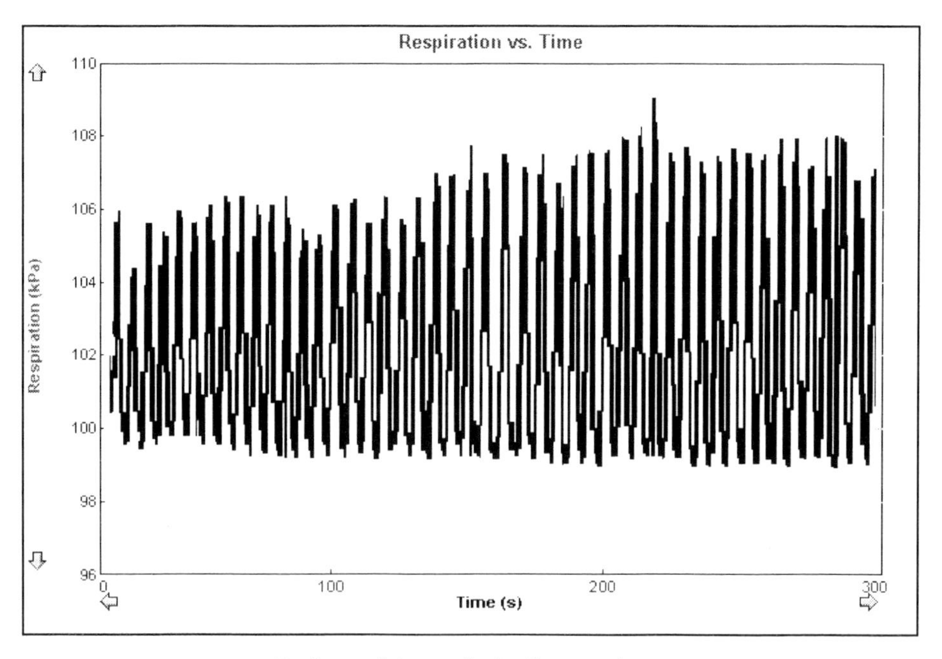

Rebreathing of air from a bag

ANSWERS TO QUESTIONS

1. Respiratory rates should increase following a person holding their breath.

2. The vertical size or amplitude of the respiratory wave will initially be greater than normal, and then return to normal after several breaths.

3. The significance of an increase in the amplitude of the waveform is that a greater volume of air is breathed during a respiratory cycle. The significance of an increase in the frequency of the waveform is that the respiratory rate has increased.

4. In most cases, the wave amplitude will increase in an attempt to compensate for the heightened carbon dioxide concentration in the blood and the lungs. An increase in wave amplitude is relative to an increase the volume of air inhaled with each respiratory cycle.

5. Based on the results of the experiment, carbon dioxide increases the volume of air inhaled during respiration in humans.

Experiment

27A

Heart Rate and Physical Fitness

The circulatory system is responsible for the internal transport of many vital substances in humans, including oxygen, carbon dioxide, and nutrients. The components of the circulatory system include the heart, blood vessels, and blood. Heartbeats result from electrical stimulation of the heart cells by the *pacemaker,* located in the heart's inner wall of the right atrium. Although the electrical activity of the pacemaker originates from within the heart, the rhythmic sequence of impulses produced by the pacemaker is influenced by nerves outside the heart. Many things might affect heart rate, including the physical fitness of the individual, the presence of drugs such as caffeine or nicotine in the blood, and the age of the person.

As a rule, the maximum heart rate of all individuals of the same age and sex is about the same. However, the time it takes individuals to reach that maximum level while exercising varies greatly. Since physically fit people can deliver a greater volume of blood in a single cardiac cycle than unfit individuals, they can usually sustain a greater work level before reaching the maximum heart rate. Physically fit people not only have less of an increase in their heart rate during exercise, but their heart rate recovers to the resting rate more rapidly than unfit people.

In this experiment, you will evaluate your physical fitness. An arbitrary rating system will be used to "score" fitness during a variety of situations. Tests will be made while in a resting position, in a prone position, as well as during and after physical exercise.

Important: Do not attempt this experiment if physical exertion will aggravate a health problem. Inform your instructor of any possible health problems that might be affected if you participate in this exercise.

OBJECTIVES

In this experiment, you will

- Determine the effect of body position on heart rates.
- Determine the effect of exercise on heart rates.
- Determine your fitness level.
- Correlate the fitness level of individuals with factors such as smoking, the amount of daily exercise, and other factors identified by students.

MATERIALS

computer
Vernier computer interface
Logger*Pro*
Vernier Exercise Heart Rate Monitor
stepping stool, 45 cm (18 inches) high
saline solution in dropper bottle

PROCEDURE

Each person in a lab group will take turns being the subject and the tester. When it is your turn to be the subject, your partner will be responsible for recording the data on your lab sheet.

1. Elastic straps, for securing the transmitter belt, come in two different sizes. Select the size of elastic strap that best fits the subject being tested. It is important that the strap provide a snug fit of the transmitter belt.
2. Wet each of the electrodes (the two grooved rectangular areas on the underside of the transmitter belt) with 3 drops of saline solution.
3. Secure the transmitter belt against the skin directly over the base of the rib cage. The POLAR logo on the front of the belt should be in line with the chest center as shown in Figure 1. Adjust the elastic strap to ensure a tight fit.

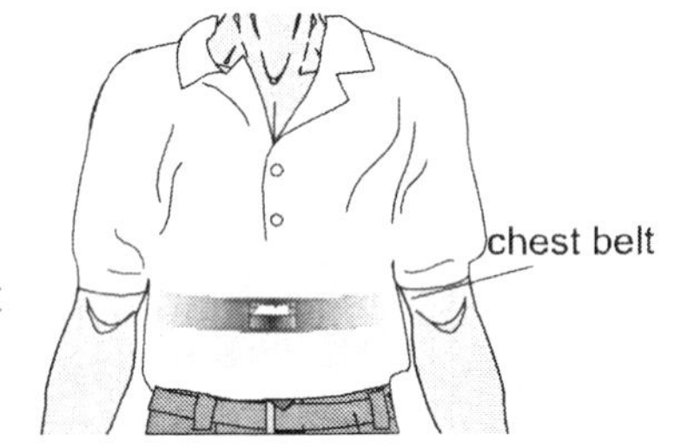

Figure 1

4. Connect the receiver module of the Exercise Heart Rate Monitor to the Vernier computer interface.
5. Have the subject hold the receiver in his right hand. Remember, the receiver must be within 80 cm of the transmitter belt while data is being collected.

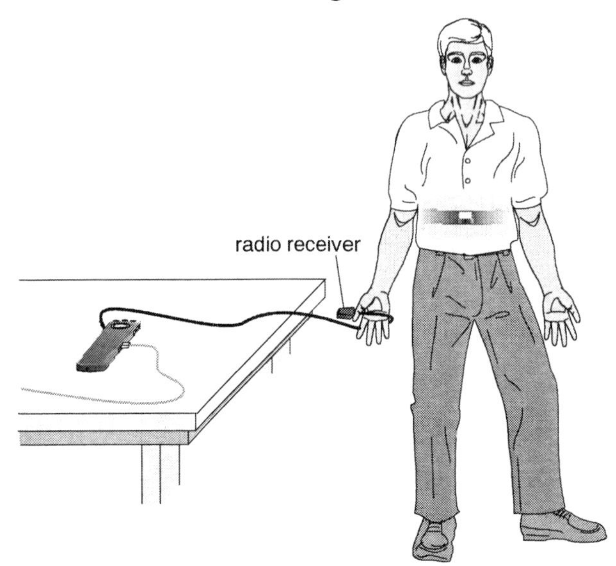

Figure 2

6. Prepare the computer for data collection by opening the file "27 Heart Rate & Fitness" from the *Biology with Computers* folder of Logger*Pro*.
7. Click [Collect] to begin monitoring heart rate.

8. Determine that the sensor is functioning correctly. The readings should be consistent and within the normal range of the individual, usually between 55 and 80 beats per minute. Click Stop when you have determined that the equipment is operating properly.

Standing heart rate

9. Click Collect to begin monitoring heart rate. Stand upright for 2 minutes.

10. Record the resulting heart rate in Table 6.

11. Use the resulting heart rate to assign fitness points based on Table 1 and record the value in Table 6.

Table 1: Standing Heart Rate			
Beats/min	Points	Beats/min	Points
60–70	12	101–110	8
71–80	11	111–120	7
81–90	10	121–130	6
91–100	9	131–140	4

Reclining heart rate

12. Recline on a clean surface or table for 2 minutes.

13. Record the resulting heart rate in Table 6.

14. Assign fitness points based on Table 2 and record the value in Table 6.

Table 2: Reclining Heart Rate			
Beats/min	Points	Beats/min	Points
50–60	12	81–90	8
61–70	11	91–100	6
71–80	10	101–110	4

Heart rate change from reclining to standing

15. Stand up next to the lab table.

16. Immediately record the peak heart rate in Table 6.

17. Subtract the reclining rate value in Step 13 from the peak heart rate after standing to find the heart rate increase after standing.

18. Locate the row corresponding to the reclining heart rate from Step 13 in Table 3.

19. Use the calculated heart rate increase after standing (from Step 17) to locate the proper column for fitness points in Table 3. Record the fitness points in Table 6

Table 3					
Reclining rate	Heart rate increase after standing				
beats/min	0–10	11–17	18–24	25–33	34+
50–60	12	11	10	8	6
61–70	12	10	8	6	4
71–80	11	9	6	4	2
81–90	10	8	4	2	0
91–100	8	6	2	0	0
101–110	6	4	0	0	0

20. Rest for 2 minutes. Click Stop to end data collection. When the rest period is over, click Collect to begin data collection.

Step test

21. Before performing the step test, record the subject's heart rate (Pre-exercise) in Table 6.
22. Perform a step test using the following procedure:
    a. Place the right foot on the top step of the stool.
    b. Place the left foot completely on the top step of the stool next to the right foot.
    c. Place the right foot back on the floor.
    d. Place the left foot completely on the floor next to the right foot.
    e. This stepping cycle should take 3 seconds to complete.
23. When five steps have been completed, record the heart rate in Table 6. Quickly move to Step 24.

Recovery rate

24. With a stopwatch or clock, begin timing to determine the subject's recovery time. During the recovery period, the subject should remain standing and relatively still. Monitor the heart rate readings and stop timing when the readings return to the pre-exercise heart rate value recorded in Step 21. Record the recovery time in Table 6.
25. Click Stop to end data collection.
26. Locate the subject's recovery time in Table 4 and record the corresponding fitness point value in Table 6. If the subject's heart rate did not return to within 10 beats/min from their pre-exercise heart rate, record a value of 6 points.

Table 4	
Time (sec)	Points
0–30	14
31–60	12
61–90	10
91–120	8

Step test for endurance

27. Subtract the subject's pre-exercise heart rate (from Step 21) from his or her heart rate after 5 steps of exercise. Record this heart rate increase in the endurance row of Table 6.

28. Locate the row corresponding to the pre-exercise heart rate in Table 5 and use the heart rate increase value to determine the proper fitness points. In Table 6, record the fitness points.

Table 5

Pre-exercise heart rate	Heart rate increase after exercise				
	0–10	11–20	21–30	31–40	41+
60–70	12	12	10	8	6
71–80	12	10	8	6	4
81–90	12	10	7	4	2
91–100	10	8	6	2	0
101–110	8	6	4	1	0
111–120	8	4	2	1	0
121–130	6	2	1	0	0
131+	5	1	0	0	0

29. Total all the fitness points recorded in Table 6. Determine the subject's personal fitness level using the scale below.

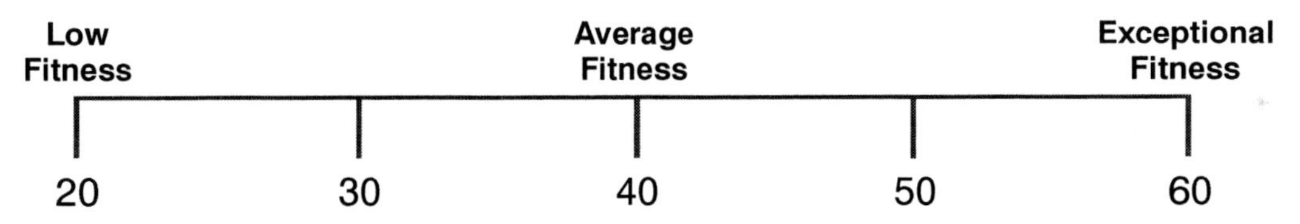

DATA

Table 6

Condition	Rate or time	Points
Standing heart rate	beats/min	
Reclining heart rate	beats/min	
Reclining to standing	beats/min	
Pre-exercise heart rate	beats/min	
After 5 steps	beats/min	
Recovery time	seconds	
Endurance	beats/min	
		Total points:

QUESTIONS

1. How did your heart rate change after moving from a standing position to a reclining position? Is this what you expected? How do you account for this?

2. How did your heart rate change after moving from a reclining position back to a standing position? Is this what you expected? How do you account for this?

3. Predict what your heart rate might be if you had exercised for twice the length of time that you actually did. Explain.

4. How does your maximum heart rate compare to other students in your group. Is this what you expected? How do you account for this?

5. Why would athletes need to work longer and harder before their heart rates were at the maximum value?

6. In examining the results of your physical fitness test, were you surprised by any of the findings? If you were, how might you explain them? Based on the results of the test, what behaviors in your life would you continue to practice? What behaviors might you think about changing?

7. Current research indicates that most heart attacks occur as people get out of bed after sleep. Account for this observation.

EXTENSION

1. Using a sphygmomanometer, learn how to measure blood pressure. Compare a person's blood pressure when reclining, to that of the same person immediately after standing from a reclined position. Relate the change in blood pressure to the heart rate values measured when going from reclining to standing.
2. Design an anonymous survey to be taken by each member of your class. In the survey, ask questions that you think might influence the test results. Examples might include:
   - Did you have more than six hours of sleep last night? Gender? Age?
   - Do you smoke? If so, how many packs per week do you smoke?
   - What was your total number of fitness points?
3. Try to determine whether any of the variables from your survey show a statistical link to fitness. You may want to use a statistical T-Tests to determine whether a relationship between the variable and physical fitness is due to chance.

Experiment

27B

Heart Rate and Physical Fitness

The circulatory system is responsible for the internal transport of many vital substances in humans, including oxygen, carbon dioxide, and nutrients. The components of the circulatory system include the heart, blood vessels, and blood. Heartbeats result from electrical stimulation of the heart cells by the *pacemaker,* located in the heart's inner wall of the right atrium. Although the electrical activity of the pacemaker originates from within the heart, the rhythmic sequence of impulses produced by the pacemaker is influenced by nerves outside the heart. Many things might affect heart rate, including the physical fitness of the individual, the presence of drugs such as caffeine or nicotine in the blood, and the age of the person.

As a rule, the maximum heart rate of all individuals of the same age and sex is about the same. However, the time it takes individuals to reach that maximum level while exercising varies greatly. Since physically fit people can deliver a greater volume of blood in a single cardiac cycle than unfit individuals, they can usually sustain a greater work level before reaching the maximum heart rate. Physically fit people not only have less of an increase in their heart rate during exercise, but their heart rate recovers to the resting rate more rapidly than unfit people.

In this experiment, you will evaluate your physical fitness. An arbitrary rating system will be used to "score" fitness during a variety of situations. Tests will be made while in a resting position, in a prone position, as well as during and after physical exercise.

Important: Do not attempt this experiment if physical exertion will aggravate a health problem. Inform your instructor of any possible health problems that might be affected if you participate in this exercise.

OBJECTIVES

In this experiment, you will

- Determine the effect of body position on heart rates.
- Determine the effect of exercise on heart rates.
- Determine your fitness level.
- Correlate the fitness level of individuals with factors such as smoking, the amount of daily exercise, and other factors identified by students.

MATERIALS

computer
Vernier computer interface
Logger*Pro*
Vernier Heart Rate Monitor
stepping stool, 45 cm (18 inches) high

PROCEDURE

1. Each partner in a team will take turns testing and being the test subject. When being tested, your partner should record your data on your lab sheet.

2. Connect the Heart Rate Monitor to the Vernier computer interface. Prepare the computer for data collection by opening the file "27 Heart Rate & Fitness" in the *Biology with Computers* folder.

3. Attach the earclip of the heart rate monitor to the ear lobe. Sit still and remain quiet during testing. Click [Collect] to begin monitoring heart rate.

4. Determine that the sensor is functioning correctly. The readings should be consistent and within the normal range of the individual, usually between 55 and 80 beats per minute. Reposition the earclip if readings appear sporadic. Click [Stop] when you have determined that the equipment is operating properly.

Standing heart rate

5. Click [Collect] to begin monitoring heart rate. Stand upright for 2 minutes.

6. Record the resulting heart rate in Table 6.

7. Use the resulting heart rate to assign fitness points based on Table 1 and record the value in Table 6.

Table 1: Standing Heart Rate

Beats/min	Points	Beats/min	Points
60–70	12	101–110	8
71–80	11	111–120	7
81–90	10	121–130	6
91–100	9	131–140	4

Reclining heart rate

8. Recline on a clean surface or table for 2 minutes.

9. Record the resulting heart rate in Table 6.

10. Assign fitness points based on Table 2 and record the value in Table 6.

Table 2: Reclining Heart Rate			
Beats/min	Points	Beats/min	Points
50–60	12	81–90	8
61–70	11	91–100	6
71–80	10	101–110	4

Heart rate change from reclining to standing

11. Stand up next to the lab table.

12. Immediately record the peak heart rate in Table 6.

13. Subtract the reclining rate value in Step 9 from the peak heart rate after standing to find the heart rate increase after standing.

14. Locate the row corresponding to the reclining heart rate from Step 9 in Table 3.

15. Use the calculated heart rate increase after standing (from Step 13) to locate the proper column for fitness points in Table 3. Record the fitness points in Table 6

Table 3					
Reclining rate	Heart rate increase after standing				
beats/min	0–10	11–17	18–24	25–33	34+
50–60	12	11	10	8	6
61–70	12	10	8	6	4
71–80	11	9	6	4	2
81–90	10	8	4	2	0
91–100	8	6	2	0	0
101–110	6	4	0	0	0

16. Rest for 2 minutes. Click Stop to end data collection. When the rest period is over, click Collect to begin data collection. Press ENTER to erase the latest data.

Step test

17. Record the current heart rate in Table 6 before performing the step test.

18. Hold the earclip to prevent it from moving while exercising.

19. Perform a step test using the following procedure:
    a. Place the right foot on the top step of the stool.
    b. Place the left foot completely on the top step of the stool next to the right foot.
    c. Place the right foot back on the floor.
    d. Place the left foot completely on the floor next to the right foot.
    e. This stepping cycle should take 3 seconds to complete.

20. Perform five steps and then stop exercising. Record the heart rate in Table 6. You may need

to allow multiple readings to obtain an accurate heart rate value, because movement can cause interference with readings. Once the heart rate has been recorded quickly move to Step 21.

Recovery rate

21. With a stopwatch or clock, begin timing to determine the recovery rate. During the recovery period, remain standing, relatively still and monitor the heart rate readings. Stop timing when the readings return to the standing heart rate value recorded in Step 6. Record the recovery time in seconds in Table 6.

22. Click Stop to end data collection.

23. Assign fitness points for the recovery time based on the information in Table 4. If the heart rate did not return to within 10 beats/min from the standing position value (from Step 6) after it stabilized, subtract two points from the point value in Table 4. Record the fitness points in Table 6.

Table 4	
Time (sec)	Points
0–30	14
31–60	12
61–90	10
91–120	8

Step test for endurance

24. Subtract the normal standing heart rate (from Step 6) from the heart rate after 5 steps of exercise (from Step 20). Record this heart rate increase in the endurance row of Table 6.

25. Assign fitness points based on the information in Table 5. Locate the row corresponding to the standing heart rate (from Step 6) in Table 5 and use the heart rate increase (from Step 24) to obtain fitness points. Record the fitness points in Table 6.

Table 5					
Standing rate	Heart rate increase after exercise				
(beats/min)	0–10	11–20	21–30	31–40	41+
60–70	12	12	10	8	6
71–80	12	10	8	6	4
81–90	12	10	7	4	2
91–100	10	8	6	2	0
101–110	8	6	4	1	0
111–120	8	4	2	1	0
121–130	6	2	1	0	0
131+	5	1	0	0	0

26. Total all the fitness points recorded in Table 6. Determine the personal fitness level using the scale below.

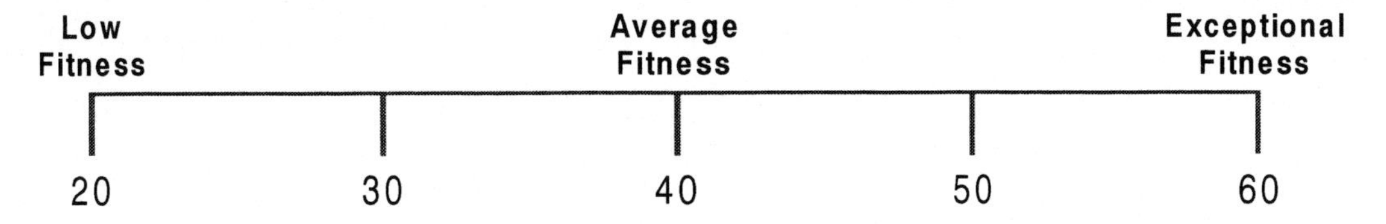

DATA

Table 6		
Condition	Rate or time	Points
Standing heart rate	beats/min	
Reclining heart rate	beats/min	
Reclining to standing	beats/min	
Before step test	beats/min	
After 5 steps	beats/min	
Recovery rate	seconds	
Endurance	beats/min	
		Total points:

QUESTIONS

1. How did your heart rate change after moving from a standing position to a reclining position? Is this what you expected? How do you account for this?
2. How did your heart rate change after moving from a reclining position back to a standing position? Is this what you expected? How do you account for this?
3. Predict what your heart rate might be if you had exercised for twice the length of time that you actually did. Explain.
4. How does your maximum heart rate compare to other students in your group. Is this what you expected? How do you account for this?
5. Why would athletes need to work longer and harder before their heart rates were at the maximum value?
6. How do you evaluate your physical fitness? Do you agree with the rating obtained from this experiment? Explain.
7. Current research indicates that most heart attacks occur as people get out of bed after sleep. Account for this observation.

EXTENSION

1. Using a sphygmomanometer, learn how to measure blood pressure. Compare a person's blood pressure when reclining, to that of the same person immediately after standing from a reclined position. Relate the change in blood pressure to the heart rate values measured when going from reclining to standing.

2. Design an anonymous survey to be taken by each member of your class. In the survey, ask questions that you think might influence the test results. Examples might include:
   - Did you have more than six hours of sleep last night? Gender? Age?
   - Do you smoke? If so, how many packs per week do you smoke?
   - What was your total number of fitness points?

3. Try to determine whether any of the variables from your survey show a statistical link to fitness. You may want to use a statistical T-Tests to determine whether a relationship between the variable and physical fitness is due to chance.

Experiment
27

TEACHER INFORMATION

Heart Rate and Physical Fitness

There are two different versions of this lab exercise. Version A was written using the Exercise Heart Rate Monitor, which consists of a chest belt and plug-in receiver. This unit measures heart rate by detecting the electrical impulses of the heart with the chest belt and relaying the information to the plug-in receiver. Version B uses the Heart Rate Monitor, which uses a photocell earclip to measure pulse rate.

1. When using the Heart Rate Monitor, the first reading displayed will often be inaccurate. It is best if students ignore this reading. If students encounter an unusually high or low reading while collecting data, it is most likely the result of movement. It is best to omit the inaccurate reading and take another.

2. When using the Heart Rate Monitor, the test subject should remain still while data are collected. This sensor is very sensitive to changes in ambient light or movement. This will help to avoid any movement artifacts being plotted on the graph.

3. With the Heart Rate Monitor, heart rate can be monitored at other points on the body. These include the finger tips and the web of skin between the thumb and index finger. Students should always monitor heart rate from the same body location for each part of the experiment.

4. The Exercise Heart Rate Monitor includes a transmitter belt, receiver module, large elastic strap, and small elastic strap.

5. It is important to have good contact between the transmitter belt and the test subject when using the Exercise Heart Rate Monitor. It is very important that the belt fit snug, but not too tight. Both electrodes should be wetted with either saline solution or contact lens solution. A 5% salt solution works well and can be prepared by adding 5 g per 100 mL of solution. Typical symptoms of inadequate contact with the electrodes are a noisy signal with erroneous peaks, missing heart beat readings, or a flat–line display. If the students receive a flat reading with no heart rate detected, have them move the transmitter and the receiver closer together. The range of the transmitter in the chest belt is 60 to 80 cm.

6. Computer monitors can be a source of electrical interference. Move the receiver module of the Exercise Heart Rate Monitor as far from the computer monitor as possible.

7. The receiver module of the Exercise Heart Rate Monitor will receive signals from the closest transmitter source. To avoid confusion or erroneous readings, have test subjects from different lab teams stay at least 2 m apart.

8. It is possible to alter your heart rate by simply decreasing your respiratory rate and relaxing. Encourage students to stay alert and breathe normally.

SAMPLE RESULTS

Sample data from two students are listed below. The first student was a 17 year old male and the second was a 16 year old female.

Table 6
Sample Student Data

Condition	Rate, Student 1 (beats/min) or time	Points	Rate, Student 2 (beats/min) or time	Points
Standing heart rate	73	11	93	9
Reclining heart rate	54	12	69	11
Reclining to standing	69	11	84	8
Pre-exercise heart rate	72		93	
After 5 steps	87		116	
Recovery rate	57 s	12	98 s	10
Endurance	14	10	23	2
Total		56		40

ANSWERS TO QUESTIONS

1. The heart rate generally lowers when a student moves from a standing position to a reclining position. The forces of gravity do not have to be overcome for blood to flow while in a reclining position.
2. The heart rate generally increases when a student moves from a reclining position to a standing position. The forces of gravity do have to be overcome while in a standing position.
3. The heart rate generally increases when a student exercises twice as long. It will not increase to twice the rate, however, because the heart will adjust to the new stress and increase the blood flow to meet the body's needs. The blood flow is proportional to the heart rate. When the blood flow is appropriate, the heart rate will no longer continue increasing.
4. Answers will vary. Factors such as weight, regular exercise, health, etc., may play a part in determining the maximum heart rate of a student.
5. An athletes heart is more efficient at moving blood through the body. Each contraction of an athlete's heart moves a greater volume of blood than an average individual. More blood per contraction means more oxygen for the body's cells. Because of this athletes must work harder to increase their heart rate to its maximum values.
6. Answers will vary.
7. The heart rate increases significantly when an individual moves from a reclining position to a standing position. The force of gravity on the blood make the heart work harder, as in Question 2. This increased stress might provoke a heart attack in susceptible people.

Experiment

28

Monitoring EKG

An electrocardiogram, or EKG, is a graphical recording of the electrical events occurring within the heart. A typical EKG tracing consists of five identifiable deflections. Each deflection is noted by one of the letters P, Q, R, S, or T. The P wave is the first waveform in a tracing and represents the depolarization of the heart's atria. The next waveform is a complex and consists of the Q, R, and S deflection. The QRS complex represents the depolarization of the heart's ventricles. The deflection that represents the repolarization of the atria is usually undetectable because of the intensity of the QRS waveform. The final waveform is the T wave and it represents the repolarization of the ventricles.

Because an EKG is a recording of the heart's electrical events, it is valuable in diagnosing diseases or ailments that damage the conductive abilities of the heart muscle. When cardiac muscle cells are damaged or destroyed, they are no longer able to conduct the electrical impulses that flow through them. This causes the electrical signal to terminate at the damaged tissue or directed away from the signal flow. The termination or redirection of the electrical signal will alter the manner in which the heart contracts. A cardiologist can look at a patient's electrocardiogram and determine the presence of damaged cardiac muscle based on the waveform as well as the time interval between electrical events.

In this activity, you will use the EKG sensor to make a five-second graphical recording of your heart's electrical events. From this recording, you will identify the previously mentioned waveform components and determine the time intervals associated with each.

OBJECTIVES

In this experiment, you will

- Use the EKG Sensor to graph your heart's electrical activity.
- Determine the time interval between EKG events.
- Calculate heart rate based on your EKG recording.

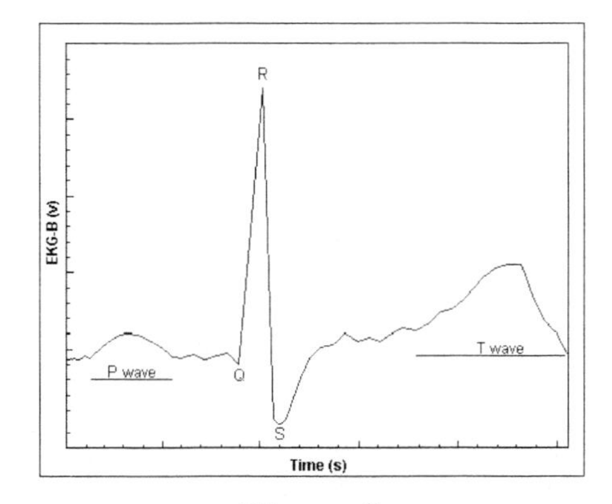

Figure 1

MATERIALS

computer
Vernier computer interface
Logger*Pro*
Vernier EKG Sensor
disposable electrode tabs

PROCEDURE

1. Connect the EKG Sensor to the Vernier computer interface. Prepare the computer for data collection by opening the "28 Monitoring EKG" file in the *Biology with Computers* folder.

2. Attach three electrode tabs to your arms, as shown in Figure 2. A single patch should be placed on the inside of the right wrist, on the inside of the right upper forearm (below elbow), and on the inside of the left upper forearm (below elbow).

3. Connect the EKG clips to the electrode tabs as shown in Figure 2. Sit in a reclined position in a chair or lay flat on top of a lab table. The arms should be hanging at the side unsupported. When everything is positioned properly, click Collect to begin data collection. If your graph has a stable baseline as shown below, continue to Step 4. If your graph has an unstable baseline, collect a new set of data by clicking Collect. Repeat data collection until your graph has a stable baseline.

Green (negative)
Red (positive)
Black (ground)

Figure 2

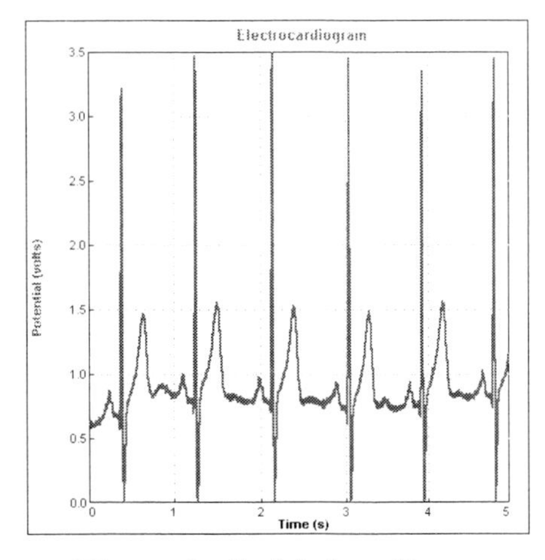

Figure 3: Stable baseline

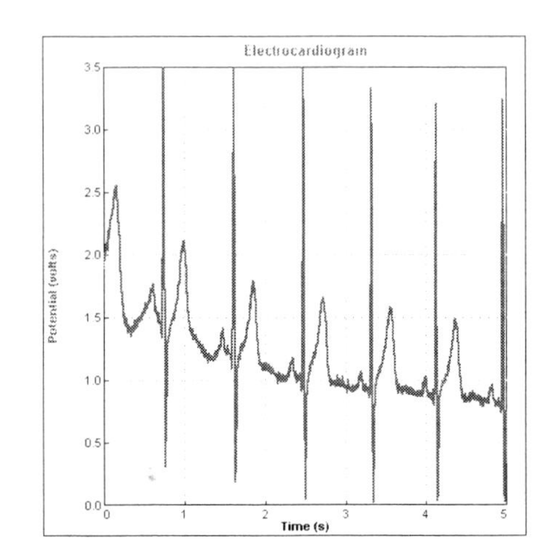

Figure 4: Unstable baseline

4. Click the Examine button, [icon], to analyze the data. As you move the mouse pointer across the screen, the x and y values are displayed in the Examine window that appears. For three heart beats, identify the various EKG waveforms using Figure 5 and determine the time intervals listed below. Record the average for each set of time intervals in Table 1.

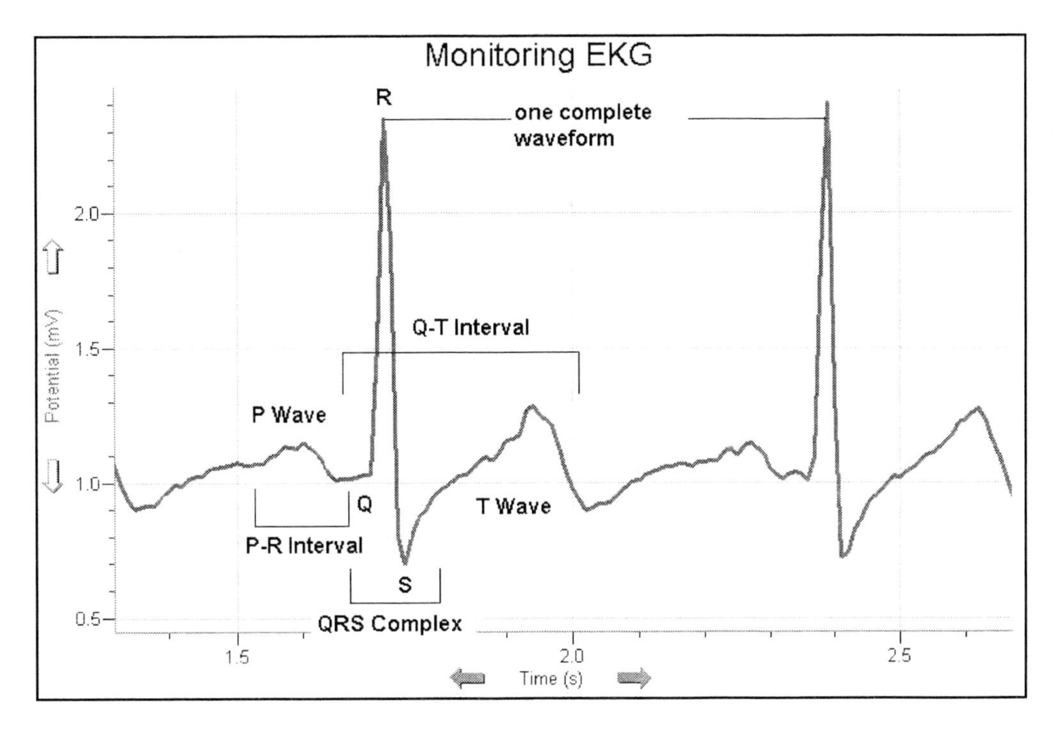

- **P-R interval:** time from the beginning of P wave to the start of the QRS complex.
- **QRS complex:** time from Q deflection to S deflection.
- **Q-T interval:** time from Q deflection to the end of the T.

5. Calculate the heart rate in beats/min using the EKG data. Remember to include the time between the end of the T Wave and the beginning of the next P Wave. Use the total number of seconds for one full heart cycle in the equation. Record the heart rate in Table 1.

$$\frac{\#\,\text{beats}}{\text{minute}} = \frac{1\,\text{beat}}{\_\_\ \text{seconds}} \times \frac{60\,\text{seconds}}{1\,\text{minute}}$$

6. Print a copy of your EKG. Identify and label the various waveforms.

DATA

Table 1	
Interval	Time (s)
P - R	
QRS	
Q - T	

Heart Rate	___________ beats/min

Table 2	
Standard Resting Electrocardiogram Interval Times	
P - R interval	0.12 to 0.20 s
QRS interval	less than 0.10 s
Q - T interval	0.30 to 0.40 s

QUESTIONS

1. The electrocardiogram is a powerful tool used to diagnose certain types of heart disease. Why is it important to look at the time intervals of the different waveforms?
2. What property of heart muscle must be altered in order for an EKG to detect a problem? Explain.
3. Based on what you have learned regarding electrocardiograms, can they be used to diagnose all heart diseases or defects? Explain.
4. Describe a cardiovascular problem that could be diagnosed by a cardiologist using an electrocardiogram.

EXTENSION

Using data collected with the EKG Sensor, it is possible to determine a more accurate maximum heart rate value for an individual. The commonly used formula for calculating maximum heart rate is:

220 bpm – Individual's Age = Max Heart Rate

While this formula is sufficient for general purposes, it fails to take into account physical differences such as size, and fitness level. For example, an individual that engages in regular exercise will likely have a heart that operates more efficiently due to the effects of athletic training.

To calculate your maximum heart rate, do the following:

a. Run in place or perform some type of exercise, such as jump-n-jacks, for 1-minute.

b. Repeat Steps 1 – 4 to collect and analyze your electrocardiogram. When analyzing the data in Step 4, only determine the average Q-T interval.
c. Divide 60 seconds by the Q-T interval to calculate your maximum heart rate.

Experiment
28

TEACHER INFORMATION

Monitoring EKG

1. Always have the test subject lay flat on a lab table or sit in a chair with his arms at his sides when data are being collected.

2. If the test subject's skin is oily or a lotion has been applied, scrub the area of skin where the electrode is to be attached with soap and water prior to applying the electrode.

3 EKG electrodes should not be placed over large muscle groups, such as those found on the upper arm. If the electrodes are placed over a large muscle, the EKG signal will be noisy due to muscle artifacts caused during muscle contraction.

4. Students will find that not all EKGs look alike. They should not be alarmed. The heart is simply a muscle and like any muscle it will differ physically from one individual to another. What they should see, is that all EKGs contain the same waveforms even though they differ in size and shape.

5. The stored calibration in this experiment works well for EKG Sensor. The calibration will convert millivolts directly to volts.

SAMPLE RESULTS

Sample data from a 28 year old male subject is shown below.

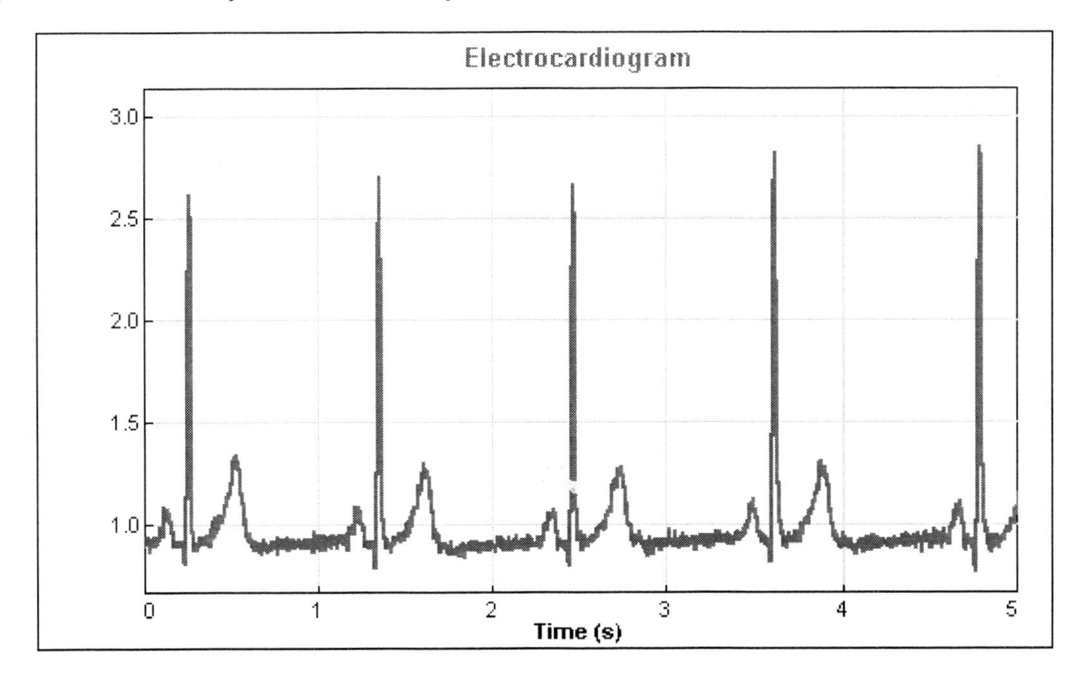

Table 1	
P-R	0.13 seconds
QRS	0.04 seconds
Q-T	0.34 seconds
Heart Rate	81 beats/min

Maximum Heart Rate (using formula)	192 beats/min
Maximum Heart Rate (using EKG)	177 beats/min

ANSWERS TO QUESTIONS

1. By looking at time intervals between cardiac events and comparing them to standard values, one can determine if a person's heart has sustained any damage that will alter the muscle cells ability to conduct an electrical signal. For example, in first degree heart block, impulse propagation through the AV node is impeded, resulting in an abnormally long P-R interval.

2. In order for a problem to be detected by an EKG, the disease must alter the cardiac muscle cell's ability to conduct an electrical impulse.

3. Many diseases affect the heart without altering its electrical properties. A physical defect such as a faulty pulmonary or aortic valve is going to disturb the mechanical action of the heart, but not the electrical action that would show up on an EKG.

4. The most common problems that can be diagnosed from an EKG are heart block, atrial fibrillation, and ventricular fibrillation. Ventricular fibrillation is the most severe and is often the result of a myocardial infarction or heart attack. During ventricular fibrillation, there is an abnormality in impulse conduction in the ventricles, resulting in continuous disorganized contractions inadequate to produce blood flow.

Experiment

29

Ventilation and Heart Rate

In this experiment, you will investigate the effect of altering the levels of oxygen and carbon dioxide on the rate at which the heart beats. Two different methods of ventilation will be used to investigate this phenomenon. The first method, *hyperventilation*, is when the breathing rate of an organism is greater than what is necessary for proper exchange of oxygen and carbon dioxide. This will be achieved by a period of rapid breathing by the test subject. The second method, *hypoventilation*, occurs when there is a decrease in ventilation without a decrease in oxygen consumption or carbon dioxide production by the body. True hypoventilation is usually the result of a disease. The test subject will simulate this condition by holding his or her breath for a period of time. The test subject's heart rate will be monitored using the Exercise Heart Rate Monitor.

OBJECTIVES

In this experiment, you will

- Monitor the heart rate of the test subject using the Exercise Heart Rate Monitor.
- Evaluate the effects of hyperventilation and hypoventilation on heart rate.

MATERIALS

computer
Vernier computer interface
Logger*Pro*
Vernier Exercise Heart Rate Monitor
saline solution in dropper bottle

PROCEDURE

Each person in a lab group will take turns being the subject and the tester. When it is your turn to be the subject, your partner will be responsible for recording the data on your lab sheet.

1. Elastic straps, for securing the transmitter belt, come in two different sizes. Select the size of elastic strap that best fits the subject being tested. It is important that the strap provide a snug fit of the transmitter belt.
2. Wet each of the electrodes (the two grooved rectangular areas on the underside of the transmitter belt) with 3 drops of saline solution.
3. Secure the transmitter belt against the skin directly over the base of the rib cage. The POLAR logo on the front of the belt should be in line with the chest center as shown in Figure 1. Adjust the elastic strap to ensure a tight fit.
4. Connect the receiver module of the Exercise Heart Rate Monitor to the Vernier computer interface.

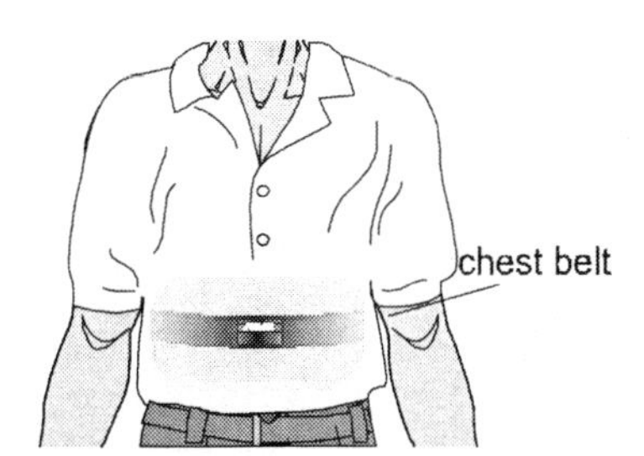

Figure 1

5. Have the subject hold the receiver in his or her right hand as shown in Figure 2. Remember, the receiver must be within 80 cm of the transmitter belt while data is being collected.

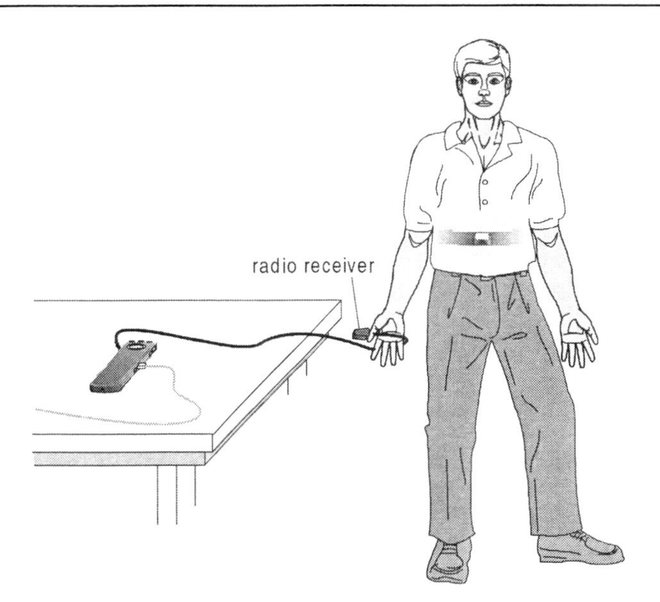

Figure 2

6. Prepare the computer for data collection by opening the file "29 Ventilation Heart Rate" from the *Biology with Computers* folder of Logger*Pro*.

7. Click Collect to begin monitoring heart rate.

8. Determine that the sensor is functioning correctly. The readings should be consistent and within the normal range of the individual, usually between 55 and 80 beats per minute. Click Stop when you have determined that the equipment is operating properly.

Part I Hyperventilation

9. Instruct the test subject to breathe normally while sitting still. Click Collect to begin monitoring heart rate. After collecting data for 30 seconds, have the test subject make rapid shallow breaths for the next 30 seconds. The test subject should breathe normally during the remainder of the data collection. Data collection will stop after 120 seconds.

10. Click the Examine button, , to examine the heart rate values plotted on the graph. Record the test subject's heart rate in Table 1 for every 10 second interval.

11. Click anywhere on the graph. Type "Hyperventilation" as the title of the graph and then press ENTER.

12. Print a copy of the graph. Enter your name(s) and the number of copies of the graph you want.

Part II Hypoventilation (simulated)

13. Click Collect to begin monitoring heart rate. After collecting data for 30 seconds, have the test subject take a large breath and hold it as long as possible. The test subject should not hold his or her breath longer than 60 seconds. The test subject should breathe normally during the remainder of the data collection after releasing his or her breath. Data collection will stop after 120 seconds.

14. Click the Examine button, , to examine the heart rate values plotted on the graph. Record the test subject's heart rate in Table 1 for every 10 second interval.

15. Click anywhere on the graph. Type "Hypoventilation" as the title of the graph and then press ENTER.

16. Print a copy of the graph. Enter your name(s) and the number of copies of the graph you want.

PROCESSING THE DATA

Table 1												
Time (sec)	10	20	30	40	50	60	70	80	90	100	110	120
Hyperventilation												
Hypoventilation												

QUESTIONS

1. What happens to the heart rate during hyperventilation?
2. What happens to the heart rate during hypoventilation?
3. List several factors that you think may have caused the test subject's heart rate to change in each of the trials.
4. What happens to the oxygen levels in your lungs during hyperventilation? Carbon dioxide levels?
5. In what way would the change in heart rate that corresponds with holding your breath be advantageous in other types of organisms? What organisms might commonly exhibit such an adaptation?

Experiment 29

TEACHER INFORMATION

Ventilation and Heart Rate

1. The Exercise Heart Rate Monitor includes a transmitter belt, receiver module, large elastic strap, and small elastic strap.
2. It is important to have good contact between the transmitter belt and the test subject when using the Exercise Heart Rate Monitor. It is very important that the belt fit snugly, but not too tight. Both electrodes should be wet with either saline solution or contact lens solution. A 5% salt solution works well and can be prepared by adding 5 g of NaCl per 100 mL of solution. Typical symptoms of inadequate contact with the electrodes are a noisy signal with erroneous peaks, missing heart beat readings, or a flat-line display. If the students receive a flat reading with no heart rate detected, have them move the transmitter and the receiver closer together. The range of the transmitter in the chest belt is 80 cm.
3. Computer monitors can be a source of electrical interference. Keep the receiver module of the Exercise Heart Rate Monitor as far as possible from any computer monitors in the class.
4. The receiver module of the Exercise Heart Rate Monitor will receive signals from the closest transmitter source. To avoid confusion or erroneous readings, have test subjects from different lab teams stay at least 2 meters apart.
5. It is possible to alter your heart rate by simply decreasing your respiratory rate and relaxing. Encourage students to stay alert and breathe normally.
6. Anyone prone to dizziness, nausea, or headaches should not be selected as the test subject.

SAMPLE RESULTS

Table 1

time (s)	10	20	30	40	50	60	70	80	90	100	110	120
Hypoventilation	82	79	79	66	67	63	65	69	98	79	82	85
Hyperventilation	79	81	78	92	91	99	103	95	88	80	82	80

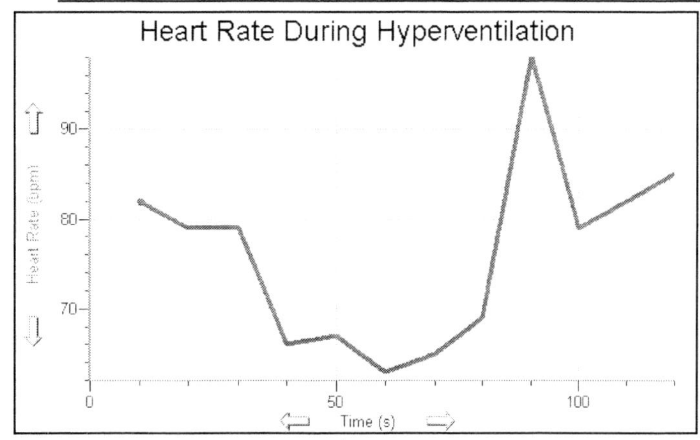

Hyperventilation

Heart Rate During Hypoventilation
100
90
80
0
50
100
Time (s)

Hypoventilation

ANSWERS TO QUESTIONS

1. Heart rate will increase during hyperventilation.

2. Heart rate will decrease during hypoventilation.

3. Answers will vary. Factors that may be listed are blood carbon dioxide concentrations, blood oxygen concentrations, blood pressure, body temperature, and hormones.

4. Hyperventilation results in an increase of oxygen levels in the alveoli of the lungs and a decrease in carbon dioxide levels. The opposite results during hypoventilation.

5. When heart rate decreases, so does the rate at which carbon dioxide levels increase in the lungs. By slowing the increase of carbon dioxide in the lungs, an organism could hold its breath for a longer period of time. Aquatic mammals are a good example of organisms that use such an adaptation. This allows whales to stay under water for as long as one hour. A seal's heart rate can change from 150 beats per minute to 10 beats per minute when it dives underneath the water to search for food.

Experiment

30

Oxygen Gas and Human Respiration

The process of breathing accomplishes two important tasks for the body. During inhalation, oxygen-rich air is brought into your lungs. During exhalation, air depleted in oxygen and rich in carbon dioxide is forced out. Oxygen is then transported to the cells where it is used in the process of respiration, yielding carbon dioxide as a product.

$$C_6H_{12}O_6 + 6\ O_2 \rightarrow 6\ CO_2 + 6\ H_2O + \text{energy}$$

glucose oxygen carbon dioxide water

Gas exchange takes place in the lungs at the membrane between the alveoli and the pulmonary capillaries. It is here that oxygen diffuses into the bloodstream and carbon dioxide diffuses out. Under normal circumstances, there is an equilibrium between the oxygen and carbon dioxide levels in the blood. Several mechanisms are involved in maintaining this balance. One such mechanism involves chemoreceptors. These specialized cells respond to changes in carbon dioxide, oxygen and H^+ concentrations and influence the body's ventilation patterns to maintain the proper balance of blood gases.

In this experiment, you will determine what factors affect how long you can hold your breath. You will be tested under two different conditions. The first condition is normal breathing. The second condition is immediately following hyperventilation. *Hyperventilation* is when your breathing rate is greater than what is necessary for proper exchange of oxygen and carbon dioxide. This will be achieved by a period of rapid breathing prior to holding your breath.

OBJECTIVES

In this experiment, you will

- Use an O_2 Gas Sensor to determine residual oxygen levels in exhaled air.
- Evaluate how internal O_2 and CO_2 concentrations influence breathing patterns.

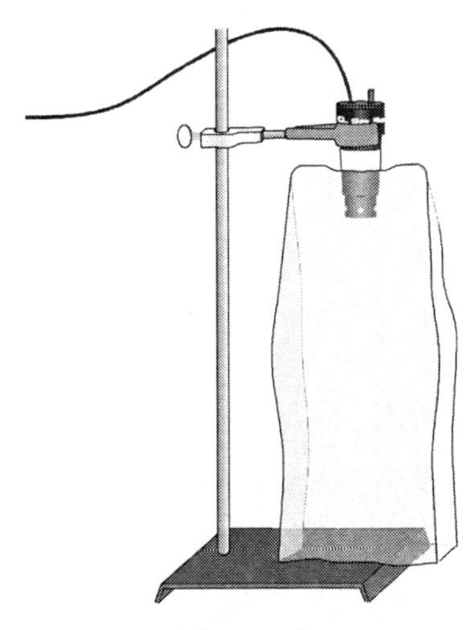

Figure 1

MATERIALS

computer
Vernier computer interface
Logger *Pro*
Vernier O_2 Gas Sensor
ring stand
test tube clamp
bread bag

PRELIMINARY QUESTIONS

1. How long do you think you can hold your breath?
2. When you hold your breath, what do you think happens to the oxygen concentration in your lungs? Explain.
3. When you hold your breath, what do you think happens to the carbon dioxide concentration in your lungs? Explain.
4. On average, people can hold their breath for a minute. What do you think prevents people from holding their breath for 2 or 3 minutes?

PROCEDURE

Each person in a lab group will take turns being the subject and the tester. When it is your turn to be the subject, your partner will be responsible for operating the equipment.

1. Secure the O_2 Gas Sensor using a test tube clamp and ring stand as shown in Figure 1. The plastic bread bag should already be taped to the sensor.
2. Connect the O_2 Gas Sensor to the Vernier computer interface.
3. Prepare the computer for data collection by opening the "30 Oxygen and Human Resp" file in the *Biology with Computers* folder.
4. When you begin collecting data, it is important that data collection begins at the same point the subject begins to hold his breath.
   a. Have the subject take a deep breath and hold it. Immediately click Collect to begin data collection. The subject should hold his breath as long as possible.
   b. When the subject can no longer hold his breath, he should blow his breath into the bread bag and twist the open end shut. This should result in the bread bag filled with the air the subject was holding in his lungs. Allow data collection to proceed for the full 120 seconds.
   c. When data collection has finished, open the bread bag and pull it back over the sensor exposing the sensor to room air. Leave the bag in that position until you are prepared to collect data again.
   d. Click the Examine button, . The cursor will become a vertical line. As you move the mouse pointer across the screen, the oxygen and time values corresponding to its position will be displayed in the box at the upper-left corner of the graph. Scroll across the data to determine how long the subject held his breath. Record the time in Table 1. Determine the maximum and minimum oxygen concentrations and record them in Table 1. To remove the examine box, click the upper-left corner of the box.
5. Move your data to a stored run. To do this, choose Store Latest Run from the Experiment menu.

6. Collect data following mild hyperventilation.
   a. Pull the bread bag back down off of the sensor in preparation for data collection.
   b. Have the subject take 10 quick deep breaths, forcefully blowing out all air after each breath. The subject should then take an 11th breath and hold it. Immediately click Collect to begin data collection. The subject should hold his breath as long as possible.
   c. When the subject can no longer hold his breath, he should blow his breath into the bread bag and twist the open end shut. This should result in the bread bag filled with the air the subject was holding in his lungs. Allow data collection to proceed for the full 120 seconds.
   d. When data collection has finished, open the bread bag and pull it back over the sensor exposing the sensor to room air.
   e. Click the Examine button, . The cursor will become a vertical line. As you move the mouse pointer across the screen, the oxygen and time values corresponding to its position will be displayed in the box at the upper-left corner of the graph. Scroll across the data to determine how long the subject held his breath. Record the time in Table 1. Determine the maximum and minimum oxygen concentrations and record them in Table 1. To remove the examine box, click the upper-left corner of the box.
7. Both runs should now be displayed on the same graph. Use the displayed graph and the data in Table 1 to answer the questions below.

DATA

Table 1

	Breath held (s)	Maximum O_2 concentration (%)	Minimum O_2 concentration (%)	Change in oxygen (%)
Normal				
Hyperventilation				

QUESTIONS

1. Did the oxygen concentration change as you expected? If not, explain how it was different.
2. Did the amount of time you held your breath change after hyperventilation (taking the 10 quick breaths)? If so, did the time increase or decrease? Explain.
3. After hyperventilation, was the resulting concentration of oxygen in your exhaled breath higher or lower than in the first attempt? How much did it change? What do you contribute this to?
4. On the first trial, what do you believe forced you to start breathing again?
5. On the second trial, what do you believe forced you to start breathing again?
6. Based on your answers to questions 4 and 5, does the concentration of oxygen or carbon dioxide have a greater influence on how long one can hold his breath?

Experiment **30**

TEACHER INFORMATION

Oxygen Gas and Human Respiration

This exercise is meant as an investigative study. Results may vary from those printed here. Variations in results may be due to different body mass, age, fitness, and sex of the students.

1. The best bag to use for this experiment is a bread bag. This is the same bag that comes with every loaf of bread you purchase from the super market. The advantages to this type of bag are that it is pliable, large, and easy to obtain.
2. To secure the bag to the O_2 Gas Sensor, cut a small hole the size of a half dollar and feed the sensor through the hole. Use tape to seal the bag to the sensor and prevent any air from escaping at that junction. Most any type of tape will work.
3. Once the sensor is mounted on the ring stand with the test tube clamp, students can rotate the sensor enough so that it is pointing more in their direction. This may prove easier to use rather than having the sensor pointing straight down.
4. Students may find it easier to hold their breath if they are facing away from the computer screen.
5. Students with asthma or other respiratory ailments should not participate as the subject in this experiment.

ANSWERS TO PRELIMINARY QUESTIONS

1. Student answers will vary as to how long they think they can hold their breath.
2. Oxygen levels in the lungs are going to drop as more oxygen moves into the bloodstream.
3. Carbon dioxide levels in the lungs are going to increase as more moves out of the bloodstream.
4. The main factor that prevents people from holding their breath longer is the increase in carbon dioxide concentration in the blood. Carbon dioxide, oxygen, and H^+ concentrations all contribute to voluntary control of ventilation. The body is far more sensitive to changes in carbon dioxide concentration than it is to changes in oxygen. Because of this, voluntary control of ventilation is primarily controlled by the concentration of carbon dioxide found in the blood.

SAMPLE RESULTS

Table 1				
	Breath held (s)	Maximum O_2 concentration (%)	Minimum O_2 concentration (%)	Change in oxygen (%)
Normal	55	20.5	15.7	4.8
Hyperventilation	89	20.4	14.2	6.2

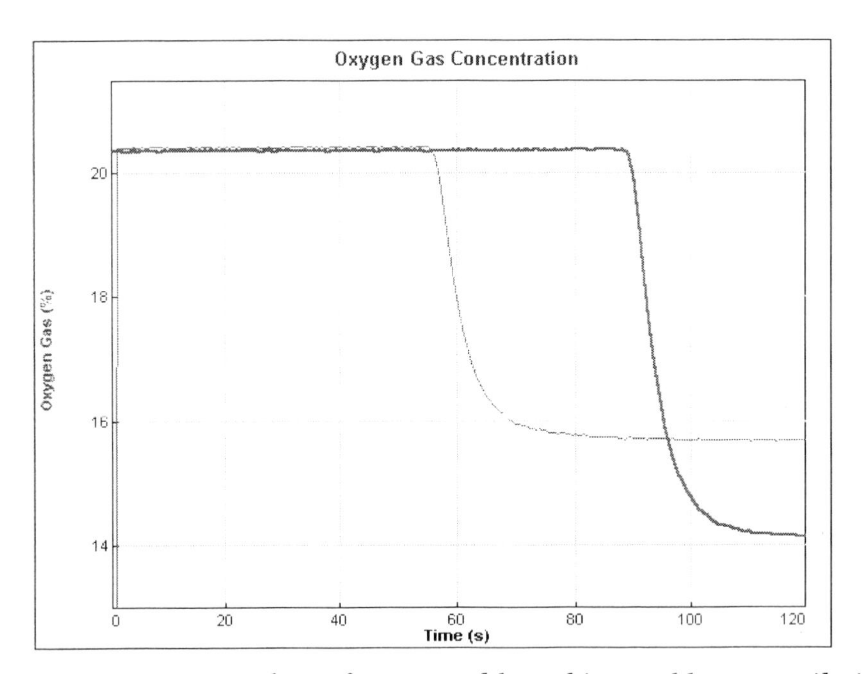

Oxygen concentration data after normal breathing and hyperventilation

ANSWERS TO QUESTIONS

1. Answers vary.
2. During hyperventilation, the carbon dioxide concentration in the body drops. This should result in a person being able to hold their breath significantly longer than normal.
3. After hyperventilation, students will likely find that the oxygen concentration of their expired air was lower than under normal breathing conditions. Since they are able to hold their breath longer, more oxygen will be taken up and less will be expired.
4. Answers vary.
5. Answers vary.
6. Initially, students will believe that they are forced to breath again because they have run out of air. After seeing that they used more air in the second trial when they hyperventilated, they should reason that oxygen was not the determining factor. This should lead them to the conclusion that carbon dioxide is the reason they were forced to breath again.

Experiment

31A

Photosynthesis and Respiration

Plants make sugar, storing the energy of the sun into chemical energy, by the process of photosynthesis. When they require energy, they can tap the stored energy in sugar by a process called cellular respiration.

The process of photosynthesis involves the use of light energy to convert carbon dioxide and water into sugar, oxygen, and other organic compounds. This process is often summarized by the following reaction:

$$6\ H_2O + 6\ CO_2 + \text{light energy} \rightarrow C_6H_{12}O_6 + 6\ O_2$$

Cellular respiration refers to the process of converting the chemical energy of organic molecules into a form immediately usable by organisms. Glucose may be oxidized completely if sufficient oxygen is available by the following equation:

$$C_6H_{12}O_6 + 6\ O_2 \rightarrow 6\ H_2O + 6\ CO_2 + \text{energy}$$

All organisms, including plants and animals, oxidize glucose for energy. Often, this energy is used to convert ADP and phosphate into ATP.

OBJECTIVES

In this experiment, you will

- Use an O_2 Gas Sensor to measure the amount of oxygen gas consumed or produced by a plant during respiration and photosynthesis.
- Determine the rate of respiration and photosynthesis of a plant.

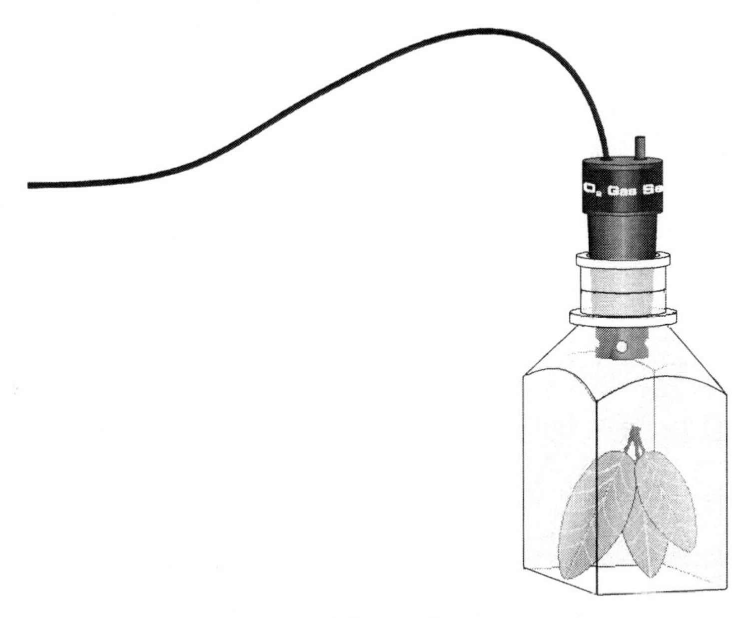

Figure 1

MATERIALS

computer
Vernier computer interface
Logger *Pro*
Vernier O_2 Gas Sensor
aluminum foil
250 mL respiration chamber
plant leaves
500 mL tissue culture flask
lamp
forceps

PROCEDURE

1. Connect the O_2 Gas Sensor to the Vernier interface.

2. Prepare the computer for data collection by opening the file "31A Photosyn-Resp (O2)" from the *Biology with Computers* folder of Logger*Pro.*

3. Obtain several leaves from the resource table and blot them dry, if damp, between two pieces of paper towel.

4. Place the leaves into the respiration chamber, using forceps if necessary. Wrap the respiration chamber in aluminum foil so that no light reaches the leaves.

5. Place the O_2 Gas Sensor into the bottle as shown in Figure 1. Gently push the sensor down into the bottle until it stops. The sensor is designed to seal the bottle without the need for unnecessary force. Wait 3 minutes before proceeding to Step 6.

6. Click Collect to begin data collection. Data will be collected for 10 minutes.

7. When data collection has finished, determine the rate of respiration:
   a. Move the mouse pointer to the point where the data values begin to decrease. Hold down the left mouse button. Drag the pointer to the point where the data ceases to decline and release the mouse button.
   b. Click on the Linear Fit button, , to perform a linear regression. A floating box will appear with the formula for a best fit line.
   c. Record the slope of the line, *m*, as the rate of respiration in Table 1.
   d. Close the linear regression floating box.

8. Move your data to a stored run. To do this, choose Store Latest Run from the Experiment menu.

9. Remove the aluminum foil from around the respiration chamber.

10. Fill the tissue culture flask with water (not the respiration chamber) and place it between the lamp and the respiration chamber. The flask will act as a heat shield to protect the plant.

11. Turn the lamp on. Place the lamp as close to the leaves as reasonable. Do not let the lamp touch the tissue culture flask. Note the time. The lamp should be on for 3 minutes prior to beginning data collection.

12. After the three-minute time period is up, click Collect to begin data collection. Data will be collected for 10 minutes.

13. When data collection has finished, determine the rate of photosynthesis:
    a. Move the mouse pointer to the point where the data values begin to increase. Hold down the left mouse button. Drag the pointer to the point where the data ceases to rise and release the mouse button.
    b. Click on the Linear Fit button, [icon], to perform a linear regression. Choose "Latest: Oxygen Gas" and a floating box will appear with the formula for a best fit line.
    c. Record the slope of the line, *m*, as the rate of photosynthesis in Table 1.
    d. Close the linear regression floating box.
14. Print a graph showing your photosynthesis and respiration data.
    a. Label each curve by choosing Text Annotation from the Insert menu. Enter "Photosynthesis" in the edit box. Repeat to create an annotation for the "Respiration" data. Drag each box to a position near its respective curve. Adjust the position of the arrow heads.
    b. Print a copy of the graph, with both data sets displayed. Enter your name(s) and the number of copies of the graph you want.
15. Remove the plant leaves from the respiration chamber, using forceps if necessary. Clean and dry the respiration chamber.

DATA

Table 1	
Leaves	Rate of photosynthesis/respiration (ppt/min)
In the Dark	
In Light	

QUESTIONS

1. Were either of the rate values a positive number? If so, what is the biological significance of this?
2. Were either of the rate values a negative number? If so, what is the biological significance of this?
3. Do you have evidence that cellular respiration occurred in leaves? Explain.
4. Do you have evidence that photosynthesis occurred in leaves? Explain.
5. List five factors that might influence the rate of oxygen production or consumption in leaves. Explain how you think each will affect the rate?

EXTENSIONS

1. Design and perform an experiment to test one of the factors that might influence the rate of oxygen production or consumption in Question 5.

2. Compare the rates of photosynthesis and respiration among various types of plants.

TEACHER INFORMATION

Experiment
31A

Photosynthesis and Respiration

1. Spinach leaves purchased from a grocery store work very well and are readily available any time of the year. For best results, keep the leaves cool until they are to be used. Just before use, expose the leaves to bright light for 5 minutes.
2. A fluorescent ring lamp works very well since it bathes the plant in light from all sides and it gives off very little heat. When using a ring lamp as shown below, it is not necessary to use a heat shield.

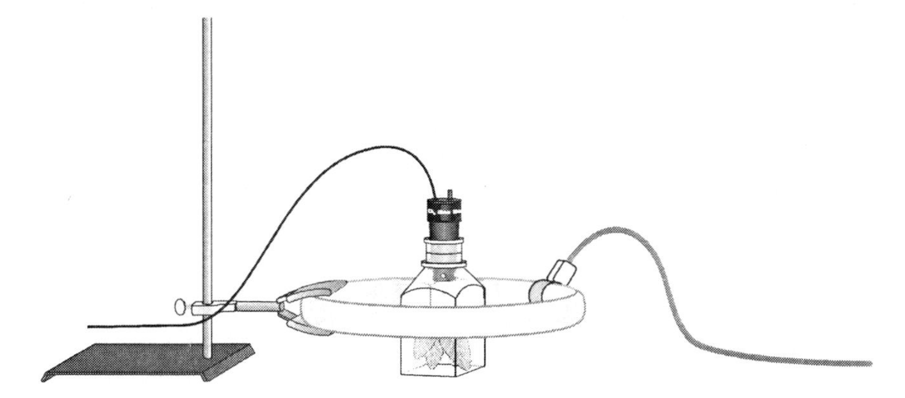

3. If tissue culture flasks are not available, a beaker or flask of water will also work as a heat shield. The tissue culture flask is very thin, however, and will allow leaves to receive much more light from the same lamp.
4. On a nice, sunny day, this experiment may be performed using sun light. If so, no heat shield is needed.
5. To extend the life of the O_2 Gas Sensor, always store the sensor upright in the box in which it was shipped.

SAMPLE RESULTS

Table 1	
Leaves	Rate of Respiration/Photosynthesis (ppt/min)
In the Dark	–0.138
In Light	0.271

ANSWERS TO QUESTIONS

1. The rate value for leaves in the light was a positive number. The biological significance of this is that O_2 is produced during photosynthesis. This causes the concentration of O_2 to increase, as the O_2 is converted into glucose.

2. The rate value for leaves in the dark was a negative number. The biological significance of this is that O_2 is consumed during cellular respiration. This causes the concentration of O_2 to decrease as glucose is oxidized for energy.

3. Yes, cellular respiration occurred in leaves, since O_2 decreased when leaves were in the dark and photosynthesis was not possible.

4. Yes, photosynthesis occurred in leaves, since O_2 increased when leaves were exposed to light.

5. Answers may vary. They might include:

   a. A greater number of leaves should increase the rate, since there are more chloroplasts to undergo photosynthesis and more cells to require energy through cellular respiration.
   b. A greater light intensity will increase the rate of photosynthesis. It may not affect the rate of cellular respiration, however.
   c. A cooler room may decrease both rates, as cellular metabolism decreases in cooler weather.
   d. Facing the top of the leaves toward the light should increase the rate of photosynthesis, since the chloroplasts are closer to the light source.
   e. If the plants overheat due to the heat from the lamp, they may wilt and stop functioning. This will decrease all rates.
   f. If there are too many leaves, diffusion may be restricted and prevent accurate readings. This may apparently decrease both rates.

Experiment

31B

Photosynthesis and Respiration

Plants make sugar, storing the energy of the sun into chemical energy, by the process of photosynthesis. When they require energy, they can tap the stored energy in sugar by a process called cellular respiration.

The process of photosynthesis involves the use of light energy to convert carbon dioxide and water into sugar, oxygen, and other organic compounds. This process is often summarized by the following reaction:

$$6\ H_2O + 6\ CO_2 + \text{light energy} \rightarrow C_6H_{12}O_6 + 6\ O_2$$

Cellular respiration refers to the process of converting the chemical energy of organic molecules into a form immediately usable by organisms. Glucose may be oxidized completely if sufficient oxygen is available by the following equation:

$$C_6H_{12}O_6 + 6\ O_2 \rightarrow 6\ H_2O + 6\ CO_2 + \text{energy}$$

All organisms, including plants and animals, oxidize glucose for energy. Often, this energy is used to convert ADP and phosphate into ATP.

OBJECTIVES

In this experiment, you will

- Use a CO_2 Gas Sensor to measure the amount of carbon dioxide consumed or produced by a plant during respiration and photosynthesis.
- Determine the rate of respiration and photosynthesis of a plant.

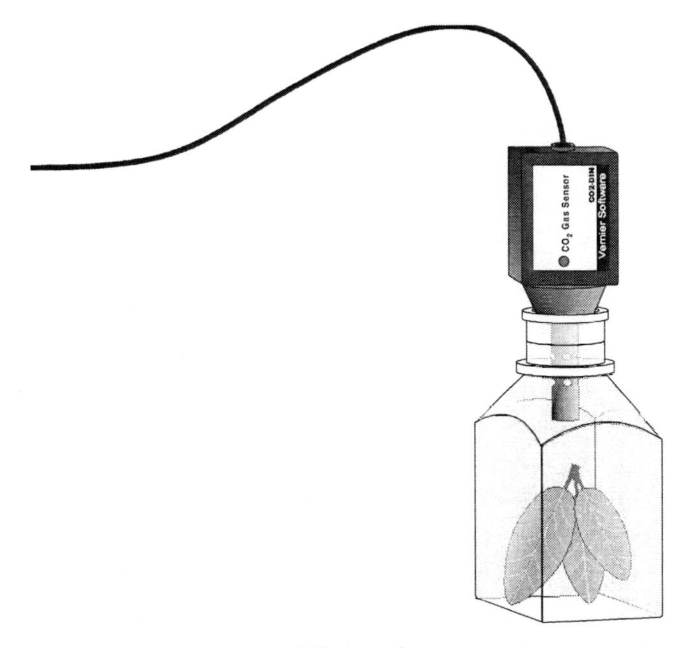

Figure 1

MATERIALS

computer
Vernier computer interface
Logger *Pro*
Vernier CO_2 Gas Sensor
aluminum foil
250 mL respiration chamber
plant leaves
500 mL tissue culture flask
lamp
forceps

PROCEDURE

1. Connect the CO_2 Gas Sensor to the Vernier interface.

2. Prepare the computer for data collection by opening the file "31B Photosyn-Resp (CO2)" from the *Biology with Computers* folder of Logger*Pro*.

3. Obtain several leaves from the resource table and blot them dry, if damp, between two pieces of paper towel.

4. Place the leaves into the respiration chamber, using forceps if necessary. Wrap the respiration chamber in aluminum foil so that no light reaches the leaves.

5. Place the CO_2 Gas Sensor into the bottle as shown in Figure 1. Gently twist the stopper on the shaft of the CO_2 Gas Sensor into the chamber opening. Do not twist the shaft of the CO_2 Gas Sensor or you may damage it. Wait 3 minutes before proceeding to Step 6.

6. Click Collect to begin data collection. Data will be collected for 10 minutes.

7. When data collection has finished, determine the rate of respiration:
   a. Move the mouse pointer to the point where the data values begin to increase. Hold down the left mouse button. Drag the pointer to the point where the data ceases to rise and release the mouse button.
   b. Click on the Linear Fit button, , to perform a linear regression. A floating box will appear with the formula for a best fit line.
   c. Record the slope of the line, *m*, as the rate of respiration in Table 1.
   d. Close the linear regression floating box.

8. Move your data to a stored run. To do this, choose Store Latest Run from the Experiment menu.

9. Remove the aluminum foil from around the respiration chamber.

10. Fill the tissue culture flask with water (not the respiration chamber) and place it between the lamp and the respiration chamber. The flask will act as a heat shield to protect the plant leaves.

11. Turn the lamp on. Place the lamp as close to the leaves as reasonable. Do not let the lamp touch the tissue culture flask. Note the time. The lamp should be on for 3 minutes prior to beginning data collection.

12. After the three-minute time period is up, click Collect to begin data collection. Data will be collected for 10 minutes.

13. When data collection has finished, determine the rate of photosynthesis:
    a. Move the mouse pointer to the point where the data values begin to decrease. Hold down the left mouse button. Drag the pointer to the point where the data ceases to decline and release the mouse button.
    b. Click on the Linear Fit button, , to perform a linear regression. Choose "Latest: CO2" and a floating box will appear with the formula for a best fit line.
    c. Record the slope of the line, *m*, as the rate of photosynthesis in Table 1.
    d. Close the linear regression floating box.

14. Print a graph showing your photosynthesis and respiration data.
    a. Label each curve by choosing Text Annotation from the Insert menu. Enter "Photosynthesis" in the edit box. Repeat to create an annotation for the "Respiration" data. Drag each box to a position near its respective curve.
    b. Print a copy of the graph, with both data sets displayed. Enter your name(s) and the number of copies of the graph you want.

15. Remove the plant leaves from the respiration chamber, using forceps if necessary. Clean and dry the respiration chamber.

DATA

Table 1	
Leaves	Rate of Respiration/Photosynthesis (ppt/min)
In the Dark	
In Light	

QUESTIONS

1. Were either of the rate values positive? If so, what is the biological significance of this?
2. Were either of the rate values negative? If so, what is the biological significance of this?
3. Do you have evidence that cellular respiration occurred in leaves? Explain.
4. Do you have evidence that photosynthesis occurred in leaves? Explain.
5. List five factors that might influence the rate of Carbon Dioxide production or consumption in leaves. Explain how you think each will affect the rate?

EXTENSIONS

1. Design and perform an experiment to test one of the factors that might influence the rate of Carbon Dioxide production or consumption in Question 5.
2. Compare the rates of photosynthesis and respiration among various types of plants.

TEACHER INFORMATION

Photosynthesis and Respiration

1. Spinach leaves purchased from a grocery store work very well and are readily available any time of the year. Do not purchase the pre-packaged spinach in a bag. For best results, keep the leaves cool until they are to be used. Just before use, expose the leaves to bright light for 5 minutes.

2. A fluorescent ring lamp works very well since it bathes the plant in light from all sides and it gives off very little heat. When using a ring lamp as shown below, it is not necessary to use a heat shield.

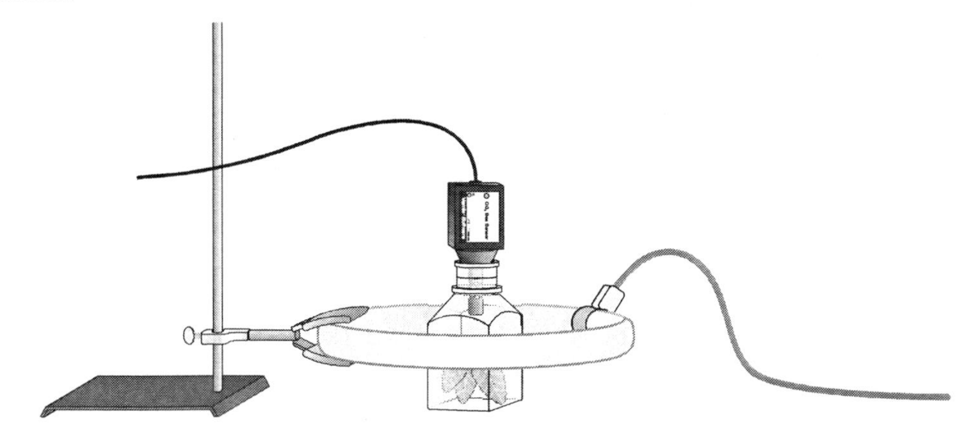

3. If tissue culture flasks are not available, a beaker or flask of water will also work. The tissue culture flask is very thin, however, and will allow leaves to receive much more light from the same lamp.

4. On a nice, sunny day, this experiment may be performed using sun light. If so, no heat shield is needed.

5. When students are placing the probe in the respiration chamber, they should gently twist the stopper into the chamber opening. Warn the students not to twist the probe shaft or they may damage the sensor.

6. To conserve battery power, we suggest that AC Adapters be used to power the interfaces rather than batteries when working with the CO_2 Gas Sensor. An AC Adapter is shipped with each LabPro interface at the time of purchase.

SAMPLE RESULTS

Table 1

Leaves	Rate of Respiration/Photosynthesis (ppt/min)
In the Dark	0.039
In Light	–0.076

ANSWERS TO QUESTIONS

1. The CO_2 rate value for leaves in the dark was a positive number. The biological significance of this is that CO_2 is produced during respiration. This causes the concentration of CO_2 to increase, as sugar is oxidized and broken into CO_2, water, and energy.

2. The rate value for leaves in the light was a negative number. The biological significance of this is that CO_2 is consumed during photosynthesis. This causes the concentration of CO_2 to decrease, as the CO_2 is converted into glucose.

3. Yes, cellular respiration occurred in leaves, since CO_2 increased when leaves were in the dark and photosynthesis was not possible.

4. Yes, photosynthesis occurred in leaves, since CO_2 decreased when leaves were exposed to light.

5. Answers may vary. They might include:
   a. A greater number of leaves should increase the rate, since there are more chloroplasts to undergo photosynthesis and more cells to require energy through cellular respiration.
   b. A greater light intensity will increase the rate of photosynthesis. It may not affect the rate of cellular respiration, however.
   c. A cooler room may decrease both rates, as cellular metabolism decreases in cooler weather.
   d. Facing the top of the leaves toward the light should increase the rate of photosynthesis, since the chloroplasts are closer to the light source.
   e. If the plants overheat due to the heat from the lamp, they may wilt and stop functioning. This will decrease all rates.
   f. If there are too many leaves, diffusion may be restricted and prevent accurate readings. This may apparently decrease both rates.
   g. If water vapor increases in the chamber, the both rates will appear to be more positive than they should be, as water vapor absorbs infrared light. Since the CO_2 Gas Sensor functions by measuring the amount of infrared light reaching a photodetector, water vapor will effect the readings.

Experiment
31C

Photosynthesis and Respiration

Plants make sugar, storing the energy of the sun into chemical energy, by the process of photosynthesis. When they require energy, they can tap the stored energy in sugar by a process called cellular respiration.

The process of photosynthesis involves the use of light energy to convert carbon dioxide and water into sugar, oxygen, and other organic compounds. This process is often summarized by the following reaction:

$$6\ H_2O + 6\ CO_2 + \text{light energy} \rightarrow C_6H_{12}O_6 + 6\ O_2$$

Cellular respiration refers to the process of converting the chemical energy of organic molecules into a form immediately usable by organisms. Glucose may be oxidized completely if sufficient oxygen is available by the following equation:

$$C_6H_{12}O_6 + 6\ O_2 \rightarrow 6\ H_2O + 6\ CO_2 + \text{energy}$$

All organisms, including plants and animals, oxidize glucose for energy. Often, this energy is used to convert ADP and phosphate into ATP.

OBJECTIVES

In this experiment, you will

- Use an O_2 Gas Sensor to measure the amount of oxygen gas consumed or produced by a plant during respiration and photosynthesis.
- Use a CO_2 Gas Sensor to measure the amount of carbon dioxide consumed or produced by a plant during respiration and photosynthesis.
- Determine the rate of respiration and photosynthesis of a plant.

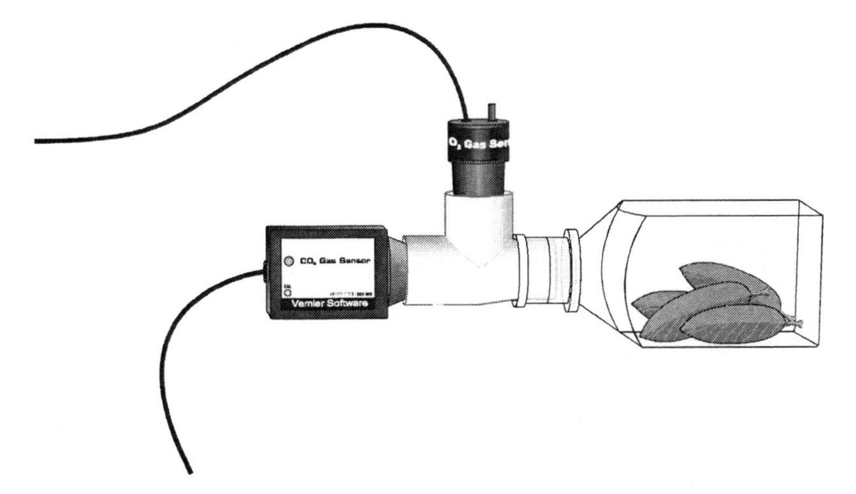

Figure 1

MATERIALS

computer
Vernier computer interface
Logger *Pro*
Vernier O_2 Gas Sensor
Vernier CO_2 Gas Sensor
CO_2–O_2 Tee
250 mL respiration chamber
plant leaves
500 mL tissue culture flask
lamp
aluminum foil
forceps

PROCEDURE

1. Connect the CO_2 Gas Sensor to Channel 1 and the O_2 Gas to Channel 2 of the Vernier computer interface.

2. Prepare the computer for data collection by opening the file "31C Photo (CO2 and O2)" from the *Biology with Computers* folder of Logger*Pro*.

3. Obtain several leaves from the resource table and blot them dry, if damp, between two pieces of paper towel.

4. Place the leaves into the respiration chamber, using forceps if necessary. Wrap the respiration chamber in aluminum foil so that no light reaches the leaves.

5. Insert the CO_2–O_2 Tee into the neck of the respiration chamber. Place the O_2 Gas Sensor into the CO_2–O_2 Tee as shown in Figure 1. Gently push the sensor down into the Tee until it fits snugly. The sensor is designed to seal the Tee without the need for unnecessary force. The O_2 Gas Sensor should remain vertical throughout the experiment. Place the CO_2 Gas Sensor into the Tee directly across from the respiration chamber as shown in Figure 1. Gently twist the stopper on the shaft of the CO_2 Gas Sensor into the chamber opening. Do not twist the shaft of the CO_2 Gas Sensor or you may damage it. Wait 10 minutes before proceeding to Step 6.

6. Click [Collect] to begin data collection. Collect data for fifteen minutes and click [Stop].

7. When data collection has finished, determine the rate of respiration:
   a. Click anywhere on the CO_2 graph. Move the mouse pointer to the point where the data values begin to increase. Hold down the left mouse button. Drag the pointer to the point where the data ceases to increase and release the mouse button.
   b. Click on the Linear Fit button, [Linear Fit], to perform a linear regression. A floating box will appear with the formula for a best fit line.
   c. Record the slope of the line, *m*, as the rate of respiration in Table 1.
   d. Close the linear regression floating box.
   e. Repeat Steps 7a – d for the O_2 graph. However, you will need to move the mouse pointer to the point where the data values begin to decrease. Hold down the mouse button and drag to the point where the data ceases to decrease.

8. Move your data to a stored run. To do this, choose Store Latest Run from the Experiment menu.

9. Remove the aluminum foil from around the respiration chamber.

10. Fill the tissue culture flask with water (not the respiration chamber) and place it between the lamp and the respiration chamber. The flask will act as a heat shield to protect the plant leaves.

11. Turn the lamp on. Place the lamp as close to the leaves as reasonable. Do not let the lamp touch the tissue culture flask. Note the time. The lamp should be on for 5 minutes prior to beginning data collection.

12. After the five-minute time period is up, click Collect to begin data collection. Collect data for 15 minutes and click Stop.

13. When data collection has finished, determine the rate of photosynthesis:

    a. Click anywhere on the CO_2 graph. Move the mouse pointer to the point where the data values begin to decrease. Hold down the left mouse button. Drag the pointer to the point where the data ceases to decrease and release the mouse button.

    b. Click on the Linear Fit button, , to perform a linear regression. Choose "Latest CO2" and a floating box will appear with the formula for a best-fit line.

    c. Record the slope of the line, *m*, as the rate of photosynthesis in Table 1.

    d. Close the linear regression floating box.

    e. Repeat steps 13a-d for the O_2 graph. However, you will need to move the mouse pointer to the point where the data values begin to increase, hold down the mouse button and drag to the point where the data ceases to increase.

14. Print a graph showing your photosynthesis and respiration data.

    a. Label each curve by choosing Text Annotation from the Analyze menu. Enter "Photosynthesis" in the edit box. Repeat to create an annotation for the "Respiration" data. Drag each box to a position near its respective curve. Adjust the position of the arrow heads.

    b. Print a copy of the graph, with both data sets displayed. Enter your name(s) and the number of copies of the graph you want.

15. Remove the plant leaves from the respiration chamber, using forceps if necessary. Clean and dry the respiration chamber.

DATA

Table 1

Leaves	CO_2 rate of production/consumption (ppt/min)	O_2 rate of production/consumption (ppt/min)
In the dark		
In the light		

QUESTIONS

1. Were either of the rate values for CO_2 a positive number? If so, what is the biological significance of this?
2. Were either of the rate values for O_2 a negative number? If so, what is the biological significance of this?
3. Do you have evidence that cellular respiration occurred in leaves? Explain.
4. Do you have evidence that photosynthesis occurred in leaves? Explain.
5. List five factors that might influence the rate of oxygen production or consumption in leaves. Explain how you think each will affect the rate?

EXTENSIONS

1. Design and perform an experiment to test one of the factors that might influence the rate of oxygen production or consumption in Question 5.
2. Compare the rates of photosynthesis and respiration among various types of plants.

TEACHER INFORMATION

Photosynthesis and Respiration

1. Spinach leaves purchased from a grocery store work very well and are readily available any time of the year. Do not purchase the pre-packaged spinach in a bag. For best results, keep the leaves cool until they are to be used. Just before use, expose the leaves to bright light for 5 minutes.
2. A fluorescent ring lamp works very well since it bathes the plant in light from all sides and it gives off very little heat. When using a ring lamp as shown below, it is not necessary to use a heat shield.

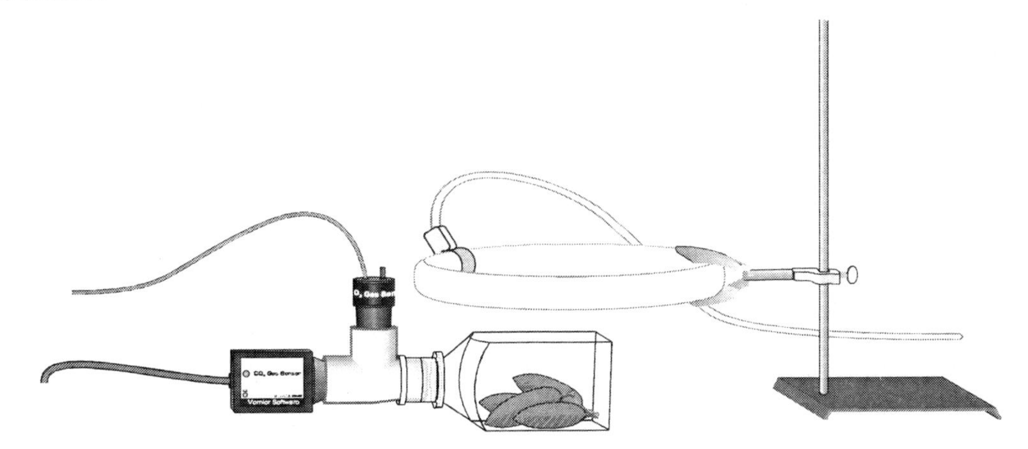

3. If tissue culture flasks are not available, a beaker or flask of water will also work. The tissue culture flask is very thin, however, and will allow leaves to receive much more light from the same lamp.
4. On a nice, sunny day, this experiment may be performed using sun light. If so, no heat shield is needed.
5. To extend the life of the O_2 Gas Sensor, always store the sensor upright in the box in which it was shipped.
6. The waiting time before taking data may need to be adjusted depending on the rate of diffusion of the oxygen gas and the carbon dioxide gas. Monitor the gas concentrations and start collecting data when the levels of gas begin to move in the correct direction. It may take up to 15 minutes for the Oxygen Gas level to begin increasing once the light is turned on.
7. When students are placing the probe in the respiration chamber, they should gently twist the stopper into the chamber opening. Warn the students not to twist the probe shaft or they may damage the sensor.
8. To conserve battery power, we suggest that AC Adapters be used to power the interfaces rather than batteries when working with the CO2 Gas Sensor. An AC Adapter is shipped with each LabPro interface at the time of purchase.

SAMPLE RESULTS

Table 1		
Leaves	CO_2 rate of production/consumption (ppm/min)	O_2 rate of production/consumption (%/min)
In the dark	39.0	–0.0138
In the light	–75.6	0.0271

ANSWERS TO QUESTIONS

1. The CO_2 rate value for leaves in the dark was a positive number. The biological significance of this is that CO_2 is produced during respiration. This causes the concentration of CO_2 to increase, as sugar is oxidized and broken into CO_2, water, and energy.

2. The O_2 rate value for leaves in the dark was a negative number. The biological significance of this is that O_2 is consumed during cellular respiration. This causes the concentration of O_2 to decrease as glucose is oxidized for energy.

3. Yes, cellular respiration occurred in leaves, since O_2 decreased when leaves were in the dark and photosynthesis was not possible.

4. Yes, photosynthesis occurred in leaves, since O_2 increased when leaves were exposed to light.

5. Answers may vary. They might include:
   a. A greater number of leaves should increase the rate, since there are more chloroplasts to undergo photosynthesis and more cells to require energy through cellular respiration.
   b. A greater light intensity will increase the rate of photosynthesis. It may not affect the rate of cellular respiration, however.
   c. A cooler room may decrease both rates, as cellular metabolism decreases in cooler weather.
   d. Facing the top of the leaves toward the light should increase the rate of photosynthesis, since the chloroplasts are closer to the light source.
   e. If the plants overheat due to the heat from the lamp, they may wilt and stop functioning. This will decrease all rates.
   f. If there are too many leaves, diffusion may be restricted and prevent accurate readings. This may apparently decrease both rates.

Appendix

A

Using the CD

Important: You do not need to use the CD included with this book unless you want to change something in the student instructions to meet your needs.

The CD located inside the back cover of this book contains four folders:

- **Biology w Computers Word** - For use with this book. Uses Logger *Pro* 3.1 or newer computer software with LabPro. Contains the word-processing files for each of the 31 student experiments in this book, *Biology with Computers, Third Edition.*
- **Biology w Calculators Word** - Not for use with this book. Supports the DataMate calculator program with LabPro or CBL 2. Contains files for each of the student experiments in the book, *Biology with Calculators*.
- **Biology w Handhelds Word** – Not for use with this book. Supports the DataPro handheld program with LabPro. Contains the files for each of the student experiments and teacher information in the book, *Biology with Handhelds*.
- **Word Files for Older Book** - Uses Logger *Pro* 2 computer software with LabPro, Serial Box Interface, or ULI. Contains the word-processing files for each of the 31 student experiments in the book, *Biology with Computers, Second Edition.*

Using the *Biology with Computers* Word-Processing Files

Start Microsoft Word, then open the file of your choice from the Biology w Computers Word folder. Files can be opened directly from the CD or copied onto your hard drive first. These files can be used with any version of Microsoft Word that is Word 97 or newer.

All file names begin with the experiment number, followed by an abbreviation of the title; e.g., 01 Energy in Food is the file name used for *Experiment 1, Energy in Food.* This provides a way for you to edit the tests to match your lab situation, your equipment, or your style of teaching. The files contain all figures, text, and tables in the same format as printed in *Biology with Computers.*

Appendix

Remote Data Collection Using LabPro

LabPro can collect data without a computer attached. Having a computer attached to LabPro is preferable, because it provides much more flexibility in your data collection and provides a screen for immediate feedback of your results. There are times, however, when disconnecting LabPro from the computer to collect data is useful. For example, remote data collection is perfect for gathering acceleration data on a roller coaster. For these times, there are three methods available.

Quick Setup Method

Using the Quick Setup Method, you can collect data with LabPro, then retrieve it using a computer running Logger *Pro*. You can collect up to 99 points without using a computer to set up data collection. This method will work only with auto-ID sensors.

1. Set up LabPro for remote data collection
   a. Put fresh batteries in the LabPro.
   b. Connect your auto-ID sensor(s) to LabPro.
   c. Press the QUICK SETUP button. A flash of the yellow LED and two beeps will verify setup.
2. Collect data.
   a. When you are ready to collect data, press the START/STOP button. You will hear a beep indicating that data collection has begun. The green LED will blink each time a reading is made.
   b. When data collection is completed, the green LED will no longer blink.[1] You do not need to press the START/STOP button to end data collection. (You can, however, stop data collection early by pressing the START/STOP button before data collection is finished.)
3. Retrieve the data.
   a. When data collection is finished, go to the computer and attach the interface.
   b. Start Logger *Pro*.
   c. If a Remote Data Available window appears, simply click the YES button and choose to retrieve remote data into the current file. If a window does not appear when the interface is reconnected, choose Remote ▸ Retrieve Remote Data from the Experiment menu.
   d. Click [OK] to retrieve data into the LabPro Remote data set. You may want to rename the time and data columns after the data are retrieved.

Set Up for Time Based Method

Using the Set Up for Time Based Method, you will use a computer to set up LabPro, detach it for a real time data collection, and then reattach it to the computer to retrieve the data. This gives you more flexibility than the Quick Setup Method. With this method, you can (a) use any sensors (not just auto-ID), (b) control the time between samples, and (c) control the number of data points collected (up to about 12,000).

1. Set up LabPro for remote data collection.
   a. Put fresh batteries in the LabPro.

[1] The duration of data collection depends on the sensor(s) connected.

b. Set up the sensors, LabPro, and Logger *Pro* just as you normally would to collect data in Time Based mode.
c. Instead of clicking the Collect button, choose Remote ▸ Remote Setup from the Experiment menu.[2] A summary of your setup will be displayed.
d. Click OK to prepare the LabPro.
e. Disconnect the LabPro from the computer.
f. If it has not already been saved, save the experiment file so it can be used to later retrieve the data from LabPro.

2. Collect data.

a. When you are ready to collect data, press the START/STOP button.
b. When data collection is complete, the yellow LED will flash briefly. You can also stop data collection early by pressing the START/STOP button before data collection is finished.

3. Retrieve the data.

a. Start Logger *Pro* if it is not already running. Choose Continue Without Interface in the Connect LabPro dialog and click OK.
b. Open the experiment file used to set up LabPro.
c. Attach LabPro to the computer.
d. If a Remote Data Available window appears, click the YES button. Click, OK which accepts the default to retrieve remote data into the current file. If a window does not appear when the interface is reconnected, choose Remote ▸ Retrieve Remote Data from the Experiment menu and follow the on-screen instructions.
e. The data will be retrieved.

Set Up for Selected Events Method

Using the Set Up for Selected Events Method, you will use a computer to set up LabPro, detach it to collect individual data points, and then reattach it to retrieve the data. With this method, you can use any sensors (not just auto-ID), and collect individual data points whenever you press the START/STOP button.

1. Set up LabPro for remote data collection

a. Put fresh batteries in the LabPro.
b. Set up the sensors, LabPro, and Logger *Pro* just as you normally would to collect data using the Selected Event mode.
c. Instead of clicking the Collect button, choose Remote ▸ Remote Setup from the Experiment menu.[2] A summary of your setup will be displayed.
d. Click OK to prepare the LabPro.
e. Disconnect the LabPro from the computer.
f. If it has not already been saved, save the experiment file so it can be used to later retrieve the data from LabPro.

2. Collect data. Every time you press the START/STOP button, LabPro will collect one data point. The green LED will remain on and the yellow LED will flash with each reading. Repeat as often as you want, for up to 99 readings.

---

[2] Sensors that require a warm-up time may have additional on-screen instructions.

3. Retrieve the data.
   a. Start Logger *Pro* if it is not already running. Choose Continue Without Interface in the Connect LabPro dialog and click OK.
   b. Open the experiment file used to set up LabPro.
   c. Attach LabPro to the computer.
   d. If a Remote Data Available window appears, click the YES button. Click, OK which accepts the default to retrieve remote data into the current file. If a window does not appear when the interface is reconnected, choose Remote ▸ Retrieve Remote Data from the Experiment menu and follow the on-screen instructions.
   e. The data will be retrieved.

Sensor Warm-up Times

Some sensors require a certain warm-up period before an accurate measurement can be taken. LabPro handles this automatically in remote modes by warming up each sensor for the appropriate amount of time. The Turbidity Sensor, for example, requires a 2-second warm-up period. Follow the on-screen instructions regarding warm-up times for sensors.

The table below lists all sensors with warm-up times of 2 seconds or longer. If a sensor is not on this list, its warm-up time is negligible.

Sensor	Warm-up Time (s)
Ammonium Ion-Selective Electrode	2
Calcium Ion-Selective Electrode	2
Chloride Ion-Selective Electrode	2
CO_2 Sensor	90
Colorimeter	300
Conductivity Probe	2
Dissolved Oxygen Probe	600
Dual-Range Force Sensor	2
Electrode Amplifier	30
Flow Rate Sensor	2
Force Plate	2
Nitrate Ion-Selective Electrode	2
Oxygen Gas Sensor	2
pH Sensor	30
Turbidity Sensor	2
UVA Sensor	2
UVB Sensor	2

The following sensors cannot be set up to collect remote data:

Digital Control Unit
Drop Counter
Heat Pulser
Photogate
Radiation Monitor
Student Radiation Monitor
Rotary Motion Sensor

Vernier Products for Biology

All software and laboratory interfacing hardware required for the experiments contained in this book are available from Vernier Software & Technology and can be found in this appendix. The purchase of Vernier programs includes a site license that permits you to make as many copies as you wish for use in your own school. You may also use the programs on networks within your school at no extra cost. A table of order codes and prices can be found on the next page.

LabPro

Vernier LabPro provides a portable and versatile data collection device for any class studying biology. A wide variety of Vernier probes and sensors can be connected to each of the four analog channels and two sonic/digital channels. LabPro is connected to a computer using a serial or USB port or to a TI graphing calculator or Palm OS Handheld Data collection with LabPro and a computer is controlled by Logger *Pro* software.

Included with the purchase of a LabPro (LABPRO) is a Voltage Probe, computer cables (USB and serial for Macintosh and Windows), calculator cradle, short calculator link cable, user's manual, and AC power supply.

Data-Collection Software for Biology

Logger *Pro* data software is the data-collection software for this *Biology with Computers* lab manual. Logger *Pro* software now comes with a free site license for both Windows and Macintosh, so you only need to order one copy of Logger *Pro* for your school or college department.

Software	Macintosh and Windows CD
Logger *Pro*	(order code: LP)

Vernier Sensors for Biology

EKG Sensor

The EKG Sensor measures electrical signals produced by the heart. It uses three disposable electrodes. An EKG graph is displayed, demonstrating to students the contraction and repolarization of the heart's chambers. A package of 100 disposable electrodes is included with each sensor.

Exercise Heart Rate Monitor

The Exercise Heart Rate Monitor is ideal for determining the heart rate of actively moving individuals. With this sensor, a person's heart rate can be monitored during, as well as after exercise. The Exercise Heart Rate Monitor consists of a wireless transmitter belt and a receiver module that plugs into a Vernier LabPro, CBL, or CBL 2. The transmitter belt senses the electrical signals generated by the heart, much like an EKG. For each heart beat detected, a signal is transmitted to the receiver module, and a heart rate is determined.

O_2 Gas Sensor

The new O_2 Gas Sensor measures oxygen concentration in air. Many of the experiments currently performed using the CO_2 Gas Sensor can be performed or complemented using the O_2 Gas Sensor. Due to its wide measurement range (0-27% oxygen by volume), it can also be used to monitor oxygen concentration during human respiration.

CO_2 Gas Sensor

The CO_2 Gas Sensor measures gaseous carbon dioxide levels in the range of 0 to 5000 ppm. This sensor is great for measuring changes in CO_2 levels during plant photosynthesis and respiration. You can easily monitor changes in CO_2 levels occurring in respiration of organisms as small as crickets or beans!

Respiration Monitor Belt

Our Respiration Monitor Belt is used with our Gas Pressure Sensor to measure human respiration. Simply strap the belt around your chest, then pump air into the belt with the hand bulb. You can then monitor the pressure associated with the expansion and contraction of your chest during breathing. Requires the Gas Pressure Sensor.

Stainless Steel Temperature Probe

The Stainless Steel Temperature Probe is an accurate, durable, and inexpensive sensor for measuring temperature. Range: –40°C to +135°C

pH Sensor

Our pH Sensor is a Ag-AgCl gel-filled combination electrode and amplifier. It includes a convenient storage solution container that can be attached directly to the electrode. Range: 0 to 14 pH units

Gas Pressure Sensor

The Gas Pressure Sensor can be used for a variety of experiments in biology where gases, such as oxygen and carbon dioxide, are either produced or consumed in a reaction. The pressure range is 0 to 2.1 atm (0 to 210 kPa). It comes with a variety of pressure-sensor accessories, including a syringe, plastic tubing with two Luer-lock connectors, two rubber stoppers with Luer-lock adapters, and one two-way valve.

Colorimeter

The Vernier Colorimeter allows you to study the light absorption of various solutions. It is great for Beer's law experiments, determining the concentration of unknown solutions, or studying changes in concentration vs. time. Fifteen 3.5 mL cuvettes are included.

Conductivity Probe

This probe is great for environmental testing for salinity, total dissolved solids (TDS), or conductivity in water samples. Biology students can use it to investigate the difference between ionic and molecular compounds, strong and weak acids, salinity, or ionic compounds that yield different ratios of ions. The Conductivity Probe can monitor concentration or conductivity at three different sensitivity settings: 0-200 μS/cm, 0-2000 μS/cm, and 0-20,000 μS/cm.

Dissolved Oxygen Probe

Use the Dissolved Oxygen Probe to determine the concentration of oxygen in aqueous solutions in the range of 0–14 mg/L (ppm). It has built-in temperature compensation and a fast response time. This probe is great for water quality, biology, or ecology. Included with the probe is a zero-oxygen solution, two membrane caps, a 100% calibration bottle, and electrode filling solution. Replacement membranes are available (order code MEM).

Ion-Selective Electrodes

The Vernier family of Ion-Selective Electrodes (ISEs) can be used to measure the concentration of a specific ion in aqueous samples. The species available include Nitrate (NO_3^-), Chloride (Cl^-), Calcium (Ca^{2+}), and Ammonium (NH_4^+).

Vernier Products for *Biology with Computers*

Item	Order Code
Vernier LabPro interface	(LABPRO)
CO_2 Gas Sensor	(CO2-BTA)
O_2 Gas Sensor	(O2-BTA)
EKG Sensor	(EKG-BTA)
Exercise Heart Rate Monitor	(EHR-BTA)
Stainless Steel Temperature Probe	(TMP-BTA)
pH Sensor	(PH-BTA)
Gas Pressure Sensor	(GPS-BTA)
Conductivity Probe	(CON-BTA)
Colorimeter	(COL-BTA)
Respiration Monitor Belt (needs GPS-BTA)	(RMB)
Biology with Computers lab manual	(BWC-LP)

Optional Vernier Products

Graphical Analysis (for Windows *and* Mac)	(GA)
CO2-O2 Tee	(CO2-TEE)
Replacement Cuvettes (package of 100)	(CUV)
pH Buffer Set (for pH 4, 7, and 10 buffers)	(PHB)
EKG Electrodes	(ELEC)
D.O. Probe Membrane Cap	(MEM)
Ion-Selective Electrodes (NO_3^-, Cl^-, Ca^{2+}, NH_4^+)	(order code varies)

Appendix

D

Safety Information

Chemical Hazard Information

The reference source for the chemical hazard information in this book is the 2002 edition of Flinn Scientific's *Chemical & Biological Catalog Reference Manual.* Flinn Scientific, Inc. is an acknowledged leader in the areas of chemical supply, apparatus and laboratory equipment supply, and chemical safety. Flinn's *Chemical & Biological Catalog Reference Manual* is an outstanding reference to be used as you order chemicals, store chemicals, mix solutions, use chemicals in you classroom, and dispose of chemicals. Most of the chemicals and the equipment used in *Biology with Computers* are available from this catalog. We strongly urge you to obtain and use a current copy of the above mentioned publication by contacting Flinn Scientific at the address below:

Flinn Scientific, Inc.
P.O. Box 219
Batavia, Illinois 60510
Telephone (800) 452-1261
www.flinnsci.com

The Flinn hazard code is used in the teacher information section of many tests in *Biology with Computers* to describe any possible hazards associated with the chemical reagents used. The Flinn hazard code (A–D) is defined as follows:

A. Extremely Hazardous. This category includes, but is not limited to, concentrated acids, severely toxic, severely corrosive, unstable and /or explosive chemicals.

B. Hazardous. This category includes, but is not limited to, chemicals that are toxic/poisons, corrosive, contain heavy metals, and/or are alleged/proven carcinogens.

C. Somewhat Hazardous. This category includes, but is not limited to, chemicals that are highly flammable/combustible, moderately toxic and/or oxidants.

D. Relatively Non-Hazardous. This category includes, but is not limited to, chemicals that are irritants and/or allergens.

Appendix

Equipment and Supplies

A list of equipment and supplies for all the experiments is given below. The amounts listed are for a class of up to 30 students working in groups of two, three, or four students in a classroom equipped with eight computers. The materials have been divided into **nonconsumables**, **consumables**, and **chemicals**. Most consumables and chemicals will need to be replaced each year. Most nonconsumable materials may be used many years without replacement. Some substitutions can be made.

Nonconsumables

Item	Amount	Experiment
balance	2	1, 22, 23a, 23b
basting bulb	8	12b, 16b, 23a-b
beaker, 100 mL	16	8, 9, 11a-d, 18, 19
beaker, 1 L	16	11a-d, 16b, 23a-b
beaker, 250 mL	16	3, 6b, 7, 14, 18, 19, 22, 25
beaker, 400 mL	8	4, 6a
beaker, 50 mL	16	3
beaker, 600 mL	24	2, 6b, 7, 12a, 16a, 17, 23a-b, 24a-b
borer, cork, 14 mm	1	22
bottle, dropper	8	12b, 16b
bottle, Nalgene, 250 mL	8	6a, 11a-b, 11d, 12a, 16a, 23a-b, 31a-c
bottle, sampling, 500 mL	8	20, 21
bottle, spray	2	10
can, small, metal	8	1
clamp, utility	16	1, 2, 4, 5, 10, 11c, 12b, 16b, 17, 18, 22, 30
clamp, dialysis tubing	16	4
clamp, plastic tubing	16	10
CO2–O2 Tee	8	11d, 31c
cup, small, plastic	8	20
cup, small, Styrofoam	8	19
cuvette, colorimeter	24	7, 8, 9, 13
fan	2	10

floodlight, 100-watt	8	7, 10
food holder	8	1
forceps	8	8, 9, 11c, 24a-b, 31a-c
glass beads	250	11c
glass-marking pencil	8	13
gloves, latex	class set	8, 9
goggles	class set	all (except 15, 20, 21, 26-30)
graduated cylinder, 100 mL	8	1, 11c
graduated cylinder, 10 mL	8	6a-b, 13, 17, 24a-b
graduated cylinder, 25 mL	8	17
graduated cylinder, 50 mL	8	3, 7
heater, small electric	2	10
lab apron	class set	3, 8, 9
lamp, desk	8	31a-c
meter stick	8	15
microplate, 24-well	8	8, 9
microscope	8	13
microscope slide w/cover slip	8	13
milk container, 1 gallon	8	19
pan, shallow, plastic or metal	8	25
pencil	8	7, 22
pipet pump (or pipet bulb)	8	7, 16b, 17
pipet, 2 mL	8	17
pipet, 5 mL	8	7, 13
plastic tubing w/Luer-lock fitting	16	6b, 11c, 12b, 16b, 22, 24b
plastic window screen, 12 cm x 12 cm	136	25
razor blade, knife or scalpel	8	2, 8, 9, 10
ring stand	8	1, 2, 4, 5, 10, 11c, 12b, 16b, 17, 18, 22, 30
ring, 10 cm	8	1
rubber bands, medium	120	25
rubber stopper assembly	16	6b, 11c, 12b, 16b, 22, 24b

ruler, metric	8	2, 7, 8, 9, 10
scissors	2	4, 7, 25
siphon tube	2	25
stepping stool, 45 cm (18-inches) high	4	27a-b
stirring rod, glass	16	1, 2, 4
stopper, cork	8	7
stopper, rubber, single-hole	16	1
stopper, rubber, split, single-hole	8	1
stopwatch	8	7, 8, 9, 11c, 24a-b
string	32 meters	15, 22
syringe, plastic	8	10
test tube rack	16	4, 6a-b, 9, 11c, 12b, 13, 14, 17, 24a-b
test tube, 10 × 100 mm	32	7, 12a, 16a, 17
test tube, 18 × 150 mm	40	4, 6a-b, 8, 9, 11c, 12b, 13, 16b, 17, 24a-b
test tube, 25 × 150 mm screw top	64	14, 25
thermometer	8	6a-b, 11a-d, 12a-b, 16a-b, 23a-b, 24a-b
tissue culture flask, 500 mL	8	31a-c
wash bottle	8	3, 17, 18
Water Depth Sampler	2	21

Consumables

Item	Amount	Experiment
algal culture	1	25
Alka Seltzer®	5 tablets	3
aluminum foil	1 roll	7, 14, 25, 31a-c
antacid tablets	20 g	3
aspirin	20 g	3
beet, red	2	8, 9
beral pipets	100	4, 9, 10, 13
beral pipets, 1 mL graduated	200	6a-b, 7, 8, 9, 12a, 16a, 24a-b
Bufferin®	20 g	3

chromatography paper	8 pcs	7 (extension)
cotton swabs	300	8, 9, 13
cotton, absorbent	1 bag	11c
cotton, non-absorbent	1 bag	11c
crickets, adult	80	23a-b
dental floss	1 roll	4
dialysis tubing, 2.5 cm × 12 cm	1/2 roll	4
disposable electrode tabs	1 pkg	28
distilled water	30 L	2, 3, 4, 5, 7, 13, 14, 25
egg white	20 g	3
fruit juice	200 mL	3
gelatin	200 mL	3
graph paper	32 pcs	2, 8, 10, 13
ice	5 bags	6a-b, 7, 11a-d, 23a-b
Lactaid (droplet form)	1 bottle	24a-b
leaves, fresh picked	25	31a-c
liver	20 g	3, 6a-b
marble (rock)	20 g	3
marshmallows	8	1
matches	8 boxes	1
nuts, various	8	1
paper bag, small	8	26
Parafilm, 5 x 5 cm	64	14
peas (garden)	400	11a-d
plant cuttings	8	10
plant, aquatic (elodea or anacharis)	16	14
plastic bag, bread or produce	16	26, 30
plastic bag, gallon	4	10
pond water	7 L	5, 14, 18, 25
popcorn, popped	8	1
potato, whole	9	3, 22

quartz (rock)	20 g	3
snails, aquatic	16	14
soda water	200 mL	3
spinach, fresh	2 bunches	7, 31a-c
starch	20 g	3
straws	30	18
tape, masking	1 roll	10
Tes-Tape or other glucose test paper	1 pkg	24a-b
tissues, lint free	80	7, 8, 9, 13
toothpicks	20	8, 9
towels, paper	30	22
vegetable oil	200 mL	12b, 16b
vitamin B	20 g	3
vitamin C	20 g	3
wooden splint	16	1
yeast	6 pkgs	3, 6a-b, 12a-b, 16a-b, 17, 24a-b

Chemicals

Item	Amount	Experiment
1-propanol	100 mL	8
2,6-Dichloroindophenol (DPIP)	1 g	7
acetone	5 mL	7 (extension)
agar	15 g	2
aluminum chloride	10 g	5
buffer solution, pH 4	500 mL	3, 6a, 6b, 9
buffer solution, pH 7	1 liter	3, 6a, 6b, 9, 18
buffer solution, pH 10	500 mL	3, 6a, 6b, 9
calcium chloride	10 g	5
ethanol	100 mL	5, 8
fructose	5 g	12a, 12b
galactose	5 g	24a, 24b

glucose	25 g	5, 12a, 12b, 17, 24a, 24b
hydrochloric acid (0.1 M)	100 mL	3
hydrogen peroxide (3%)	750 mL	6a, 6b
lactose	10 g	12a, 12b, 24a, 24b
methanol	100 mL	8
petroleum ether	45 mL	7 (extension)
potassium hydroxide	75 g	11c
potassium phosphate, dibasic	200 g	7
potassium phosphate, monobasic	150 g	7
sodium chloride (table salt)	350 g	2, 3, 4, 5, 9, 27a, 29
sodium hydroxide (0.1 M)	100 mL	3
sodium lauryl sulfate	1 g	9
sucrose	600 g	4, 5, 7, 12a, 12b, 16a, 16b, 22
sulfuric acid (0.2 M)	25 mL	18

Suppliers

Carolina Biological Supply Co.
2700 York Road
Burlington, NC 27215
1-800-334-5551
www.carolina.com

Frey Scientific
100 Paragon Parkway
Mansfield, OH 44903
1-800-225-FREY
www.freyscientific.com

Fisher Science Education
485 South Frontage Rd.
Burr Ridge, IL 60521
1-800-955-1177
www.fisheredu.com

Flinn Scientific Inc.
PO Box 219
Batavia, IL 60510
1-800-452-1261
www.flinnsci.com

NASCO
901 Janesville Ave.
Fort Atkinson, WI 53538
1-800-558-9595
www.homeschool-nasco.com

Sargent-Welch Scientific Co.
PO Box 5529
Buffalo Grove, IL 60089
1-800-727-4368
www.sargentwelch.com

Science Kit and Boreal Labs
PO Box 5003
Tonawnada, NY 14150
1-800-828-7777
www.sciencekit.com

Ward's Natural Science Establishment
PO Box 1712
Rochester, NY 14603
1-800-962-2660
www.wardsci.com

Index

(by Experiment Number)

O

P

R

S

T

V

W

X

Y